T0139570

FINE-TUNING IN THE PHYSICAL UN[

Is the Universe fine-tuned for complexity, life, or something else? [
overview of fine-tuning arguments in physics, with contributions from
in their fields, sheds light on this often-used but seldom-understood [
reviews a specific subject in modern physics – such as dark energy,
system formation – and discusses whether any parameters in our curr[
to be fine-tuned and, if so, to what degree. Connections and differe[
fine-tuning arguments are made clear, and detailed mathematical der[
fine-tuned parameters are given. This accessible yet precise introducti[
physics will aid students and researchers across astrophysics, atomic ar[
and cosmology, as well as all those working at the intersections of physi[

DAVID SLOAN is a lecturer in physics at Lancaster University. Followin[
State, he has held positions at Utrecht, Oxford and Cambridge Universiti[

RAFAEL ALVES BATISTA is a research fellow at Radboud University. He [
from the University of Hamburg. He was previously a postdoctoral [
Universities of São Paulo and Oxford.

MICHAEL TOWNSEN HICKS is a researcher in the philosophy of science a[
physics at the University of Birmingham. He completed his PhD at Rutge[
has held postdoctoral positions at the Universities of Oxford and Cologne[

ROGER DAVIES is the Philip Wetton Professor of Astrophysics at the Univ[
and Director of the Hintze Centre for Astrophysical Surveys. He is forr[
Physics Department at Oxford, a past President of the Royal Astronomi[
current President of the European Astronomical Society.

FINE-TUNING IN THE PHYSICAL UNIVERSE

Edited by

DAVID SLOAN
Lancaster University

RAFAEL ALVES BATISTA
Radboud University, Netherlands

MICHAEL TOWNSEN HICKS
University of Birmingham

ROGER DAVIES
University of Oxford

CAMBRIDGE
UNIVERSITY PRESS

A careful inspection of the values of the different parameters has led to the suggestion that at least a few of those constants of nature must be fine-tuned if life is to emerge. That is, relatively small changes in their values would have resulted in a universe in which there would be a blockage in one of the stages in emergent complexity that lead from a 'big bang' to atoms, stars, planets, biospheres, and eventually intelligent life (e.g., [2, 3, 6, 25]).

We can easily imagine laws that were not all that different from the ones that actually prevail but would have led to a rather boring universe – laws which would have led to a universe containing dark matter and no atoms; laws where there were hydrogen atoms but nothing more complicated and, therefore, no chemistry (and no nuclear energy to keep the stars shining); laws where there was no gravity; laws where there was a universe where gravity was so strong that it crushed everything; laws where the cosmic lifetime was so short that there was no time for evolution; or laws where the expansion was too fast to allow gravity to pull stars and galaxies together.

Some physicists regard such apparent fine-tunings as nothing more than statistical flukes. They would claim that we should not be surprised that nature seems 'tuned' to allow intelligent life to evolve – we would not exist otherwise. This attitude has been countered by John Leslie, who gives a nice metaphor. Suppose you were up before a firing squad. A dozen bullets are fired at you, but they all miss. Had that not happened, you would not be alive to ponder the matter. But your survival is still a surprise – one that it's natural to feel perplexed about.

Other physicists are motivated by this perplexity to explore whether 'fine-tuning' can be better understood in the context of parallel universe models. In this connection, it's important to stress that such models are consequences of several much-studied physical theories – for instance, cosmological inflation and string theory. The models were not developed simply to remove perplexity about fine-tuning.

Before we explore some prerequisites for complexity, it is instructive to examine a pedagogical diagram that demonstrates in a simple way the properties of a vast range of objects in our Universe. This diagram (Figure 1.1), adapted from Carr and Rees [5], shows the mass vs size (on a logarithmic scale) of structures from the subatomic scale to the cosmic scale. Black holes, for example, lie on a line of slope 1 in this $\log M - \log R$ plot. A black hole the size of a proton has a mass of some 10^{38} protons, which simply reflects how weak the force of gravity is. Solid objects such as rocks or asteroids, which have roughly the atomic density, lie along a line of slope 3, as do animals and people. Self-gravity is so weak that its effects are unnoticeable up to objects even the size of most asteroids. From there on, however, gravity becomes crucial – causing, for instance, planets to be spherical – and by the time objects reach a mass of about $0.08 M_{\odot}$, they are sufficiently squeezed by gravity to ignite nuclear reactions at their centres and become stars. The bottom-left corner of Figure 1.1 is occupied by the subatomic quantum regime. On the extreme left is the 'Planck length' – the size of a black hole whose Compton wavelength is equal to its Schwarzschild radius. Classical general relativity cannot be applied on scales smaller than this (and indeed may break down under less extreme conditions). We then need a quantum theory of gravity. In the absence of such a theory, we cannot understand

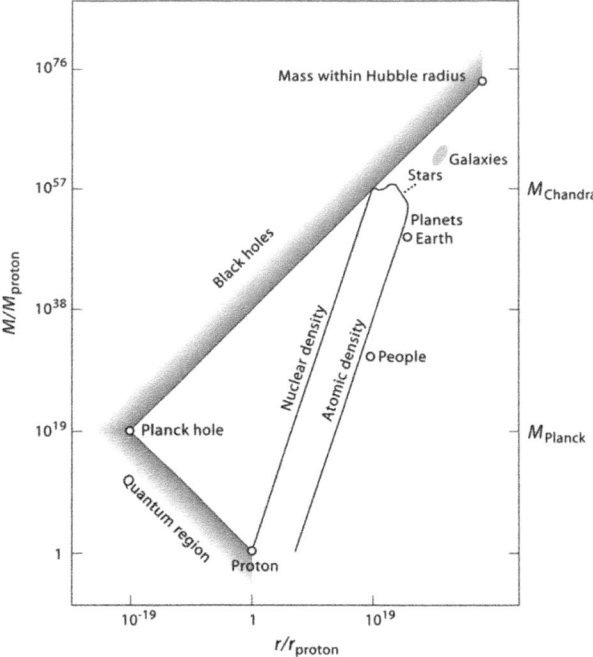

Figure 1.1 This diagram summarises the scales of stars, planets, black holes, and other bodies in a log-log plot of mass against radius. Ordinary lumps lie on the line of slope 3. The mass, in units of the proton mass, scales roughly as the cube of the radius. That line would eventually cross the black hole line (of slope one) at a mass of about 100 million suns. However, it is curtailed before it can do so. The reason is that any mass above about that of Jupiter (containing more than 10^{54} atoms) would be crushed by gravity to a higher density than an ordinary solid. If G were different, the shape of the diagram would not change much, but the number of powers of 10 between the scale of stars and of atoms would scale as the inverse 3/2 power.

the Universe's very beginnings (i.e., what happened at eras comparable to the Planck time of 10^{-43} seconds).

Despite this unmet challenge, it is impressive how much progress has been made in cosmology. In the early 1960s, there was no consensus that our Universe had expanded from a dense beginning. But we have now mapped out, at least in outline, the evolution of our Universe, and the emergence of complexity within it, from about a nanosecond after the Big Bang. At that time, our observable Universe was roughly the size of the solar system, and characterised by energies of the order of those currently realised at the Large Hadron Collider (LHC) near Geneva. Nucleosynthesis of the light elements gives us compelling corroboration of the hot and dense conditions in the first few seconds of the Universe's existence (see Chapter 7; see also, e.g., [8] for a recent review).

The cosmic microwave background (CMB) provides us with not only an astonishingly accurate proof for a black-body radiation state that existed when the Universe was

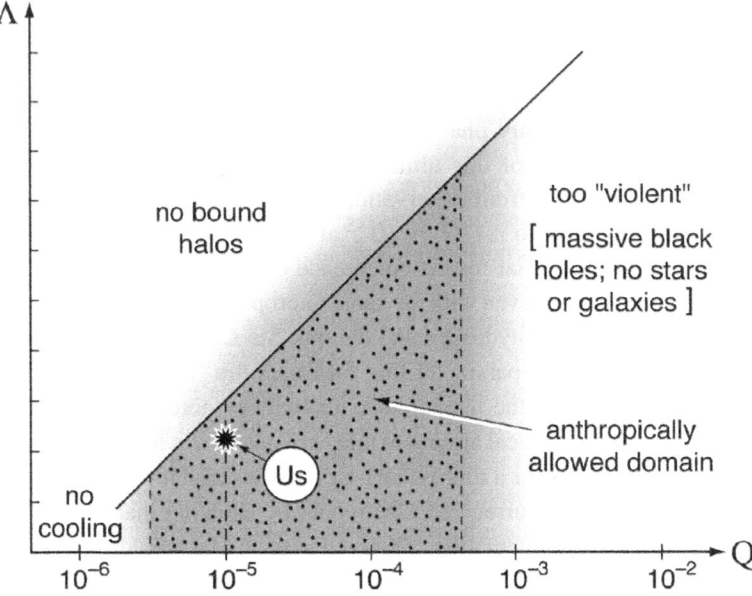

Figure 1.3 Plot of the cosmological constant Λ versus amplitude of fluctuations in cosmic microwave background Q. The shaded-dotted region shows conditions that allow for the existence of complexity.

to the distance of the horizon. In contrast, an astronomer in a high-Q universe would resemble a climber in a mountain landscape, where one peak could dominate the view, and averages are not well defined.)

What about the other extreme, a 'smoother' universe with $Q \lesssim 10^{-6}$? In this case, the disruptive dark energy would push protogalaxies apart before they had a chance to collapse. Even if the dark energy were not there, any galaxies that formed in a lower-Q universe would be small and rather loosely bound (and forming later than in our actual Universe). At $Q \lesssim 10^{-6}$, stars would still form, but material enriched in heavy elements and ejected via stellar winds or supernovae may escape from the shallow gravitational potential wells, not allowing for second-generation stars and planetary systems to form. For values of Q that are significantly smaller than 10^{-6}, there would be inefficient radiative cooling, and therefore, stars would not form within a Hubble time. The conclusion from this discussion (summarised also in [25]; see Figure 1.3) is that for a universe to be conducive for complexity and life, the amplitude of the fluctuations should best be between 10^{-6} and 10^{-4} and, therefore, not particularly finely tuned.

1.2.4 Non-Trivial Chemistry

For life to emerge, the Universe requires nuclear fusion. Fusion not only powers the stars; nucleosynthesis at the hot stellar centres also forges elements such as carbon, oxygen, iron, and phosphorus, all of which are essential for life as we know it. In general, many of the elements in the periodic table participate in the complex chemistry required for the formation of planets and the evolution of their biospheres.

To obtain the nuclear fusion reactions that lead to the creation of the periodic table requires a certain balance between the strength of the electromagnetic force (that repels two protons from each other) and the strong nuclear force (that attracts them). This balance, in our Universe, where the strong nuclear force is about a hundred times stronger than the electromagnetic force, is responsible for the fact that we do not have atomic numbers higher than 118. Had the ratio of the two interactions been much smaller, carbon and heavier elements could not have formed, but the necessary tuning is not excessive.

Similarly, much has been written about Fred Hoyle's prediction of the existence of a 7.65 MeV resonant level of ^{12}C [14, 16]. However, while the prediction itself was indeed remarkable, the degree of fine-tuning required for the energy of that level is not fantastic (e.g., [27, 33]; see [10] for a recent study of this energy level).

The topic of chemistry actually allows us to examine a much more extreme counterfactual universe – a 'nuclear-free universe' – in which hydrogen is the only element that exists. Surprisingly, on the large scale, such a universe would not look much different from ours. Gravity would ensure that galaxies would still form, and even stars would shine (albeit generally for shorter times) by releasing their gravitational energy as they contract to form white dwarfs and black holes. Even Jupiter-like planets composed of solid hydrogen could exist. Of course, no complexity or life of the types we are familiar with will emerge in such a universe (only perhaps something similar to Fred Hoyle's science fiction concept of *The Black Cloud* [15]).

1.2.5 'Tuned' Cosmic Expansion Rate

The results from the Planck satellite depicted in Figure 1.2 (in combination with observations of baryon acoustic oscillations, lensing reconstruction and a prior on the Hubble constant) give for the cosmic energy budget $\Omega_m \sim 0.3$, $\Omega_\Lambda \sim 0.7$, with baryons making less than 5% of this budget [30]. If the cosmic acceleration is indeed driven by a cosmological constant (energy of the physical vacuum, with an equation of state parameter $w = P/\rho = -1$), then the acceleration will continue forever (see Chapter 3). It is clear, however, that if the dark-energy density would have dominated over the matter density (dark matter + baryons) much earlier in the life of our Universe, galaxies would never have formed (this is also dependent on the value of Q; see discussion in the next section). This means that for complexity to arise, some constraints are needed on the ratios of Ω_m/Ω_Λ and Ω_b/Ω_{DM} (where Ω_b denotes the baryon fraction and Ω_{DM} the dark matter fraction). The second ratio is crucial because even though dark matter dominates over baryonic matter in our Universe, without the latter, there would be no stars, no planets, and no life.

As an aside we should note that the nature of the dark energy that propels the cosmic acceleration is one of the most fascinating puzzles in modern cosmology (and one that may not be solved until we have a better understanding of the granular structure of space-time on the Planck scale). Despite its importance for fundamental physics, the dark energy hardly affected any astrophysical phenomena in our Universe; in contrast, the evolution of our Universe so far – the emergence of and morphology of galaxies, clusters, and so forth – has been dominated by the effects of dark matter.

1.3 The Multiverse

As far as we can tell, the laws of physics and the values of the cosmological parameters are the same throughout our entire observable Universe. But the observable Universe is limited by the horizon, which is determined by the finite age of our Universe. What lies beyond this Hubble volume? The homogeneity and isotropy of our observable Universe, with the absence of any perceptible gradient across it (to the 10^{-5} level) suggest (though, of course, do not prove) that the same laws continue to apply thousands of times further. Indeed, many arguments suggest that galaxies beyond the horizon outnumber those we see by a vast factor – perhaps so vast that all combinatorial options would occur repeatedly, and we'd all, far beyond the horizon, have avatars.

Furthermore, some models for the inflationary phase lead to what has been dubbed 'eternal inflation' [24, 39]. According to these models, our Big Bang could be just one 'pocket universe' in a huge ensemble – one island of space-time in a vast archipelago. This scenario also fits well with the 'landscape' concept of string theory, in which there are some 10^{500} metastable vacua solutions, of which our Universe is but one [4, 19]. So the question arises: how large is physical reality?[1]

The first thing to realise is that because we live in an accelerating universe, galaxies are disappearing over an 'event horizon', so we will not observe their far future (rather, as we cannot observe the fate of an object that falls into a black hole after it has crossed the horizon). If the acceleration continued, then after about a trillion years, observers in the remnant of the Milky Way (or its merged product with the Local Group) would not be able to see (again, even in principle) any galaxy other than their own. This does not mean that those galaxies whose light would have been stretched beyond the cosmic scale would not exist.

Moreover, galaxies that are already beyond our current horizon will never become observable, even in principle. Yet most researchers would be relaxed about claims that these galaxies exist in the same way that, in the middle of the ocean, you expect that an ocean extends beyond the terrestrial horizon. These never-observable galaxies would have emerged from the same Big Bang as we did. But suppose that we imagine separate Big Bangs. Are space-times completely disjoint from ours any less real than regions forever

---

[1] It's perhaps necessary, especially in addressing philosophical readers, to inject a clarification at the start. Many would define 'the Universe' as 'everything there is' – and if that's the definition, then there plainly cannot be more than one. If there are other domains (perhaps originating in other Big Bangs, and perhaps differing from our observable domain in size, content, or dimensionality), then we should really define the whole enlarged ensemble as 'the Universe'. We then need a new word – 'metagalaxy', for instance – to denote the domain to which cosmologists and astronomers have observational access. However, so long as this whole idea remains speculative, it is probably best to leave the term 'universe' undisturbed, with its traditional connotations, even though this then demands a new term, 'multiverse', for the whole (still hypothetical) ensemble.

unobservable which are the aftermath of 'our' Big Bang? Surely not – so these other universes, too, should count as parts of physical reality.

Similarly, while we cannot observe any free quarks, we believe that quarks exist, because the Standard Model of particle physics has successfully passed many experimental tests. Likewise, we are disposed to believe in what Einstein's theory tells us about the metric within black holes (inside the event horizon), because general relativity has gained high credibility by being tested in numerous observations and experiments.

If we can develop a theory that makes numerous predictions that are testable (and confirmed) in the observable part of the Universe, then we should be prepared to accept its predictions in unobservable parts.

We currently have no theories of microphysics that are 'battle tested' above the energies reachable in the biggest particle accelerators. These energies are exceeded throughout the first nanosecond after the Big Bang. Theorists who model the inflation era therefore make assumptions about the physics. Some such assumptions predict eternal inflation; others do not (see Chapter 4). Some predict the landscape scenario; others do not. The details of this physics are already somewhat constrained (by, for instance, the observed properties of the fluctuations in the CMB), but we are still far from being able to prove or disprove a model like Linde's 'eternal inflation' [24]. We should therefore be open-minded about how far the aftermath of 'our' Big Bang extends beyond our horizon and also about whether other Big Bangs exist as part of physical reality. Once we are willing to entertain the notion that a multiverse may exist, an even more intriguing question arises: are the laws of physics and the values of the physical constants the same in other members of this ensemble of universes, or are they different?

If they are different, then what we call 'laws of nature' may be no more than local by-laws governing just our cosmic patch. Moreover, many of these pocket universes could be stillborn or sterile. That is, the physical laws prevailing in them (or the values of the parameters) may preclude the emergence of any kind of complexity, and life, in particular. They simply may not satisfy one or more of the prerequisites we discussed in Section 1.2.

The mere possibility that physical reality can encompass such a multiverse provides a strong motivation to develop the lines of thought outlined in Section 1.2 to explore a variety of counterfactual universes, with different values of physical constants, to ascertain which ranges of parameters would allow complexity. The identification and selection of such 'biophilic' universes constitutes what has been dubbed anthropic reasoning (e.g., [2, 12, 25, 34, 41]). Obviously, we humans find ourselves not in a typical member of the multiverse but in a typical domain in the subset of universes that allows complexity and life to emerge and evolve [12]. Copernican humility can only be taken so far. We live on an ordinary terrestrial planet orbiting an ordinary star in its habitable zone. The observable Universe may contain as many as two trillion galaxies [7], and our universe may be only one member of an ensemble of some 10^{500} universes. But our Universe is not 'typical'.

To give a simple analogy (which we believe was first suggested by physicist Leonard Susskind), suppose you wake up in the morning and think, 'What am I?' It seems that a natural answer may be 'I am an insect', since insects have the largest biomass of terrestrial animals. It is estimated that at any time, there are some 10^{19} insects alive. The reason that this argument is false is that by being able to wonder 'what am I?' we have already selected a small subset of the animal kingdom. On the other hand, we can argue that the probability that the answer to 'what am I?' is 'I am Leonardo da Vinci' is still very small.

The ability to actually determine the ranges of all the parameters that allow complexity to develop and the probability for its emergence is currently beyond what physics can achieve, since it requires a knowledge of all the probability distributions and the correlations among them. What we can currently do is a 'poor man's' simplified version of this daunting task – going beyond the discussion of Section 1.2, where parameters were varied one at a time, and analysing the 'anthropic' domain in a two-parameter diagram. For instance, we can depict different values of the dark energy (assumed to be due to a cosmological constant Λ) and the amplitude of the fluctuations Q (Figure 1.3; Ref. [25]). Structures can only form so long as gravity overwhelms the repulsive effect of Λ. A higher value of Q implies earlier formation of structure, and therefore, higher values of Λ would still be anthropically allowed in such a universe.

Another two-parameter example (see [38]) takes Q and the density of dark matter as two parameters that could vary. If the dark matter density were higher than in our actual Universe, the cosmic expansion would become matter dominated at an earlier stage, allowing more time for the growth of structures from initial fluctuations. So the anthropically allowed values of Q would then extend downward. (This contrasts with the effect of higher values of Λ, which extend the allowable Q upwards).

We are currently far from having any theory that determines the values of Λ or Q or the dark matter density (and we know even less about the relative likelihood of various combinations of these constants or how they might be correlated). We are even further from having a cosmological model that can put a 'measure' on the probability density of various combinations. But if we did, we would then have another way of testing – and, in principle, refuting – whether the 'fine-tuning' was due to anthropic selection. We could do this by examining whether we existed in a 'typical' part of the anthropically allowed multiverse or whether the tuning was even more 'special' than anthropic constraints required. This line of reasoning can be illustrated by a simple analogy:

Even if we knew nothing about how stars and planets formed, we would not be surprised to find that our Earth's orbit was fairly close to circular: had it been highly eccentric, water would boil when the Earth was at perihelion and freeze at aphelion – a harsh environment unconducive to our emergence. However, a modest orbital eccentricity, up to 0.1 or so, is plainly not incompatible with life. But suppose it had turned out that the Earth moved in a near-perfect circle with an eccentricity of 0.000001. Some quite different explanation would then be needed: anthropic selection from orbits whose eccentricities had a 'Bayesian prior' that was uniform in the range 0–1 could plausibly account for an eccentricity of 0.1, but not for one as tiny as this.

The methodology requires us to decide what range of values is compatible with our emergence. It also requires a specific theory that gives the relative Bayesian priors for any particular value within that range. With this information, one can then ask if our actual

Universe is 'typical' of the subset in which we could have emerged. If it is a grossly atypical member even of this subset (not merely of the entire multiverse), then we would need to abandon our hypothesis that the numbers were anthropically selected.

Most physicists would consider the 'natural' value of 'dark energy' in the 'landscape' to be large, because it is a consequence of a very complicated microstructure of space. Current evidence suggests that the 'dark energy' has an actual value 5–10 times below the anthropically allowed maximum (other parameters being constrained to their actual values). That would put our Universe between the 10th or 20th percentile of universes in which galaxies could form. In other words, our Universe is not significantly more special, with respect to Λ, than our emergence demanded. But suppose that (contrary to current indications), observations showed that Λ made no discernible contribution to the expansion rate and was thousands of times below the threshold, not just 5–10 times. This 'overkill precision' would (like the precisely circular orbit in the analogy given earlier) raise doubts about the hypothesis that Λ was equally likely to have any value and suggest that it was zero for some fundamental reason (or that it had a discrete set of possible values, and all the others were well above the threshold).

In principle, we could, when theoretical models were more advanced, analyse other important parameters of physics in the same way, to test whether our Universe is typical of the habitable subset that could harbour complex life. The methodology requires us to decide what values are compatible with our emergence. It also requires a specific theory that gives the probability of any particular value. For instance, in the case of Λ, is there a set of discrete vacua or a continuum of values? In the latter case, we need to know whether all values are equally probable or whether the probability density is clustered at a low value.

1.4 Conceptual Shifts

The introduction of the multiverse and of anthropic reasoning has generated considerable controversy, sometimes even accompanied by passionately negative reactions from a number of physicists. We have already discussed the first main objection – the sentiment that envisaging causally disconnected, unobservable universes is in conflict with the traditional 'scientific method'. We have emphasised that modern physics already contains many unobservable domains (e.g., free quarks, interiors of black holes, galaxies beyond the cosmological event horizon). If we had a theory that applied to the ultra-early Universe but gained credibility because it explained, for instance, some features of the microphysical world (the strength of the fundamental forces, the masses of neutrinos, and so forth) we should take seriously its predictions about 'our' Big Bang and the possibility of others.

We are far from having such a theory, but the huge advances already made should give us optimism about new insights in the next few decades. Indeed, even at this early stage, eternal inflation and the landscape scenario already make some predictions that are, in principle, testable. For example, in eternal inflation, our Universe is expected to have a very small (10^{-4}) negative curvature (a 'bubble'). Therefore, future measurements of spatial curvature (including measurements of the 21 cm transition) could falsify eternal inflation (e.g. [21]). Similarly, accelerator experiments can (in principle) generate conditions in which a number of metastable vacuum solutions are possible, thereby testing the premises of the landscape scenario. It is also possible (although the probability for such an event is very low), for another inflating bubble to pop close and collide with our bubble Universe, leaving an imprint in our CMB (e.g., [40]). These simple examples demonstrate that even though the multiverse idea is still in its infancy, this scenario constitutes a bona fide topic (albeit speculative) of scientific discourse, rather than metaphysics.

We have also pointed out that an anthropic explanation can be refuted, if the actual parameter values are far more 'special' than anthropic constraints require.

Many physicists still hope that a unique, self-consistent theory of the Universe will unambiguously determine the values of all the physical parameters. This is a lofty goal, but the history of science has already demonstrated that a quest for first-principle explanations for everything can fail. The great astronomer Johannes Kepler tried to find answers to two questions: (1) Why were there precisely six planets (only six were known at his time) and (2) What was it that determined the spacings among planetary orbits? Eventually, he thought he found the answer, and he published it in his book *Mysterium Cosmographicum* (originally published in 1597; [20]). Kepler's answer was impressive and at the same time absolutely wrong. He constructed a model of the solar system in which the five Platonic solids were embedded one inside the other and together in a surrounding sphere. This created exactly six spaces (like the number of planets), and by choosing the order of the solids in a particular way, the spacing agreed with the relative sizes of the orbits to within 10%. The model was impressive because Kepler used mathematics to explain observed phenomena. It was completely wrong because Kepler did not understand at the time that neither the number of the planets nor their orbits were fundamental phenomena that required first-principles explanations. Rather, we understand today that both the number of planets and their orbits are accidental variables whose values are determined by the environmental conditions in which the planetary system formed. Earth's orbit is special only insofar as it is in the habitable zone around the Sun.

The same may be true for at least some of the parameters of our Universe, such as the values of Λ and Q. These may be random variables in the multiverse, whose only 'explanations' are offered by anthropic arguments. In view of our current ignorance as to what is truly fundamental and what is not, we should keep an open mind to all options.

More specifically, some 'constants' may be truly universal and others not. As an analogy (which we owe to Paul Davies), consider the form of snowflakes. Their ubiquitous sixfold symmetry is a direct consequence of the shape of water molecules. But snowflakes display an immense variety of patterns because each is moulded by its micro-environments: how each flake grows is sensitive to the fortuitous temperature and humidity changes during its growth as it falls towards the ground. If physicists achieved a fundamental theory, it would tell us which aspects of nature were direct consequences of the bedrock theory (just as the symmetrical template of snowflakes is due to the basic structure of a water molecule) and which (like the distinctive pattern of a particular snowflake) were contingencies, taking many values across the multiverse.

If we indeed live in a multiverse, this would be a fifth (and, in some sense, the grandest) Copernican revolution. First, Copernicus showed that we are not at the centre of the solar system; Harlow Shapley showed that the solar system is not at the centre of our galaxy; the Kepler Space Observatory showed that there are billions of planetary systems in the Milky Way; Edwin Hubble and his namesake telescope have shown that there are trillions of galaxies in the observable Universe; and now we realise that our observable domain may be only a tiny part of an unimaginably large and diverse ensemble. The next few decades will hopefully shed some light on the reality of the multiverse.

One thing, however, is clear. Our cosmic horizons have expanded precisely as fast as human knowledge. Every one of the five Copernican revolutions marked an incredible human achievement. In that sense, we remain of central significance to our Universe.

References

[1] Bahé, Y. M., *et al.* 2017. 'The Hydrangea Simulations: Galaxy Formation in and around Massive Clusters'. *Monthly Notices of the Royal Astronomical Society*, **470**(Oct.), 4186–4208.

[2] Barrow, J. D., and Tipler, F. J. 1986. *The Anthropic Cosmological Principle*. Oxford University Press.

[3] Bostrom, N. 2002. *Anthropic Bias: Observation Selection Effects in Science and Philosophy*. Routledge.

[4] Bousso, R., and Polchinski, J. 2000. 'Quantization of Four-Form Fluxes and Dynamical Neutralization of the Cosmological Constant'. *Journal of High Energy Physics*, **6**(June), 006.

[5] Carr, B. J., and Rees, M. J. 1979. 'The Anthropic Principle and the Structure of the Physical World'. *Nature*, **278**(Apr.), 605–612.

[6] Carter, B. 1974. 'Large Number Coincidences and the Anthropic Principle in Cosmology'. Pages 291–298 of Longair, M. S. (ed), *Confrontation of Cosmological Theories with Observational' Data*. IAU Symposium, vol. 63.

[7] Conselice, C. J., *et al.* 2016. 'The Evolution of Galaxy Number Density at $z < 8$ and Its Implications'. *Astrophysical Journal*, **830**(Oct.), 83.

[8] Cyburt, R. H., *et al.* 2016. 'Big Bang Nucleosynthesis: Present Status'. *Reviews of Modern Physics*, **88**(1), 015004.

[9] Dressing, C. D., and Charbonneau, D. 2015. 'The Occurrence of Potentially Habitable Planets Orbiting M Dwarfs Estimated from the Full Kepler Dataset and an Empirical Measurement of the Detection Sensitivity'. *Astrophysical Journal*, **807**(July), 45.

[10] Epelbaum, E., *et al.* 2013. 'Viability of Carbon-Based Life as a Function of the Light Quark Mass'. *Physical Review Letters*, **110**(11), 112502.

[11] Freese, K. 2017. 'Status of Dark Matter in the Universe'. *International Journal of Modern Physics D*, **26**(Jan.), 1730012.

[12] Garriga, J., Livio, M., and Vilenkin, A. 2000. 'Cosmological Constant and the Time of Its Dominance'. *Physical Review D*, **61**(2), 023503.

[13] Hinshaw, G., *et al.* 2013. 'Nine-Year Wilkinson Microwave Anisotropy Probe (WMAP) Observations: Cosmological Parameter Results'. *Astrophysical Journal Supplements*, **208**(Oct.), 19.

[14] Hoyle, F. 1954. 'On Nuclear Reactions Occuring in Very Hot STARS. I. the Synthesis of Elements from Carbon to Nickel'. *Astrophysical Journal Supplements*, **1**(Sept.), 121.

[15] Hoyle, F. 2010. *The Black Cloud*. Penguin Modern Classics. Penguin Books Limited.

[16] Hoyle, F., *et al.* 1953. 'A State in C^{12} Predicted from Astrophysical Evidence'. *Physical Review*, **92**, 1095.

[17] Iben, Jr., I. 2013a. *Stellar Evolution Physics, Volume 1: Physical Processes in Stellar Interiors*. Cambridge University Press.

[18] Iben, Jr., I. 2013b. *Stellar Evolution Physics, Volume 2: Advanced Evolution of Single Stars*. Cambridge University Press.

[19] Kachru, S., *et al.* 2003. 'De Sitter Vacua in String Theory'. *Physical Review D*, **68**(4), 046005.

[20] Kepler, J., and Aiton, E. J. 1981. *Mysterium Cosmographicum*. Abaris Books.

[21] Kleban, M., and Schillo, M. 2012. 'Spatial Curvature Falsifies Eternal Inflation'. *Journal of Cosmology and Astroparticle Physics*, **6**(June), 029.

[22] Kobayashi, M., and Maskawa, T. 1973. 'CP-Violation in the Renormalizable Theory of Weak Interaction'. *Progress of Theoretical Physics*, **49**(Feb.), 652–657.

[23] Linde, A. 2017. 'A Brief History of the Multiverse'. *Reports on Progress in Physics*, **80**(2), 022001.

[24] Linde, A. D. 1986. 'Eternally Existing Self-Reproducing Chaotic Inflanationary Universe'. *Physics Letters B*, **175**(Aug.), 395–400.

[25] Livio, M., and Rees, M. J. 2005. 'Anthropic Reasoning'. *Science*, **309**(Aug.), 1022–1023.

[26] Livio, M., and Silk, J. 2017. 'Where Are They?' *Physics Today*, **70**(3), 50–57.

[27] Livio, M., *et al.* 1989. 'The Anthropic Significance of the Existence of an Excited State of C-12'. *Nature*, **340**(July), 281–284.

[28] Mortonson, M. J., Weinberg, D. H., and White, M. 2013. 'Dark Energy: A Short Review'. *arXiv e-prints*, Dec., arXiv:1401.0046.

[29] Perlmutter, S., *et al.* 1999. 'Measurements of Ω and Λ from 42 High-Redshift Supernovae'. *Astrophysical Journal*, **517**(June), 565–586.

[30] Planck Collaboration *et al.* 2016. 'Planck 2015 Results. XIII. Cosmological Parameters'. *Astronomy and Astrophysics*, **594**(Sept.), A13.

[31] Riess, A. G., *et al.* 1998. 'Observational Evidence from Supernovae for an Accelerating Universe and a Cosmological Constant'. *Astronomical Journal*, **116**(Sept.), 1009–1038.

[32] Sakharov, A. D. 1967. 'Violation of CP Invariance, C Asymmetry, and Baryon Asymmetry of the Universe'. *Soviet Journal of Experimental and Theoretical Physics Letters*, **5**(Jan.), 24.

[33] Schlattl, H., *et al.* 2004. 'Sensitivity of the C and O Production on the 3α Rate'. *Astrophysics and Space Science*, **291**(Apr.), 27–56.

[34] Smolin, L. and Susskind, L. Smolin vs. Susskind Debate. Edge. www.edge.org/conversation/lee_smolin-leonard_susskind-smolin-vs-susskind-the-anthropic-principle.

[35] Snellen, I. A. G., *et al.* 2013. 'Finding Extraterrestrial Life Using Ground-Based High-Dispersion Spectroscopy'. *Astrophysical Journal*, **764**(Feb.), 182.

[36] Stark, C. C., *et al.* 2015. 'Lower Limits on Aperture Size for an ExoEarth Detecting Coronagraphic Mission'. *Astrophysical Journal*, **808**(Aug.), 149.

[37] Tegmark, M. and Rees, M. J., 1998. 'Why Is the Cosmic Microwave Background Fluctuation Level 10^{-5}?' *Astrophysical Journal*, **499**(May), 526–532.

[38] Tegmark, M., *et al.* 2006. 'Dimensionless constants, Cosmology, and Other Dark Matters'. *Physical Review D*, **73**(2), 023505.

[39] Vilenkin, A. 1983. 'Birth of Inflationary Universes'. *Physical Review D*, **27**(June), 2848–2855.

[40] Wainwright, C. L., *et al.* 2014. 'Simulating the Universe(s) II: Phenomenology of Cosmic Bubble Collisions in Full General Relativity'. *Journal of Cosmology and Astroparticle Physics*, **10**(Oct.), 024.

[41] Weinberg, S. 1987. 'Anthropic Bound on the Cosmological Constant'. *Physical Review Letters*, **59**(Nov.), 2607–2610.

2

Hierarchy of Fine-Structure Constants

Abstract

Modern physics describes the vast range of scales of structure in the Universe, with a hierarchy of forces providing connections between the microscopic and macroscopic domains. It also predicts various natural relationships between these scales which might otherwise be regarded as coincidental. However, there are numerous other relationships or 'fine-tunings' between the constants of physics – including the dimensionless coupling constants – which are unexplained by conventional physics and seem necessary for the emergence of observers. This chapter distinguishes between (1) natural tunings which arise between various scales of structure as a result of standard physics; (2) weak anthropic tunings, such as Dicke's argument for the age of the Universe, which regard the constants as given but require that we observe at a special time and place; and (3) strong anthropic tunings, which postulate relationships between the physical constants themselves. The last arise from both astrophysical and cosmological considerations and seem to be necessary for the development of complexity in the Universe. We consider possible interpretations of these fine-tunings, including the multiverse proposal, and some associated philosophical issues.

2.1 Preface

Nearly 40 years ago, I wrote an article in the journal *Nature* with Martin Rees [22], bringing together all of the known constraints on the physical characteristics of the Universe – including the fine-tunings of the physical constants – which seemed to be necessary for the emergence of observers. Such constraints had been dubbed 'anthropic' by Brandon Carter [24] – after the Greek word for 'human' – although it is now appreciated that this is a misnomer, since there is no reason to associate the fine-tunings with humankind in particular. We considered both the 'Weak' Anthropic Principle (WAP) – which accepts the laws of nature and physical constants as given and claims that the existence of observers then imposes a selection effect on where and when we observe the Universe – and the 'Strong' Anthropic Principle (SAP) – which (in the sense we used the term) suggests that the existence of observers imposes constraints on the physical constants themselves.

Anthropic claims – at least in their strong form – were regarded with a certain amount of disdain by physicists at the time and in some quarters still are. Although we took the view that any sort of explanation for the observed fine-tunings was better than none, many regarded anthropic arguments as going beyond legitimate science. The fact that some people of a theological disposition interpreted them as evidence for a Creator – attributing teleological significance to the Strong Anthropic Principle – doubtless enhanced that reaction. However, attitudes have changed considerably since then. This is not so much because the status of the anthropic arguments themselves have changed – as we will see, some of them have become firmer and others weaker. Rather, it is because there has been a fundamental shift in the epistemological status of the Anthropic Principle. This arises because cosmologists have come to realise that there are many contexts in which our Universe could be just one of a large ensemble of 'parallel' universes in which the physical constants vary. This ensemble is sometimes described as a 'multiverse'.

The multiverse proposals have not generally been motivated by an attempt to explain the anthropic fine-tunings; most of them have arisen independently out of developments in cosmology and particle physics. Nevertheless, it now seems clear that the two concepts are interlinked. For if there *are* many universes, this begs the question of why we inhabit this particular one, and – at the very least – one would have to concede that our own existence is a relevant selection effect. Indeed, since we necessarily reside in one of the life-conducive universes, the multiverse picture reduces the Strong Anthropic Principle to an aspect of the weak one. For this reason, many physicists would regard the multiverse as providing the most natural explanation of the anthropic fine-tunings

Many of the arguments were summarised in the book *Universe or Multiverse?* [21], which I edited in 2007. This grew out of a series of conferences which were sponsored by the Templeton Foundation. The first was entitled 'Anthropic Arguments in Fundamental Physics and Cosmology' and held at the home of Martin Rees, in Cambridge in 2001. It was funded by a grant awarded to myself, Robert Crittenden, Martin Rees, and Neil Turok for a project entitled 'Fundamental Physics and the Problem of Our Existence' as part of the Templeton 'Cosmology & Fine-Tuning' programme. The second meeting – with the same title as the book – was held at Stanford University in 2003 and came at a critical point in the development of the subject. It included contributions from some of the key players in the field and these provided the main part of the book. The third meeting was in 2005 and entitled 'Expectations of a Final Theory'. It was again held at Cambridge but this time hosted by Trinity College. Most of the focus was on the exciting developments in particle physics – in particular, M-theory and the string landscape scenario, which perhaps provide a plausible theoretical basis for the multiverse paradigm.

It should be stressed that *Universe or Multiverse?* was not a proselytising work and this is signified by the question mark in the title. The proponents predominated numerically, but many of the contributors were sceptical. Perhaps the most remarkable aspect of this book is that it testified to the large number of eminent physicists at the time who found the subject interesting enough to write about. It is unlikely that such a volume could have been produced even a decade earlier. The shift in attitude in the five years between the first and the third meetings is reflected in a quote from Frank Wilczek's contribution to the book [92]:

The previous gathering had a defensive air. It prominently featured a number of physicists who subsisted on the fringes, voices in the wilderness who had for many years promoted strange arguments about conspiracies among fundamental constants and alternative universes. Their concerns and approaches seemed totally alien to the consensus vanguard of theoretical physics, which was busy successfully constructing a unique and mathematically perfect Universe. Now the vanguard has marched off to join the prophets in the wilderness.

The focus of the present volume is fine-tuning, so the main purpose of this chapter is to update the discussion of my 1979 paper with Martin Rees. There will be rather little discussion of the multiverse proposal, even though this has played such an important role in establishing the respectability of anthropic arguments, since this is covered in other chapters. The first sections will discuss the Cosmic Uroborus (to put the topic in historical context), the physical constants (including the hierarchy of fine-structure constants of the title), and the different types of tunings. The middle sections will describe these tunings in more detail – the natural tunings connecting the various scales of structure in the Universe, the weak anthropic tunings and strong anthropic tunings arising in both astrophysics, and cosmology. The final sections will discuss possible interpretations of Anthropic Principle and some philosophical issues (including whether anthropic arguments qualify as legitimate science). Despite the negative response, it may induce in some quarters, the A word ('anthropic') will be used throughout this chapter.

2.2 Cosmic Uroborus

The history of physics might be regarded as a process in which the development of new instruments, like the telescope and the microscope, has allowed us to extend observations outwards to progressively larger scales and inwards to successively smaller ones. The outward journey explores the macroscopic domain and is associated with astronomy, while the inward journey explores the microscopic domain and is associated with particle physics. This process has revealed ever larger and smaller levels of structure in the Universe. Of course, a lot of interesting physics – including the whole domain of biophysics – is associated with complex structures in the intermediate mesoscopic domain, and that will be particularly relevant to this chapter.

The journey has also led to the discovery of the forces which determine the nature of these structures – gravity and electromagnetism on large scales, the weak and strong forces on short scales – and the associated laws of nature. These forces link the macroscopic and microscopic domains so that the outward and inward journeys are not disconnected but constantly throw light on each other. Both journeys have also led to new conceptual ideas and changes in our world view. The outer one has led to the shifts from geocentric to heliocentric to galactocentric to cosmocentric world views and to the radical change of view of space and time entailed in relativity theory. The inner one has led to atomic theory, quantum theory, and a progressively unified view of the forces of nature and the fundamental constituents of matter. Any final paradigm of physics must amalgamate relativity and quantum theory in some way.

So physics has revealed a unity about the Universe which makes it clear that everything is connected in a way which would have seemed inconceivable a few decades ago. This unity is succinctly encapsulated in the image of the Cosmic Uroborus (the snake eating its own tail) shown in Figure 2.1. The pictures drawn around the snake represent the different

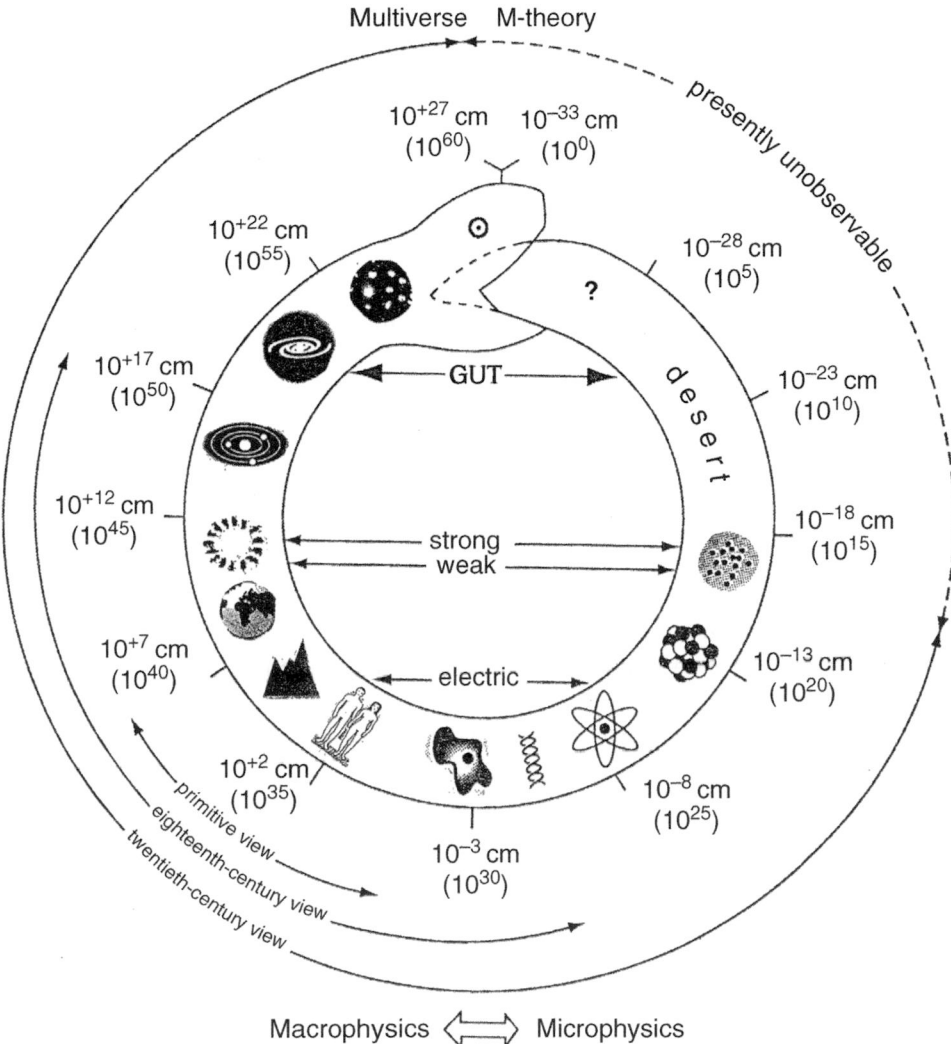

Figure 2.1 The image of the Cosmic Uroborus summarises the different levels of structure in the physical world, the intimate link between the microphysical and macroscopic domains, and the evolution of our understanding of this structure. From Reference [21].

types of structure in the Universe. Near the bottom are humans. As we move to the left, we encounter successively larger objects: a mountain, a planet, a star, a solar system, a galaxy, a cluster of galaxies, and, finally, the entire observable Universe. As we move to the right, we encounter successively smaller objects: a cell, a DNA molecule, an atom, a nucleus, a quark, the GUT scale, and, finally, the Planck length (the scale at which quantum gravity effects become important). The numbers at the edge indicate the scale of these structures in centimetres and also in units of the Planck length (in brackets). As we move clockwise from the tail to the head, the scale increases through 60 decades: from the smallest meaningful scale allowed by quantum gravity (10^{-33}cm) to the scale of the observable Universe (10^{27} cm). So one can regard the Uroborus as a clock in which each minute corresponds to a decade in scale.

The horizontal lines in Figure 2.1 correspond to the various interactions and illustrate the subtle connections between microphysics and macrophysics. For example, the electric line connects an atom to a planet because the electric force binds the electron to the nucleus in an atom and also determines the structure of solid objects. The strong and weak lines connect a nucleus to a star because the strong force which holds nuclei together also provides the energy released in the nuclear reactions which power a star, and the weak force which causes nuclei to decay also prevents stars from burning out too soon. The line associated with the grand unified theory (GUT) connects with large-scale structure because the density fluctuations which led to this originated when the temperature of the Universe was high enough for GUT interactions to be important.

The Big Bang might be regarded as the ultimate micro-macro link since it implies that the entire observable Universe was once compressed to a tiny volume. This is why the head of the snake meets the tail. Since light travels at a finite speed, we can never see further than the distance light has travelled since the Big Bang; this is about 40 billion light years, three times the age of the Universe times the speed of light because the cosmic expansion helps light travel further. More powerful telescopes probe to earlier times rather than larger distances. This is why early universe studies have led to an exciting collaboration between particle physicists and cosmologists.

As discussed by Kolb (Chapter 9), cosmologists now have a fairly complete picture of the history of the Universe. As one goes back in time, galaxy formation occurred at a billion years after the Big Bang, the background radiation last interacted with matter at 400,000 years, the Universe's energy was dominated by its radiation content before about 60,000 years, light elements were generated through cosmological nucleosynthesis at around 3 minutes, antimatter annihilated with matter at about 10^{-5} s (before which there was just a tiny fractional excess of matter), electroweak unification occurred at 10^{-9} s, the highest energy which can be probed experimentally was reached at 10^{-12} s, grand unification and inflation occurred around 10^{-35} s, and the quantum gravity era (the smallest meaningful time) was at 10^{-43} s.

The last few decades have seen two important developments on the outer front, and these are described in other chapters. First, the detection of temperature anisotropies in the cosmic background radiation and ever more precise measurements of their dependence on angular scale have confirmed the quantum origin of the density fluctuations. Second, although one expects the expansion of the Universe to slow down because of gravity, observations of distant supernovae suggest that it is accelerating. We do not know for sure what is causing this, but it must be some exotic form of dark energy, most probably related to the cosmological constant. These discoveries have led to the concordance ΛCDM model and the popularity of the inflationary scenario. Another idea that has become topical is that our entire Universe may be just one member of huge ensemble of universes called the multiverse, a notion which is particularly relevant to the theme of this book.

On the inner front, we have learnt that it may be possible to incorporate gravity into the unification of forces, leading some physicists to proclaim that we are on the verge of obtaining a Theory of Everything (TOE). However, in order to describe all the subatomic interactions, this requires extra wrapped-up dimensions of the kind proposed by Kaluza and Klein to explain electromagnetism. For example, superstring theory suggests there could be six additional spatial dimensions, and the way they are compactified is described by the Calabi-Yau group. There were originally five different superstring theories, but it was later realised that these are all parts of a single, more embracing model called M-theory, which has seven extra dimensions [93]. In one particular variant of M-theory, proposed by Randall and Sundrum [63], the 11th dimension is extended so that the physical world is viewed as a four-dimensional brane in a higher-dimensional bulk. We do not experience these extra dimensions directly – their effects only become important on the smallest and largest scales (i.e., at the top of the Uroborus).

Taken together, scientific progress on both the outer and inner fronts – culminating in the Big Bang model – can be regarded as a triumph. Indeed, as indicated by the arcs in Figure 2.1, the history of science might be regarded as an expansion of our awareness to ever larger and smaller scales. However, this achievement has come at a price. The anthropocentric view which prevailed at the start of the journey has been demolished, and the more we probe the Universe, the more irrelevant humans seem to become. According to the Newtonian paradigm, the cosmos operates likes a giant machine, oblivious to whether life or any form of consciousness is present, so the laws of physics and the characteristics of the Universe are independent of whether anybody actually observes them. Modern developments have reinforced this notion. We are completely insignificant not only as judged by scale but also in terms of duration. If the history of the Universe were compressed into a year, Homo sapiens would have existed for only the last few minutes.

Curiously, in recent decades, cosmology has brought about a reversal in this trend. This is because it seems that, in some respects, the Universe has to be the way that it is because, otherwise, it could not produce life, and we would not be here speculating about it. Indeed, since my paper with Rees, this idea has been explored in numerous works (e.g., [11, 15, 45, 54, 64]), and further references will be given in the chapter. Although this notion is referred to as the Anthropic Principle, it is not inevitable that it relates to the presence of observers, and it might be better described as the Complexity Principle. However, my personal hunch is that it does relate to observers and that the ultimate explanation will relate

On scales smaller than atoms, two more fundamental interactions come into play: the strong and the weak force. Although these are many orders of magnitude stronger than the gravitational force, they are both short range, becoming negligible beyond distances of 10^{-13} cm and 10^{-15} cm, respectively. For this reason, they do not play an important role in determining the structure of objects larger than atoms. The associated coupling constants are also energy dependent, so Reference [22] only indicates their low-energy values. The weak force has a (low-energy) dimensionless coupling constant

$$\alpha_W \equiv (g m_e^2 c/\hbar^3) \sim 10^{-11}, \tag{2.5}$$

where $g \sim 10^{-49}$ erg cm^3 is the Fermi constant. Thus, its interaction strength is intermediate between that of gravity and electricity. Reference [22] describes the strong force by the (scalar) coupling constant

$$f^2 \approx 15. \tag{2.6}$$

However, it must be stressed that the strong coupling constant is strongly energy dependent:

$$\alpha_S(E) = \frac{12\pi}{(33 - 2n_f) \ln(E^2/\Lambda_{\text{QCD}}^2)}, \tag{2.7}$$

where n_f is the number of quarks and Λ_{QCD} is the QCD scale. This goes to zero for $E \gg \Lambda_{\text{QCD}}$, corresponding to asymptotic freedom. It diverges as $E \to \Lambda_{\text{QCD}}$, but QCD theory breaks down there, and lattice gauge theory gives $\alpha_S \approx 0.1$ at the appropriate energy.

2.4 Types of Fine-Tuning

The AP claims that there are various tunings between the physical constants – including the dimensionless coupling constants – which are necessary for the emergence of observers. Indeed, there could be enough of these to determine *all* the physical constants, at least to some precision. Current physics does not explain their values, but even if it did, it would be remarkable that the values predicted turned out to be the ones required anthropically. However, the term 'tuning' is used in different senses, so we first clarify the distinction between them.

- *Natural tunings.* Standard physics implies that the mass- and length-scales of many natural objects depend on α and α_G. This implies many apparent coincidences between these scales, which might be surprising if one did not understand the underlying physics. For example, the size of cell is 10^{-3} cm, which is roughly the geometric mean of the R_P and R_U.
- *Non-anthropic tunings.* Some tunings in particle physics have no obvious anthropic aspects. For example, the Higgs mass (250 GeV) is roughly the geometric mean of the dark-energy mass and the Planck mass, but this does not seem to have anthropic significance. Wilczek's classification of fundamental parameters distinguishes between four types of tunings; he terms these enlightenment, knowledge, ignorance, and temptation, with only the last being anthropic [91].

- *Weak anthropic tunings.* Given the constants of physics, there is an inevitable selection effect on when and where observers can exist, and this may involve the coupling constants. For example, an argument of Robert Dicke [40] suggests that observers can only exist when the age of the Universe is roughly the lifetime of a main-sequence star, $t_{MS} \sim \alpha_G^{-1} t_p \sim 10^{10}$ yr. The WAP is a logical necessity, although the nature of the tunings may still be a surprise.

- *Strong anthropic tunings.* There are also coincidences between the physical constants themselves which seem necessary for observers. However, most of these coincidences merely involve prerequisites for observers (such as planets, stars, galaxies, and chemistry), so the description 'anthropic' is misleading. They are better regarded as conditions for *complexity*. The SAP is much more controversial than the WAP because it refers to counterfactual universes in which the constants are different. However, if one accepts the existence of a multiverse, the SAP essentially becomes an example of the WAP.

2.5 Scales of Structure and Natural Tunings

Straightforward physics shows that, to an order of magnitude, α and α_G determine the mass and the size of nearly every macroscopic object in the Universe. This is illustrated in Figure 2.2, from which one can read off the scales associated with the Universe itself, galaxies, stars, planets, asteroids, exploding black holes, humans, atoms, protons, and the Planck length. The scales are given in grams and centimetres and also in terms of α, α_G, and the size and mass of the atom. A derivation of some of these results is given later.

Some of these dependencies are also summarised in Table 2.2. The proton and Planck scales trivially follow from Eq. (2.3) and (2.4), but the others are less obvious. For example, stars have a mass of roughly $\alpha_G^{-3/2} \sim 10^{60}$ and a radius of roughly $\alpha_G^{-1/2} \sim 10^{20}$ in atomic units (see also Chapter 10); the largest planets (like Jupiter) have a mass and radius smaller than this by factors of $\alpha^{3/2} \sim 10^{-3}$ and $\alpha^{1/2} \sim 10^{-1}$, respectively; the mass and size of a living creature – if we assume that it must live on a planet with a suitable temperature and a life-supporting atmosphere and does not shatter whenever it falls down – must be of order $(\alpha/\alpha_G)^{3/4} \sim 10^{27}$ and $(\alpha/\alpha_G)^{1/4} \sim 10^9$. Some of the scales also depend on the proton-to-electron mass ratio, but this can be written as

$$\beta \equiv m_e/m_p \approx 10\alpha^2, \tag{2.8}$$

due to a coincidence in nuclear physics (discussed later).

It should be stressed that the dependencies on α and α_G summarised in Figure 2.2 and Table 2.2 are all consequences of standard physics. No anthropic argument has been introduced, except in deriving the scale of the Universe (which we discuss later). At first sight, the dependencies might seem surprising, but the qualitative reason can be understood as follows: any stable structure in the Universe reflects a balance between either gravity or intermolecular forces, which are trying to hold it together, and various other effects – such as pressure or quantum interactions – which are trying to blow it apart. Since gravity depends on G, quantum effects on $\hbar$, and electrical effects on e, a balance between these

A gravitationally bound cloud represents a balance between gravity and thermal and electron degeneracy pressure. The virial theorem therefore implies that the temperature T is given by

$$kT + \frac{\hbar^2}{2m_e d^2} \sim \frac{GMm_p}{R} \sim \left(\frac{N}{N_o}\right)^{2/3} \frac{\hbar c}{d}, \tag{2.14}$$

where N is the number of protons, $N_o \equiv \alpha_G^{-3/2}$, and d is the mean particle separation. As the cloud collapses, T first increases as d^{-1}, being given by

$$kT \sim \frac{GMm_p}{R}, \tag{2.15}$$

but it then decreases due to the degeneracy term, reaching a maximum

$$kT_{\text{max}} \sim \left(\frac{N}{N_o}\right)^{4/3} m_e c^2. \tag{2.16}$$

The cloud forms a star if kT_{max} exceeds the nuclear ignition threshold, $\chi m_e c^2$ with $\chi \sim 10^{-2}$, so this implies $N > 0.1 N_o$. On the other hand, the star will be unstable if it is radiation-pressure dominated. Using Eq. (2.14), the ratio of the radiation and matter pressures is

$$\frac{P_{\text{rad}}}{P_{\text{mat}}} \sim \frac{aT^4}{NkT/R^3} \sim 0.01 \left(\frac{N}{N_o}\right)^2, \tag{2.17}$$

so the upper limit is around $10\,M_C$, although a more precise calculation gives a limit of $50 M_C$. More details can be found in Chapter 10.

During its main-sequence phase, T adjusts itself so that the nuclear energy generation rate balances the luminosity, which is the radiative energy content divided by the photon leakage time. This corresponds to

$$L \sim \frac{acT^4 R^2}{\kappa M}, \tag{2.18}$$

where κ is the opacity. After its nuclear-burning (main-sequence) phase, the star resumes its collapse, and so Eq. (2.14) predicts zero temperature (corresponding to an electron-degeneracy-supported white dwarf) when the particle separation reaches

$$d_{\text{min}} \sim \left(\frac{N}{N_o}\right)^{-2/3} r_e \quad \Rightarrow \quad R \propto N^{1/3} d_{\text{min}} \propto M^{-1/3}. \tag{2.19}$$

There is no stable white dwarf configuration if the electrons become relativistic (i.e., for $kT_{\text{max}} \sim m_e c^2$) since the degeneracy term in Eq. (2.14) then goes like d^{-1} rather than d^{-2}. Therefore, M_C also gives the upper limit on the mass of a white dwarf (the Chandrasekhar mass).

Note that the mass-scale at which a collapsing cloud stops fragmenting is

$$M_{\text{frag}} \sim q^{-1/2} \left(\frac{kT}{m_p c^2}\right)^{1/4} M_C \sim 10^{-2} q^{-1/2} M_C, \tag{2.20}$$

where T is the cloud's temperature and q is the ratio of its luminosity to that of a black body with this temperature. The important point is that M_{frag} has only a weak dependence on T and is at least an order of magnitude smaller than M_C.

Solid objects represent a balance between electron degeneracy energy and electrostatic binding energy. By comparing the degeneracy term in Eq. (2.14) with e^2/d, one obtains the atomic density line

$$\rho_o \sim m_p/a_o^3 \sim e^6 m_e^3 m_p \hbar^{-6} \sim 1\,\text{g cm}^{-3}, \tag{2.21}$$

which meets the white dwarf line at

$$M \sim \left(\frac{\alpha}{\alpha_G}\right)^{3/2} m_p \sim \alpha^{3/2} M_C \sim 10^{30}\text{g}, \quad R \sim \left(\frac{\alpha}{\alpha_G}\right)^{1/2} a_o \sim 10^{10}\ \text{cm}. \tag{2.22}$$

This gives the maximum size of a planet and also lies on the Rydberg ($kT \sim \alpha^2 m_e c^2$) isotherm. The minimum mass of a planet, the mass range of a habitable planets and the maximum scale of a living creature (prescribed by the requirement that it does not fall apart when it falls through its own height) can be derived with similar arguments but is not given here.

We next discuss the galaxy scale [67, 71]. Let us assume that galaxies originate from over-dense regions in the gaseous primordial material, and that they have a mass M and radius R_B when they stop expanding and become bound. After binding, protogalaxies will virialise at a radius $\sim R_B/2$. Thereafter, they will deflate on a cooling timescale, with the virial temperature (2.15). Providing kT exceeds the Rydberg energy, the dominant cooling mechanism is bremsstrahlung with the associated cooling time

$$t_{\text{cool}} \sim \frac{1}{n \alpha \sigma_T c} \left(\frac{kT}{m_e c^2}\right)^{1/2} \sim \frac{G^{1/2} m_e^{3/2} m_p R}{\alpha^3 n^{1/2}}, \tag{2.23}$$

using $\sigma_T \sim \alpha^2 r_e^2$ and $n \sim M/m_p R^3$. The free-fall timescale, on the other hand, is

$$t_{ff} \sim (GM/R^3)^{-1/2} \sim (Gnm_p)^{-1/2}, \tag{2.24}$$

and this exceeds t_{cool} when R falls below a mass-independent value

$$R_g \sim \alpha^4 \alpha_G^{-1} \left(\frac{m_p}{m_e}\right)^{1/2} a_o \sim \alpha^3 \alpha_G^{-1} a_o \sim 100\,\text{kpc}, \tag{2.25}$$

where we have used Eq. (2.8). Until a massive cloud gets within this radius, it will contract quasi-statically and cannot fragment into stars.

This argument applies only if $kT > E_o \sim \alpha^2 m_e c^2$ at the radius R_g, so one also needs the mass to exceed

$$M_g \sim \alpha_G^{-2} \alpha^5 (m_p/m_e)^{1/2} m_p \sim \alpha_G^{-2} \alpha^4 m_p \sim 10^{12} M_\odot. \tag{2.26}$$

Gas clouds with mass below M_g cool more efficiently, owing to recombination and can never be pressure supported. On the other hand, clouds with mass above M_g are inhibited is the Eddington luminosity. A characteristic timescale, first discussed by Salpeter [69], is that over which an object of luminosity L_E would radiate away its entire rest mass:

$$t_E = \frac{c\sigma_T}{4\pi Gmp} \sim 0.1 \left(\frac{\alpha^2}{\alpha_G}\right) \left(\frac{m_p}{m_e}\right)^2 t_p. \tag{2.29}$$

If $\eta \sim 10^{-2}$ is the fraction of a star's rest mass that can be released through nuclear burning, the main-sequence lifetime is, thus,

$$t_{MS} \sim \eta \left(\frac{p_{rad}}{p_{mat}}\right)^{-1} t_E \sim 10 \left[\eta\alpha^2 \left(\frac{m_p}{m_e}\right)^2\right] \left(\frac{M}{M_C}\right)^{-2} \alpha_G^{-1} t_p, \tag{2.30}$$

the quantity in square brackets being of order unity. For a radiation-dominated star ($M > 50M_\odot$), the M-dependence disappears, and t_{MS} levels off with the value of $10^{-2}\alpha_G^{-1} t_p \sim 2 \times 10^{13}$ s. Since the characteristic mass of a star is M_C, the Dicke argument requires

$$t_o > t_{MS} \sim 10\alpha_G^{-1} t_p \sim 2 \times 10^{16} \text{s}. \tag{2.31}$$

This is a factor of 20 shorter than the current age of the Universe $t_o \approx 4 \times 10^{17}$s. However, Eq. (2.30) is really only appropriate for an upper main-sequence star because the opacity is increased at lower masses. For a solar-mass star, t_{MS} exceeds $\alpha_G^{-1} t_p$ by a factor of around $m_p/m_e \sim 10^3$, so it is better approximated as $\alpha_G^{-1} t_e$.

Since t_{MS} is very sensitive to the value of M, Dicke's argument is not very precise. It would be more convincing if one could show that life requires stars of around a solar mass. For example, it might be difficult for life-bearing planets to evolve around much more massive stars, because they are too short-lived or because one needs some lower mass convective stars to have planets. Alternatively, even if the *first* stars were massive, perhaps some of the elements vital for life must be generated through the s-processes associated with later-forming, less massive stars. Whatever the appropriate value of M, t_o cannot be much bigger than observed, else most of the Universe would have been processed into stellar remnants. Therefore, observers are most likely to exist at an epoch $t_o \sim t_{MS}$.

2.7.2 Cosmological Consequences of Dicke's Argument

Dicke's argument for solar-mass stars implies the approximate relation

$$ct_o \sim \alpha_G^{-1}(\hbar/m_e c) \sim (\alpha/\alpha_G)a_o, \tag{2.32}$$

so the ratio of the size of the observable Universe to the size of an atom is comparable to the ratio of the electric and gravitational forces between protons. There is no other explanation for this well-known coincidence within conventional physics, but Dirac [32] suggested the unconventional explanation that α_G is *always* given by

$$\alpha_G \sim \hbar/(m_e c^2 t) \sim (t/t_e)^{-1}. \tag{2.33}$$

Assuming that $\hbar$, c, and m_e are constant in time, this requires that G decreases like t^{-1}, so Dirac invoked Eq. (2.33) as the basis for a new cosmology. However, such a variation of G is inconsistent with observation.

The relation (2.32) also implies that the number of protons in the observable Universe must be of order α_G^{-2}, thereby explaining another well-known coincidence. We now describe this argument in more detail. The total mass associated with the observable Universe (i.e., the mass within the horizon volume) is $\sim \rho_o c^3 t_o^3$, where ρ_o is the present matter density. This is given by the Friedmann equation:

$$\rho_o = \frac{3H_o^2}{8\pi G} + \frac{Kc^2}{16\pi G}, \qquad (2.34)$$

where K is the scalar curvature of the Universe. Providing the K term is smaller than the others, we deduce that the mass of the Universe is

$$M_U \sim c^3 t_o^3 G^{-1} H_o^2 \sim \alpha_G^{-2} \left(\frac{m_p}{m_e} \right) m_p \sim \alpha^{-2} \alpha_G^{-2} m_p, \qquad (2.35)$$

where we have used $t_o \sim H_o^{-1}$ and Eq. (2.8). The fact that the number of protons in the Universe is of order α_G^{-2} is thus explained, providing one can neglect the K term in Eq. (2.34). Some people have argued that K must be zero by appealing to Mach's principle, but – as discussed later – there may also be anthropic reasons for expecting that the K term is small.

Given the expression for the size of the Universe, the dependencies derived in Section 2.6 allow one to predict amusing relationships between the different mass-scales discussed earlier:

- Human $\sim$ (planet $\times$ atom)$^{1/2}$
- Planck $\sim$ (exploding hole $\times$ proton)$^{1/2}$
- Exploding hole $\sim$ (Universe $\times$ proton)$^{1/2}$.
- The number of stars in the galaxy $\sim \alpha^4 \alpha_G^{-1/2}$.
- The number of galaxies in the Universe $\sim \alpha^{-6}$.

These relationships are all consequences of standard physics, providing one accepts the WAP prediction for the size of the Universe. Of course, they are only order-of-magnitude relationships, since the objects involved (humans, stars, galaxies, etc.) all span a range of sizes and masses.

2.7.3 Does WAP Suffice?

Dicke's argument helps us to understand why the preceding large number coincidences prevail. It accepts the constants of Nature as given and then shows that our existence imposes a selection effect on when we observe the Universe. As such, it is no more than a logical necessity: saying that we have to exist at a certain time is no more surprising than saying we have to exist in a certain place (e.g., close to a star). It might be surprising to find

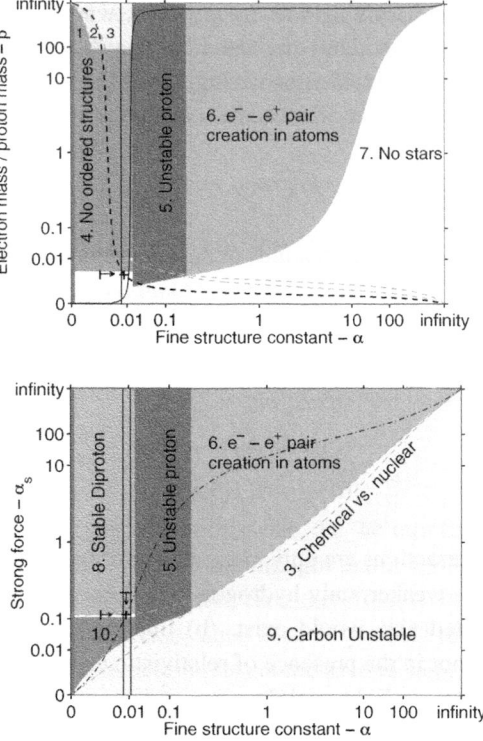

Figure 2.5 Fine-tuning of α, α_S, and m_e/m_p; from Reference [8].

Figure 2.6 Constraints on electron mass and the down-up quark mass difference from the stability of hydrogen and deuterium. The thin vertical line shows the constraint for a particular SO(10) grand unified scenario. From Reference [46].

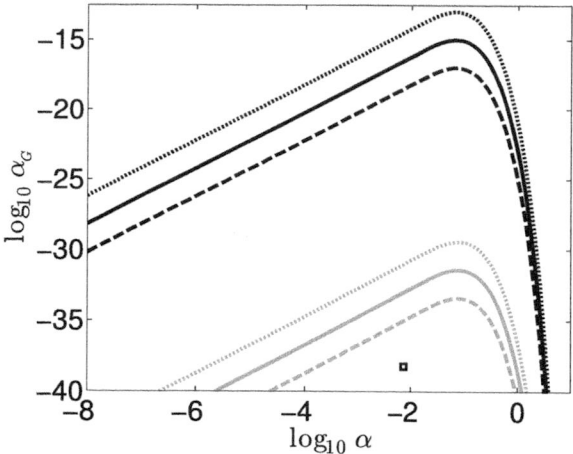

Figure 2.7 Limits on stable stars in (α_G, α) space for H-burning stars (black lines) and D-burning stars (grey lines) from Reference [9]. This shows that D-burning stars are stable in a much larger region of parameter space.

Since life requires stable nuclei other than protons and neutrons, these fundamental parameters of the Standard Model are good candidates for quantities whose values are determined anthropically.

Recent literature has focused on a fine-tuning associated with the diproton [33]. Although hydrogen burns slowly in the Sun (because it is moderated by weak inteactions), it would burn explosively if the Sun were made of heavy hydrogen (as in a nuclear bomb), and increasing f by 6% would bind the diproton [26, 62]. Barr and Khan [10] calculate the equivalent condition on the light quarks masses as

$$m_u + m_d < 0.75 \, (m_u + m_d)_{obs} = 5.3 \text{ eV}, \tag{2.48}$$

On the other hand, the deuteron would be strongly unbound for

$$m_u + m_d < 1.4 \, (m_u + m_d)_{obs} = 9.9 \text{ eV}, \tag{2.49}$$

in which case one would need much higher central temperatures in stars. Although Barrow and Tipler argue that a bound diproton would result in all the hydrogen being consumed at Big Bang nucleosynthesis, more detailed calculations show that this is not the case [57]. There could also be long-lived stars [14], although Barnes argues [9] that the strongest anthropic bound on stars in such a universe still comes from their lifetime, all stars burning out within 10^6 yr unless $\alpha_G < 10^{-30}$. Figure 2.7 is taken from his paper.

2.9 Strong Anthropic Principle: Cosmological Coincidences

The second set of fine-tunings is associated with Big Bang nucleosynthesis and the formation of galaxies and their subsequent fragmentation into stars. They involve various cosmological parameters, such as the current matter density in units of the critical density Ω_o, the amplitude of the density fluctuations Q on entering the cosmological horizon and the photon-to-baryon ratio S. Although most cosmologists might prefer to believe that these parameters were determined by processes in the early Universe rather than being prescribed freely as part of the initial conditions, even small deviations from the observed values would exclude the formation of cosmic structures.

2.9.1 Cosmological Nucleosynthesis

Let us first recall the process of cosmological nucleosynthesis, as discussed by Uzan in this volume (see Chapter 7). The prediction that one turns 24% of the mass of the Universe into helium is one of the great triumphs of the Big Bang picture. However, the only reason one gets an interesting amount of helium is because the neutron-to-proton ratio freezes out with an interesting value. The freeze-out occurs at 1 s when $T \sim 10^{10}$ K because the weak interactions become slower than the cosmological expansion rate then. However, it is only because the freeze-out temperature and neutron-proton mass difference are comparable that one gets the observed amount of helium production. It turns out that the condition for this is roughly $\alpha_G \sim \alpha_W^4$, precisely the condition required for supernovae.

Let us now describe this argument in more detail. The weak interactions proceed at a rate $n\sigma_W c$, where $n \propto T^3$ is the particle number density and $\sigma \propto T^2$ is given by Eq. (2.42). Therefore, the weak rate scales as T^5, whereas the cosmic expansion rate $\sim (GaT^4/c^2)^{1/2}$ scales as T^2. So freeze-out occurs at a temperature

$$kT_F \sim \alpha_G^{1/6} \alpha_W^{-2/3} c^2 m_e^{4/3} m_p^{-1/3}. \tag{2.50}$$

Since virtually all the frozen-out neutrons burn into helium, the resultant helium abundance is

$$Y \approx \frac{2n_N/n_P}{n_N/n_P + 1} \quad \text{where} \quad \frac{n_N}{n_P} \approx \exp\left(-\frac{\Delta c^2}{kT_F}\right). \tag{2.51}$$

Y is 24% rather than 0% or 100% only because $kT_F \approx \Delta c^2 \approx 2m_e c^2$, where we have used Eq. (2.46b). From Eq. (2.50) this derives from the 'coincidence'

$$\left(\frac{Gm_e^2}{\hbar c}\right)^{1/4} \sim \alpha_W. \tag{2.52}$$

The number on the left is the quarter power of the electron gravitational fine structure constant, so this is precisely equivalent to condition (2.44). In the Weinberg-Salam theory of weak and electromagnetic interactions, α_W to α are related by

$$\alpha_W \sim \alpha \left(\frac{m_e}{m_W}\right)^2, \tag{2.53}$$

so Eq. (2.52) may also be written in the form

$$\alpha_G \sim \left(\frac{m_e}{m_W}\right)^8 \left(\alpha^2 \frac{m_p}{m_e}\right)^2 \sim \left(\frac{m_e}{m_W}\right)^8, \tag{2.54}$$

where we have used Eq. (2.8).

It is not clear to what extent this coincidence can be interpreted anthropically. Life could probably not exist if Y were 100% (as would be the case if α_W were slightly smaller), since there would be no water. Also, the lifetime of a helium star is less than that of a hydrogen star and might not be long enough to permit evolution of life. However, it is not clear that a universe with no primordial helium would preclude life. On the other hand, the same constraint in the supernova context applies in both directions.

2.9.2 Density Parameter

Before the discovery of dark energy, there were well-known anthropic reasons for why the total matter density parameter Ω_o had to lie within an order of magnitude of 1 in order for the geometry of the universe to be nearly flat. If Ω_o were much larger than 1, the Universe would recollapse in a time $\Omega_o^{-1/2} H_o^{-1}$, and this would be less than the main-sequence time of a star. On the other hand, if Ω_o were much smaller than 1, density fluctuations would stop growing at the time $\Omega_o H_o^{-1}$, and this means that – for reasonable initial density fluctuations – galaxies would never bind at all. This argument requires that Ω_o be in the range 0.01–100. Similar arguments apply in a universe with dark energy, except that there are additional anthropic constraints on the value of the effective cosmological constant.

This is only a weak constraint on Ω_o, but in 1973, Collins and Hawking [27] gave a more precise argument, based on the observed isotropy of the microwave background. They showed that the set of spatially homogeneous cosmological models which approach isotropy at late times is of measure zero in the space of all spatially homogeneous models. So only a small set of initial conditions could give rise to a universe as isotropic as observed today, and these correspond to models which are spatially flat. Only such models would expand long enough for galaxies and intelligent life to form, suggesting that the observed isotropy might be a reflection of our own existence.

In the 1980s, early universe studies were revolutionised by the introduction of the inflationary scenario [39, 55, 85]. This *requires* that Ω_o be very close to 1 and – as discussed by Martin in this volume (Chapter 4) – observations seem to support this model. Therefore, anthropic considerations may no longer seem relevant. However, in the simplest model – with a single scalar field – inflation only works if the form of the vacuum potential $V(\phi)$ allows a sufficient number of expansion e-folds, which means that $V(\phi)$ must itself be fine-tuned [80]. Similar considerations apply in more complicated inflationary models, including quantum cosmological models [42], where universes are expected to collapse very quickly unless one imposes anthropic selection effects [44].

2.9.3 Photon-to-Baryon Ratio

As discussed by Steigman and Scherrer in this volume (Chapter 5), another important cosmological parameter is the entropy per baryon. Since the entropy is dominated by the microwave background, this is equivalent to the photon-to-baryon ratio S, which is of order $10^8 \Omega_B^{-1}$, where $\Omega_B \approx 0.05$ is the baryon density parameter. The value of S is associated with an interesting coincidence: if $S \sim 10^9$, the matter and radiation densities are comparable at the time they thermally decouple. We first examine this coincidence in more detail.

According to the hot Big Bang model, the background radiation dominated the density of the Universe until a redshift z_{eq}, which depends on S and the current CMB temperature T_o:

$$1 + z_{eq} \sim \frac{m_p c^2}{S f_B k T_o}, \quad T_o \sim \frac{\hbar c}{k} \left(\frac{S f_B}{G m_p t_o^2} \right)^{1/3}, \tag{2.55}$$

where $f_B \equiv \Omega_B / \Omega_M \approx 0.2$ and $\Omega_M \approx 0.25$ is the total matter density. In the matter-dominated era, $t \sim t_o (1 + z)^{-3/2}$, and so matter-radiation equality occurs at

$$t_{eq} \sim S^2 f_B^2 \alpha_G^{-1/2} t_p \sim 10^{12} \text{ s}. \tag{2.56}$$

On the other hand, the CMB thermally decoupled from the matter when T fell below $T_{dec} \sim 0.1 \alpha^2 m_e c^2 / k$, which corresponds to the somewhat later time

$$t_{dec} \sim t_o \left(\frac{\alpha^2 m_e}{k T_o} \right)^{-3/2} \sim S^{1/2} f_B^{1/2} \alpha_G^{-1/2} \alpha^{-6} t_p \sim 10^{13} \text{ s}. \tag{2.57}$$

(These equations use the fact that the radiation density and temperature fall like R^{-4} and R^{-1}, respectively.) The requirement that t_{eq} and t_{dec} be close to each other therefore corresponds to the coincidence

$$S \sim \alpha^{-2} f_B^{-1} \left(\frac{m_p}{m_e} \right) \sim \alpha^{-4}, \tag{2.58}$$

where we use Eq. (2.8) in the last step.

It is not clear that this coincidence has any anthropic significance, but there are certainly anthropic aspects to the values of t_{eq} and t_{dec}. For example, there is an upper limit on S if one requires that the Universe be radiation dominated at cosmological nucleosynthesis to avoid all the hydrogen going into helium. From Eq. (2.50), the weak freeze-out time is

$$t_F \sim \hbar c^2 \alpha_G^{-1/2} (k T_F)^{-2} m_p \sim \alpha_G^{-5/6} \alpha_W^{4/3} \left(\frac{m_p}{m_e} \right)^{8/3} t_p, \tag{2.59}$$

so the condition $t_{eq} > t_F$ corresponds to

$$S > f_B^{-1} \left(\frac{m_p}{m_e} \right)^{4/3} \left(\frac{\alpha_W^4}{\alpha_G} \right)^{1/6} \sim 10^4. \tag{2.60}$$

This does not pin down the value of S very precisely, but it does impose an interesting lower limit.

Other anthropic constraints on S are associated with galaxy formation (see also Chapter 6). In the standard Big Bang model, galaxies cannot form until the background radiation density falls below the matter density at t_{eq}, but this occurs before the Dicke timescale (2.30) (i.e., the main-sequence lifetime of a star) only if

$$S < \alpha^{-1}\alpha_G^{-1/4} f_B^{-1} \sim 10^{11}. \tag{2.61}$$

Since the Jeans mass in the period t_{eq} to t_{dec} has a value $M_J \sim \alpha_G^{-3/2} S^2 f_B^2 m_p$, this exceeds the galaxy mass given by Eq. (2.26) only for

$$S > \alpha_G^{-1/4}\alpha^{5/2}\left(\frac{m_p}{m_e}\right)^{1/4} \sim \alpha_G^{-1/4}\alpha^2 \sim 10^6. \tag{2.62}$$

Eqs. (2.61) and (2.62) constrain S rather tightly.

One can strengthen Dicke's weak anthropic argument for the age of the Universe by the general requirement – independent of considerations of stellar physics – that observers can exist only for $t_o > t_{dec}$, when thermodynamic disequilibrium is possible. This corresponds to the condition

$$S < 10^4 \alpha_G^{-1}\alpha^{12} f_B^{-1} \sim 10^{17} \tag{2.63}$$

but this only gives only a very weak upper limit on S. An interesting twist on this argument has been provided by Aguirre [7], who describes anthropic constraints on 'cold' cosmological models (i.e., models with an initial photon-to-baryon ratio much smaller than currently observed). He points out that such models could provide life-supporting conditions with very different values of the cosmological parameters.

In the context of limits (2.61) and (2.62), we note that there are a number of scenarios which predict $S \sim \alpha_G^{-1/4}$. This could apply, for example, if the radiation was generated by a first generation of large pregalactic objects [18]. Such objects have a characteristic lifetime t_{MS} given by Eq. (2.30), so after t_{MS}, one would expect the radiation density to be ξF times the matter density, where F is the fraction of the Universe which goes into the stars and ξ is the fraction of each star's rest mass released through nuclear energy generation. The value of t_{eq} associated with the generated S is thus $(\xi F)^{3/2} t_{MS}$, and Eqs. (2.56) and (2.30) imply

$$S \sim \left[\alpha\left(\frac{m_p}{m_e}\right) F^{3/4}\xi^{3/4}\right]\alpha_G^{-1/4}. \tag{2.64}$$

Since the term in square brackets is of order unity, one would expect $S \sim \alpha_G^{-1/4}$, as observed. The same relationship between S and α_G would apply if the radiation were generated by black holes accreting at the Eddington limit. Such scenarios are no longer mainstream, but they illustrate that the S coincidences may be explained naturally, without recourse to anthropic considerations.

Nowadays, most cosmologists believe the value of S results from of baryon-violating processes in the early Universe, most probably at the GUT epoch, around 10^{-34} s after the Big Bang. However, in most GUT models, S is predicted to be of order α^{-n}, where n is an integer, so the anthropic constraint $S < \alpha_G^{-1/4}$ merely translates into the constraint $\alpha_G < \alpha^{4n}$. If $n = 5$, this just gives the convective star condition [15].

2.9.4 Cosmological Constant

As discussed by Peacock in this volume (Chapter 3), another striking feature of the Universe is that its expansion appears to be accelerating under the influence of some form of 'dark energy'. The source of this energy is uncertain, but it may be associated with a cosmological constant, denoted by Λ in Table 2.1. One possibility is that Λ arises through quantum vacuum effects. We do not know how to calculate these, but the most natural value would be the Planck density (which is 120 orders of magnitude larger than observed value). For example, in the 'string lanscape' variant of M-theory there could be 10^{500} vacuum states [12], with the associated Λ covering the full range from minus to plus the Planck value [77, 78]. There is also the remarkable coincidence that the vacuum and matter densities are comparable at the present epoch, even though their ratio is time dependent.

As first emphasised by Weinberg [88, 89] and later studied by Efstathiou [35], Vilenkin [86], and Peacock [61], this may provide the strongest anthropic fine-tuning of all, since a priori Λ could be 120 orders of magnitude larger than observed. This is because the growth of density perturbations is quenched once Λ dominates the cosmological density, so if bound systems have not formed by then, they never will. This is not the only possible explanation for the smallness of Λ, but it may be the most plausible one. The crucial issue is whether the number of vacuum states is sufficiently large and their spacing sufficiently small to satisfy the anthropic constraints [51], but this is still unresolved. As discussed in the next section, the precise form of the Λ constraint depends on the amplitude of the primordial density fluctuations at the horizon epoch.

It should be stressed that a cosmological constant is not the only possible explanation for the cosmic acceleration. An alternative explanation is to invoke dark energy in the form of a scalar field – termed 'quintessence' [75] – and this may better explain the near equality of the vacuum and matter densities. However, some anthropic fine-tuning may be required even in this case. Kallosh [50] gives some examples from her studies of M/string theory where anthropic reasoning helps to shed light on the properties of dark energy. These issues are discussed in more detail by Peacock in Chapter 3.

2.9.5 Density Fluctuations

The precise form of the anthropic upper limit on Λ depends on the amplitude of the density fluctuations, Q, but this also has anthropic aspects. If Q is too low, no galaxies form because cooling is ineffective. If Q is too high, there is excesssive black hole formation and galaxies

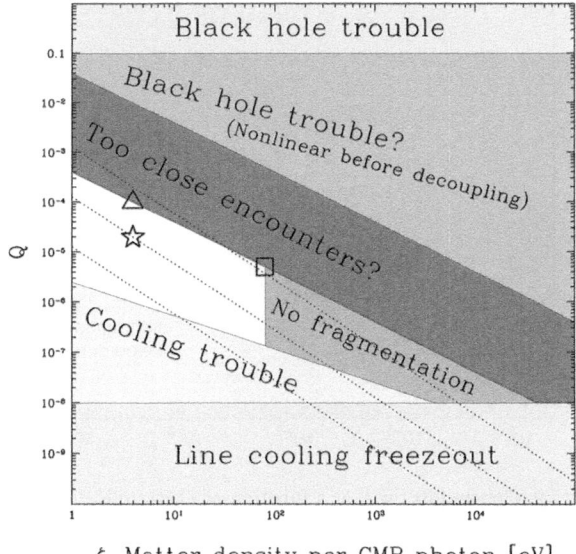

ξ, Matter density per CMB photon [eV]

Figure 2.8 Constraints on Q as a function of ξ; from Reference [82].

are too dense for long-lived planetary systems because of disruption by neighbouring stars. For $\Lambda = 0$, Tegmark and Rees showed that this leads to a constraint of the form [81]

$$\alpha^{-1}\ln(\alpha^{-2})^{-16/9}\alpha_G(\beta/\xi)^{4/3}\Omega_B^{-2/3} < Q < \alpha^{16/7}\alpha_G^{4/7}\beta^{12/7}\xi^{-8/7}, \qquad (2.65)$$

where $\xi = f_B^{-1}\alpha_G^{-1/2}S^{-1}$ is the mass of non-relativistic matter per photon in Planck units. As illustrated in Figure 2.8, this corresponds roughly to the range $10^{-6} < Q < 10^{-4}$. Later, Tegmark *et al.* [82] expanded this analysis to consider variations of Q and Λ together, and the results are illustrated in Figure 2.9. Several other authors have also considered this problem [1, 12, 13, 38].

2.9.6 Dark Matter

As discussed by Kolb in this volume (Chapter 9), the existence of dark matter with 25% of the total cosmological density is now firmly established. The anthropic significance of this is unclear. One needs dark matter in order to amplify the fluctuations at decoupling enough to provide galaxies, but one could also achieve that by increasing the value of Q. There are still many possible dark matter candidates, but several of them are associated with anthropic constraints.

If WIMPs provide the dark matter, then their density will be comparable to the baryon density provided the coincidence $\alpha_G \sim \alpha_W^4$ is satisfied [84]. This is the same relationship which arises in the context of supernovae and cosmological nucelosynthesis, essentially

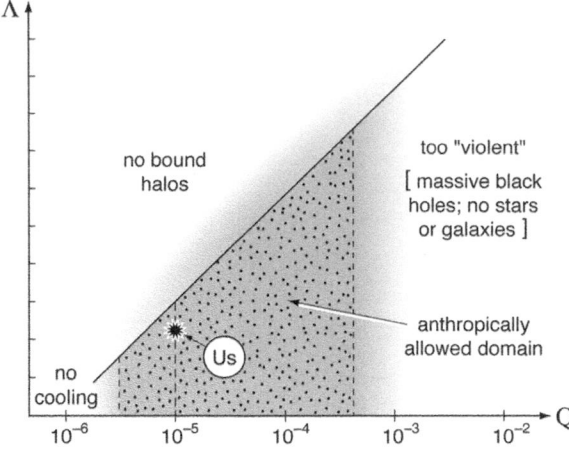

Figure 2.9 Constraints on Λ as a function of Q; from Reference [66].

because all three arguments involve the weak freeze-out condition. So there is at least an indirect sense in which this condition is anthropic.

If axions provide the dark matter, then anthropic arguments may also explain why their density is comparable with the baryon density [91]. The axion is a dark matter particle associated with the breaking of Peccei-Quinn (i.e., strong charge-parity – CP) symmetry at a time of order 10^{-30} s after the Big Bang. Large values of the symmetry-breaking energy scale, associated with large values of the Peccei-Quinn 'misalignment' angle, are forbidden in conventional axion cosmology. However, if inflation occurs after the breaking of Peccei-Quinn symmetry, then the CP-violating factor θ may vary across the different inflationary domains. So large values are permitted, providing we inhabit a domain where θ is small. Although such regions may occupy only a small volume of the multiverse, they contain a large fraction of potential observers.

In recent years, primordial black holes (PBHs) have become a popular dark matter candidate [17], partly because of the failure to find more conventional candidates but also because of the possibility that the coalescing black holes detected by LIGO could be primordial. PBHs are a natural dark matter candidate because they form in the radiation-dominated era and are therefore unrestricted by the nucleosynthesis bound on the baryonic density. However, it does require fine-tuning of the collapse fraction. In the standard scenario, the fraction of the Universe going into PBHs of mass M is only $\beta \sim 10^{-9}(M/M_\odot)^{1/2}$ at formation. Reference [20] discusses a possible anthropic resolution of this problem by invoking PBH formation at the QCD epoch. Since the horizon mass then is around M_C, this would explain why dark matter and stars have comparable masses. The collapse fraction is $\beta \sim S^{-1}$ because a baryon asymmetry of $O(1)$ is generated in the hot expanding shell around each PBH and this is then diluted by a factor β when the baryons disperse, the factor being tuned so that the times of matter-radiation equality and decoupling are comparable.

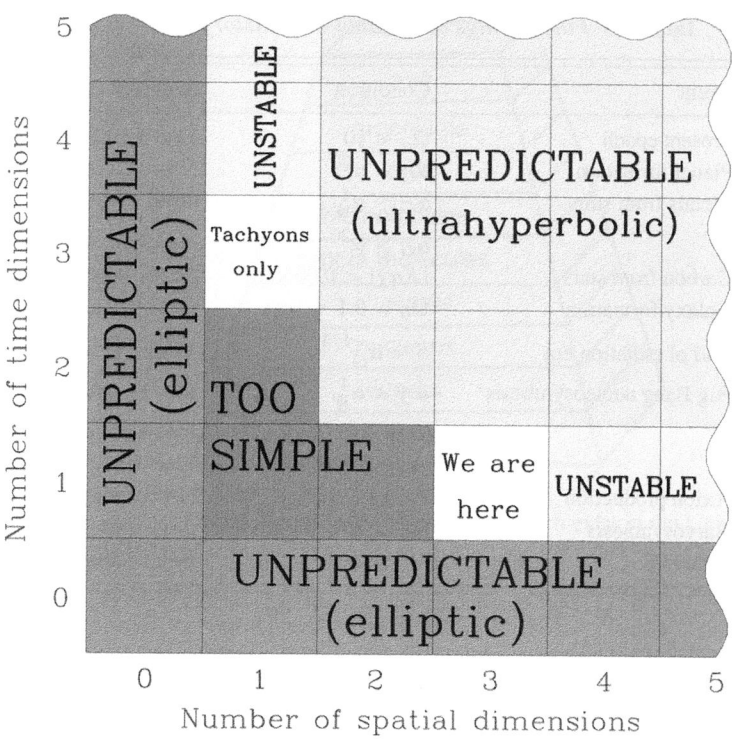

Figure 2.10 Illustrating why we live in $3 + 1$ dimensions. When the partial differential equations of nature are elliptic or ultrahyperbolic, physics has no predictive power. In the remaining (hyperbolic) cases, $n > 3$ admits no stable atoms and $n < 3$ may lack sufficient complexity for observers (e.g., no gravity). From Reference [79].

2.9.7 Number of Dimensions

Since many models of particle physics invoke extra spatial dimensions, the issue of why we live in a world with three spatial dimensions naturally arises. While there may well be some physical explanation for this, it clearly has anthropic aspects. For example, there would be no gravity with two spatial dimensions, and planetary orbits would be unstable with four of them. There are also constraints on the number of time dimensions, associated with causality. Other arguments for the number of space and time dimensions have been given by Tegmark [79], and Figure 2.10 is taken from his paper.

In the context of higher-dimensional models, it is interesting that brane cosmology [63] may give a natural explanation for the sort of power-law relations between the coupling constants which arise in the anthropic arguments. This is because the variation in the gravitational coupling constant would be associated with the change in the volume of the bulk, whereas the variation in the other coupling constants would be depend on the change in the volume of the brane or some manifold of intermediate dimensionality. In this

On the other hand, by invoking a Copernican perspective, Smolin has argued that *most* of the universes should have properties like our own, so that we are typical [72, 73]. His approach invokes a form of cosmological natural selection: the formation of black holes is supposed to generate new baby universes in which the constants are slightly mutated. In this way, after many generations, the parameter distribution will be peaked around those values for which black hole formation is maximised. This proposal involves very speculative physics since we have no understanding of how the baby universes are born, but it has the virtue of being *testable* since one can calculate how many black holes would form if the parameters were different. Note that Smolin's proposal makes no reference to observers, so it would not be regarded as anthropic in the usual sense of the term, but it still invokes a multiverse.

But how legitimate is it to invoke the existence of other universes for which there may never be any direct evidence? Smolin stresses [74] that the multiverse proposal is legitimate only if one has a theory which independently predicts it, and such a theory, to be scientific, must be falsifiable. He argues that the notion of a multiverse is neither falsifiable nor testable. However, not everybody concedes this point. For example, Rees points out [66] that one way of testing the multiverse proposal is to calculate the probability distribution for various parameters across the different universes. One would then be surprised if the value of some constant was on the tail of the distribution. In particular, if Weinberg's explanation for the value of Λ is correct, one would be surprised if its value was *much* less than the anthropic limit.

Even if one accepts that a multiverse exists and gives rise to anthropic selection effects, there is still considerable ambiguity as to how one interprets the selection effect. If the Anthropic Principle can be interpreted as a Complexity Principle, what qualifies as an observer in anthropic considerations? It would be most natural to associate the anthropic constraints with life in general rather than humans in particular. In fact, Davies explicitly associates them with a 'life principle' [28]. But in this case, does the mere existence of consciousness suffice, or is some minimum threshold of intelligence required? We have seen that this may not be so crucial since – whatever threshold one selects – it may be attained relatively quickly once the first signs of life arise [52].

2.12 Concluding Remarks

In concluding, I will make a few philosophical points, most of which relate to the common criticisms of anthropic arguments.

- In so much as the WAP is a selection effect, it is uncontroversial. It is clear that the presence of observers implies a non-random sample of the underlying population, and one needs to allow for this to avoid spurious correlations. However, one should not infer from this that the WAP is trivial, since some of the selection effects are unexpected and depend upon subtle aspects of physics. But what is being selected for? Even if it is life, one must avoid being too anthropocentric since it is clearly not restricted to humans

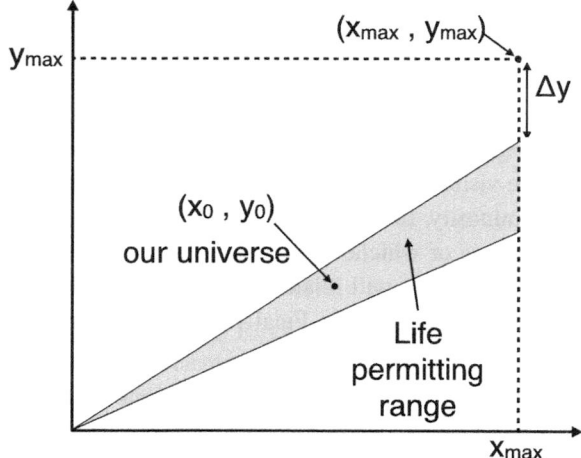

Figure 2.13 Illustrating that anthropic constraints may specify a wedge in parameter space if two constants (x, y) are allowed to vary. From Reference [8].

and may not even be carbon based. According to Smolin [72, 73], it just relates to the abundance of black holes and is irrelevant to life.

- A common objection to the SAP is that the apparent tunings involved could just be coincidental. In this context, one might compare the contrasting views of Victor Stenger [76] and Luke Barnes [8]. Clearly, this depends on the number of tunings and their precision, but it is not straightforward to assess the probabilities involved, especially since anthropic arguments do not explain exact values. In assessing the weight of the evidence, it is also relevant to know whether the tunings are post hoc or predicted. Most are post hoc, but it could be argued that the triple-α and Λ tunings were predictions.

- Many of the strong tunings discussed in this chapter are treated in isolation (i.e., with only one constant being varied at a time). But what happens if one allows some constants to covary? Stenger argues that, for two parameters, the constraints correspond to a 'wedge', as illustrated in Figure 2.13, which means that the probability is not just the product of the one-dimensional probabilities. However, Barnes points out that the wedge description does not always apply, as illustrated by Figure 2.8. In general, when considering multiple constraints, one needs to consider the intersection rather than the union of wedges. Indeed, if there were too many constraints, life would presumably be excluded altogether.

- Anthropic arguments usually only consider small variations of the constants in the 'island' of parameter space around the observed values. However, there could be other islands of life, so this may not suffice. This does not invalidate the Anthropic Principle, providing the total area of the islands is small. As Richard Dawkins remarks [29], 'However many ways there may be of being alive, there are vastly more ways of being dead'. On the other hand, there are no small islands in Smolin's proposal since our Universe should be representative of the most populated region of parameter space.

[70] Sandvik, B., Barrow, J. D., and Magueijo, J. 2002. 'A Simple Cosmology with a Varying Fine Structure Constant'. *Physical Review Letters*, **88**(3), 031302.

[71] Silk, J. 1977. 'On the Fragmentation of Cosmic Gas Clouds. I. The Formation of Galaxies and the First Generation of Stars'. *The Astrophysical Journal*, **211**(Feb.), 638–648.

[72] Smolin, L. 1992. 'Did the Universe Evolve?'. *Classical and Quantum Gravity*, **9**(Jan.), 173–191.

[73] Smolin, L. 1997. *The Life of the Cosmos*. Oxford University Press.

[74] Smolin, L. 2007. *Scientific Alternatives to the Anthropic Principle*, in B. J. Carr, ed. *Universe or Multiverse?* Cambridge University Press. Pages 323–366.

[75] Steinhardt, P. J., and Turok, N. 2006. 'Why the Cosmological Constant Is Small and Positive'. *Science*, **312**(May), 1180–1183.

[76] Stenger, V. J. 2011. *The Fallacy of Fine-Tuning: Why Universe Is Not Designed for Life*. Prometheus Books.

[77] Susskind, L. 2005. *The Cosmic Landscape: String Theory and the Illusion of Intelligent Design*. Little, Brown and Company.

[78] Susskind, L. 2007. 'The Anthropic Landscape of String Theory', in B. J. Carr, ed. *Universe or Multiverse?*. Cambridge University Press. Pages 247–266.

[79] Tegmark, M. 1997. 'On the Dimensionality of Spacetime'. *Classical and Quantum Gravity*, **14**(Apr.), L69–L75.

[80] Tegmark, M. 2005. 'What Does Inflation Really Predict?' *Journal of Cosmology and Astroparticle Physics*, **4**(Apr.), 001.

[81] Tegmark, M., and Rees, M. J. 1998. 'Why Is the Cosmic Microwave Background Fluctuation Level 10^{-5}?'. *The Astrophysical Journal*, **499**(May), 526–532.

[82] Tegmark, M., *et al.* 2006. 'Dimensionless Constants, Cosmology, and Other Dark Matters'. *Physical Review D*, **73**(2), 023505.

[83] Tegmark, M. 2007. 'Parallel Universes', in B. J. Carr, ed. *Universe or Multiverse?*. Cambridge University Press. Pages 459–491.

[84] Turner, M. S., and Carr, B. J. 1987. 'Why Should Baryons and Exotic Relic Particles Have Comparable Densities?'. *Modern Physics Letters A*, **2**, 1–7.

[85] Vilenkin, A. 1983. 'Birth of Inflationary Universes'. *Physical Review D*, **27**(June), 2848–2855.

[86] Vilenkin, A. 1995. 'Predictions from Quantum Cosmology'. *Physical Review Letters*, **74**(Feb.), 846–849.

[87] Webb, J. K., *et al.* 2001. 'Further Evidence for Cosmological Evolution of the Fine Structure Constant'. *Physical Review Letters*, **87**(9), 091301.

[88] Weinberg, S. 1987. 'Anthropic Bound on the Cosmological Constant'. *Physical Review Letters*, **59**(Nov.), 2607–2610.

[89] Weinberg, S. 1989. 'The Cosmological Constant Problem'. *Reviews of Modern Physics*, **61**(Jan.), 1–23.

[90] Wheeler, J. A. 1977. 'Genesis and Observership'. in R. Butts and J. Hintikka, eds. *Foundational Problems in the Special Sciences*. Reidel. Pages 3–33.

[91] Wilczek, F. 2007a. 'A Model of Anthropic Reasoning: The Dark to Ordinary Matter Ratio'. in B. J. Carr, ed. *Universe or Multiverse?*, Cambridge University Press. Pages 151–162.

[92] Wilczek, F. 2007b. 'Enlightnement, Knowledge, Ignorance, Temptation'. in B. J. Carr, ed. *Universe or Multiverse?*, Cambridge University Press. Pages 43–54.

[93] Witten, E. 1995. 'String Theory Dynamics in Various Dimensions'. *Nuclear Physics B*, **443**, 85–126.

Part II
Cosmological Fine-Tunings

well known that many physicists (and not a few astronomers) show a strong allergy to 'the A word'. To some extent, one can sympathise with this attitude, since the tone of some older discussions of this topic was not always helpful. Therefore, we will proceed at first with some broader terminology, attempting to define naturalness and fine-tuning initially using the viewpoint of particle physics, before setting up a more general Bayesian framework within which such issues can be discussed. With this background, we will then give a list of unexplained cosmological problems, focusing on strange parameter coincidences and, in particular, the challenge of explaining the observed level of the vacuum density.

This will lead to a survey of approaches to the observed properties of 'dark energy', especially the modelling of dark energy by considering dynamical signatures that go beyond a simple cosmological constant, or by modifying the theory of gravity. The conclusion will be that there are problems with most of these approaches, leading to a focus on observer selection as a possible solution. This leads to consideration of the possible existence of a physical ensemble of causally distinct universes, as can arise in some models of inflation (see Chapter 4 for a detailed discussion). Within such an ensemble, many elements of physics can, in practice, be different in different members of the ensemble. We begin with the possible physics of a variable cosmological constant, moving on to evidence that other pieces of 'fundamental' physics may also not be immutable. In all these cases, there are two practical difficulties with converting these general ideas into quantitative testable science. One is the 'measure problem' – the appropriate Bayesian prior for a given parameter – but the other is the observer weighting, where astronomers risk being dragged into biological issues far beyond their sphere of competence. One way to avoid this peril is to consider the impact on the history and efficiency of star formation as pieces of fundamental physics are altered.

3.2 Defining Fine-Tuning

3.2.1 Naturalness in Particle Physics

When considering the credibility of a given cosmological theory, the field borrows from particle physics. One of the most powerful principles in that subject is naturalness, which amounts to a statement that all dimensionless parameters in a theory should be of order unity. Indeed, it is common to use dimensional analysis in all areas of physics to guess the form of physical laws up to some dimensionless factor, which is then assumed to be of order unity – provided this term is interpreted generously enough to cover factors such as $(2\pi)^3$. An exception to this is that a parameter that might be expected to be of order unity can be forced to zero by means of some symmetry and a small non-zero value possibly restored by means of breaking that symmetry. For example, the fact that the electron mass is tiny compared to the > 100-GeV scale of the weak interaction can be understood via an approximate chiral symmetry (see, e.g., [59]). In non-dimensionless terms, this hypothesis implies that all particles should have roughly the same mass (or zero, like the photon) – whatever fundamental value sets the scale. It is common to assert that this should be the

Planck scale, and so the relative lightness of all known particles makes particle physics uniformly unnatural. This is known as the hierarchy problem.

A related but deeper definition of naturalness concerns quantum corrections: particle masses are not fixed at some bare value but will run according to the energy scale of observation. The interesting question is whether the resulting value is dominated by the bare value or by the correction. Consider the self-energy of the electron, and contrast the corrections to the mass up to some cut-off Λ as calculated using non-relativistic or relativistic quantum theory:

$$\frac{\delta m_e}{m_e} = \frac{4\alpha}{3\pi} \left(\frac{\Lambda}{m_e}\right) \text{ (non - relativistic)} \tag{3.1}$$

$$= \frac{3\alpha}{2\pi} \ln\left(\frac{\Lambda}{m_e}\right) \text{ (relativistic),} \tag{3.2}$$

where α is the fine-structure constant. The latter correction is natural, so the measured electron mass is never very different from its bare value. But this is not always true: for example, the Higgs mass has a correction that is quadratic in Λ, meaning that the hierarchy problem is particularly severe in this case. The need to keep the Higgs at its observed light value is one of the arguments often advanced for new physics such as supersymmetry.

Barring new physics, unnaturally small observed masses can only be accounted for by a distasteful fine-tuning, in which a hypothetical bare mass is adjusted so that it almost cancels the correction, leaving a residual small mass that is much less than either term. Cosmology has a similar problem with the value of the cosmological constant. In both subjects, fine-tuning formally deals with the observed data – but in a way that convinces no one, and which cries out for an explanation.

3.2.2 Unnatural Aspects of Cosmology

Cosmology contains a number of puzzles that disturb us at the same level as the issues in particle physics, but sometimes phrased rather differently. As discussed more fully in this section, there is undoubtedly a huge fine-tuning issue with the cosmological constant. But there are also issues that are best phrased as coincidences, and these come in two kinds:

Coincidences of Value

Here, certain parameters of the Universe take very similar values, even though they are apparently physically unrelated. The most striking puzzles here are the rough equality in density of baryons and dark matter, and the fact that recombination occurs roughly when the Universe has just become matter dominated.

Regarding the matter budget, it is normally assumed that the baryon density arises via some CP-violating process so that the numbers of baryons and photons reflect an asymmetry factor: $n_B/n_\gamma = \epsilon \sim 10^{-9}$. If the dark matter consists of incompletely annihilated particles and antiparticles, then the relic density can be written as

$$\frac{n_{DM}}{n_\gamma} \simeq \left(\sigma \frac{mc}{\hbar} \frac{m_P c}{\hbar} \right)^{-1}, \tag{3.3}$$

where σ is the annihilation cross section, m is the particle mass, and m_P is the Planck mass. Thus, the ratio of baryon and dark matter densities is

$$\frac{\rho_B}{\rho_{DM}} \simeq \epsilon \left(\sigma \frac{m_p c}{\hbar} \frac{m_P c}{\hbar} \right)^{-1}, \tag{3.4}$$

where m_p is the proton mass. This ratio is known to be close to 0.2 but is composed of physical quantities that have no known relation to each other.

As for recombination, this occurs roughly when thermal photons can ionise hydrogen: $kT \sim \alpha^2 m_e c^2$, where α is the fine-structure constant. Matter-radiation equality requires $n_\gamma kT \sim \rho_B c^2 = n_B m_p c^2$. These two temperatures will coincide if

$$\frac{n_B}{n_\gamma} \sim \alpha^2 (m_e/m_p) = 3 \times 10^{-8}. \tag{3.5}$$

So the observed near equality of the two eras arises because of the numerological coincidence between this critical value and the actual baryon-to-photon ratio. But no known physics explains why the matter-antimatter asymmetry should take this particular value (see Chapter 5).

Coincidences of Time

There are also certain puzzles that are more closely tied to our existence as observers. Although the ratio between the matter density and vacuum density varies hugely with time,

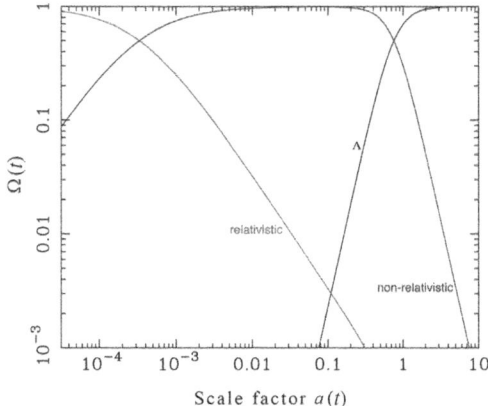

Figure 3.1 The density fraction in various components of the Universe as a function of time. The Universe is only strongly matter dominated for a little more than a single decade of expansion.

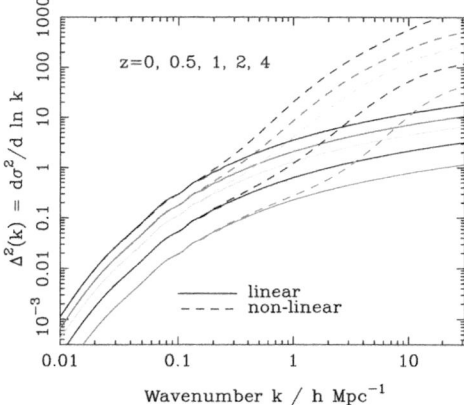

Figure 3.2 The evolution of the dimensionless mass power spectrum in the standard ΛCDM model. As the Universe expands, the high-k plateau value predicted in linear perturbation theory becomes close to unity, so non-linear effects largely erase the initial conditions up to the break in the spectrum. But as the Universe becomes Λ dominated, this evolution switches off. A small reduction in the normalisation of the spectrum would have preserved the linear information forever, but this did not happen. The curves correspond to the indicated redshifts, from left to right, respectively.

it is presently observed to be about unity – even though the history of the Universe is partitioned into many decades of expansion during which it was mostly dominated by either radiation (at early times) or vacuum energy (at late times); see Figure 3.1. A similar puzzle exists in large-scale structure, where the linear power spectrum has a break on 100-Mpc scales, to a nearly flat plateau in the dimensionless power $d\sigma^2/d\ln k \sim k^3 P(k)$ at large wave numbers. On very small scales, non-linear evolution erases the signature of the linear initial conditions, but in practice, this erasure has proceeded close to the point where all cosmological information below the break scale has been removed (see Figure 3.2). Neither of these coincidences applied in the distant past, so we have a *'why now'* problem.

3.3 Probabilistic Framework

3.3.1 Bayesian Approach

In grappling with the aforementioned puzzles, we are almost certain to use probabilistic language: 'it is very unlikely that the bare Higgs mass and its quantum corrections could cancel so precisely'; 'the vacuum density has a very large natural level, so the small observed value seems highly improbable'. Is this legitimate? The Higgs mass is what it is, so how can we talk about there being any probability that it could ever take any different value? The answer comes in the Bayesian approach to statistics. Rather than the simple view of probabilities as representing relative frequencies in an ensemble of trials, Bayesians define probability simply as a degree of belief in a proposition. Consider a theory that contains some parameter θ: if this theory is a correct description of reality, then θ has some definite value and is not a stochastic quantity. But the value of θ is unknown, and we can use the probability of different possible values as a way of quantifying our ignorance. This is done through Bayes's theorem (almost universally written as Bayes' theorem, although this is not correct English):

$$P(\theta|D) = \frac{P(\theta) \, P(D|\theta)}{P(D)}, \tag{3.6}$$

i.e., our belief about θ given data D (the posterior probability) is proportional to the prior on θ times the likelihood, $L = P(D|\theta)$. If $P(\theta|D)$ is to be normalised, then integrating over θ should give unity, requiring $P(D) = \int P(\theta) \, P(D|\theta) \, d\theta$. All of this generalises in an obvious way when θ becomes a vector of multiple parameters.

This approach has many appealing features. It focuses directly on what we care about, which is the probability of various hypotheses given the data to hand. This is in contrast to the frequentist approach, which calculates the probability of various possible outcomes on a given hypothesis, but clearly it is better to focus on what we learn from events that actually happened rather than having to think about all the other possible outcomes to the experiment that might have occurred. In particular, there is no need to have an ensemble of repeated experiments, and the Bayesian approach can happily handle unique events. The need to have a prior is a difficult and controversial part of the apparatus, however, and many are uneasy at the idea of admitting an individual's subjective degree of belief into scientific discussions.

One answer to this is that one of the main applications of the Bayesian formula is to perform inference; i.e., to estimate parameter(s) θ, together with a measure of uncertainty on them. As the quantity of data increases, the posterior probability for θ becomes concentrated around the true value, and the form of the prior becomes unimportant, provided it can be treated as constant over some small range of θ. More deeply, the prior can be considered to result from previous experimental knowledge. Consider what happens when we take two sets of data, D_1 and D_2. The posterior distribution in the face of the totality of data is

$$p(\theta|D_1, D_2) \propto p(D_1, D_2|\theta)p(\theta). \tag{3.7}$$

But the likelihood will factor into the likelihoods for the two data sets separately: $p(D_1, D_2|\theta) = p(D_1|\theta)p(D_2|\theta)$, in which case the posterior can be written as

$$p(\theta|D_1, D_2) \propto p(D_2|\theta) \left[p(D_1|\theta)p(\theta) \right]. \tag{3.8}$$

Now we see that the posterior for experiment 1 functions as the prior for experiment 2, so that Bayes's theorem neatly expresses the process of updating our belief about θ in the light of new information – although it does not remove the basic worry of how we should set a prior before we have any data at all. For further food for Bayesian thought, see, e.g., [36].

For inference – i.e., evaluating the relative probabilities of various different values of θ – the normalisation constant $P(D)$ is irrelevant, but sometimes it can be useful. This can be seen by realising that we might want to consider different models, M_i, with different parameter sets θ_i. In that case, Bayes's theorem would be written in a fuller form:

$$P(M_i, \theta_i | D) \propto P(M_i, \theta_i) \, P(D|M_i, \theta_i) \propto P(M_i) P(\theta_i | M_i) \, P(D|M_i, \theta_i), \qquad (3.9)$$

where the latter case expands the prior into a prior probability for the model itself, times a prior for the parameters in the case where that model applies. So if we want to know the probability that the *class* of model M_i applies without caring about the exact value of its parameters, then we would integrate to obtain $P(M_i|D) \propto P(M_i) \int P(\theta_i|M_i) P(D|M_i, \theta_i) \, d\theta_i$. Apart from the prior probability of the model, $P(M_i)$, this is just proportional to the earlier normalisation constant, $P(D)$, known as the Bayesian evidence. Thus, we can assess the plausibility of various models by calculating the evidence ratio between them, which allows the data to update whatever previous ratio of $P(M_i)$ values we had assigned. This is a powerful methodology, which has received considerable application in cosmology (e.g., [54]). But it is not without its difficulties, stemming from the choice of prior. This is always an issue at some level in Bayesian analyses but is often not so important in inference applications. Suppose we have a single parameter, where the prior is uniform over some range Δ: $P(\theta) = 1/\Delta$. When we compute the ratio of probabilities of two values of θ, this constant divides out as long as both values are in the allowed range, and we simply have the likelihood ratio. But in evidence ratios, the prior is more prominent. Suppose we have two models, each with a single parameter, with allowed ranges Δ_1 and Δ_2. If the likelihood has a Gaussian form, then the integral over the Gaussian is $L^{\max}\sqrt{2\pi}\sigma$, where σ is the effective precision with which θ is measured. In this case, the evidence ratio is

$$\frac{E_2}{E_1} = \left(\frac{L_2^{\max}}{L_1^{\max}}\right)\left(\frac{\sqrt{2\pi}\sigma_2}{\Delta_2}\right)\left(\frac{\sqrt{2\pi}\sigma_1}{\Delta_1}\right)^{-1}. \qquad (3.10)$$

Here, we can have the strange situation that the likelihood ratio might strongly favour model 2, but overall model 2 could be disfavoured if its prior is too wide so that the small value of σ_2/Δ_2 dominates. Things can become worse as the data improve, since σ will shrink. For example, imagine an experiment where we try to test if the vacuum energy is a cosmological constant; as discussed later, this is equivalent to introducing a parameter, w, and asking if it is consistent with -1. Suppose we measure -1.003 ± 0.001 – a 'three-sigma' detection. This corresponds to a likelihood ratio of $\exp(9/2) \simeq 90$ in favour of the more complex hypothesis. But if we had been willing to contemplate values of w between -1.5 and -0.5 ($\Delta = 1$), then the overall evidence ratio would be $1000 : 90\sqrt{2\pi} \simeq 4.4 : 1$ in favour of $w = -1$, despite the 'detection' of a deviation (the last $\sqrt{2\pi}\sigma_1/\Delta_1$ factor is absent as the cosmological constant case has no free parameter).

This is in contrast to inference, where the prior becomes progressively less important as the data improve and the likelihood becomes narrower. Indeed, studies in parameter inference often deliberately choose uninformative priors that are much broader than any reasonable value so that the conclusions are driven by the data. This is fine so far as it goes, but such priors would be completely inappropriate for evidence calculations. Here, excellent example of important astronomical conclusions being drawn from observations of life on Earth.

The Weak Anthropic Principle

But it is not possible to go very far in discussing the astronomical consequences of our local planetary observations before we run into the question of typicality: whether the conditions we observe can be extrapolated into properties characteristic of the Universe as whole. The danger in such reasoning was pointed out by Carter [10] in what he termed the Weak Anthropic Principle: we 'must be prepared to take account of the fact that our location in the Universe is necessarily privileged to the extent of being compatible with our existence as observers'. In other words, there may be times and locations in the Universe where life is impossible, or at least highly improbable, and the allowance for such censoring may condition cosmological observables or even allow us to predict them. The outstanding success of this reasoning concerns Dirac's large-number hypothesis. Dirac noted that very large dimensionless numbers often arise in both particle physics and cosmology. The ratio of the electrostatic and gravitational forces experienced by an electron in a hydrogen atom is

$$\frac{e^2}{4\pi\epsilon_0 G m_e m_p} \simeq 10^{39.4};\tag{3.11}$$

one of the problems of unifying gravity and other forces is understanding how such a vast dimensionless number can be generated naturally. In a cosmological context, the weakness of gravity manifests itself in the fact that the Hubble radius is enormously greater than the Planck length:

$$\frac{c/H_0}{\sqrt{\hbar G/c^3}} \simeq 10^{61}.\tag{3.12}$$

This number is very nearly the 1.5 powers of the previous large number: it is as if these large numbers were quantised in steps of 10^{20}. Dirac proposed that the coincidence must indicate a causal relation; requiring the proportionality to hold at all times then yields the radical consequence

$$G \propto t^{-1}\tag{3.13}$$

(because H_0 declines as $\sim t^{-1}$ as the Universe ages). See also Chapter 2.

Geological evidence shows that this prediction is not upheld in practice since the Sun would have been hotter in the past. A less radical explanation of the large-number coincidence uses the anthropic idea that life presumably requires the existence of elements other than hydrogen and helium. The Universe must therefore be old enough to allow typical stars to go through their life cycle and produce 'metals'. This condition can be expressed in terms of fundamental constants as follows. Stars have masses of the order of the Chandrasekhar mass

$$M_{\text{Chandra}} \equiv \left(\frac{\hbar c}{G m_p^2}\right)^{3/2} m_p,\tag{3.14}$$

where m_p is the proton mass. The luminosity of a star dominated by electron-scattering opacity is

$$L \sim \frac{G^4 m_p^5 M^3}{\hbar^3 c^2 \sigma_{\mathrm{T}}}. \tag{3.15}$$

The characteristic lifetime of a star, $M_{\mathrm{Chandra}} c^2 / L$, can thus also be expressed in fundamental constants and is

$$t_* \sim \frac{c\,\sigma_{\mathrm{T}}}{G\,m_p} = 5.7 \times 10^9 \text{ years.} \tag{3.16}$$

Comparing with the preceding, we see that the large-number coincidence is just $t_* \sim H_0^{-1}$; i.e., the Universe must be old enough for the stars to age. The fact that the Universe is not very much older than this may tell us that we are privileged to be in the first generation of intelligence to arise after the Big Bang: civilisations arising in $\gg 10^{10}$ years time will probably not spend their time in cosmological enquiry, as they will be in contact with experienced older races who know the answers already. Lastly, the coincidence can be used to argue for the fundamental correctness of the Big Bang as against competitors such as the steady-state theory; if the Universe is, in reality, very much older than t_*, there is no explanation for the coincidence between t_* and H_0^{-1}.

In short, the Weak Anthropic Principle states that, because intelligent life is necessary for cosmological enquiry to take place, this already imposes strong selection effects on cosmological observations. Note that, despite the name, there is no requirement for the life to be human, or even carbon based. All we say is that certain conditions can be ruled out if they do not lead to observers, and this is one of the weaknesses of the principle: are we really sure that life based on elements less massive than carbon is impossible? The whole point of these arguments is that it only has to happen once. It is nevertheless at least plausible that the 'anthropic' term may not be a complete misnomer. It has been argued (see [11]) that intelligent life may be intrinsically an extremely unlikely phenomenon, where the mean time for development could be very long:

$$\bar{t}_{\mathrm{Intelligence}} \gg t_*. \tag{3.17}$$

If the inequality was sufficiently great, it would be surprising to find even one intelligent system within the current horizon. The Anthropic Principle provides a means for understanding why the number is non-zero, even when the expectation is small, but there would be no reason to expect a second system; humanity would then probably be alone in the Universe. On the other hand, $\bar{t}_{\mathrm{Intelligence}}$ may be shorter, and then other intelligences would be common. Either possibility is equally consistent with the selection effect imposed by our existence, although the fact that life on Earth took billions of years to develop is consistent with the former view; if life is a rare event, it would not appear early in the allowed span of time.

At the outset, we should be very clear that the deduced existence of dark energy depends on the correctness of the Friedmann equation, and this is not guaranteed. We possibly have the wrong theory of gravity, and we have to replace the Friedmann equation with something else. Alternative models do exist, particularly in the context of extra dimensions, and these must be borne in mind. Nevertheless, as a practical framework, it makes sense to stick with the Friedmann equation and see if we can get consistent results. If this programme fails, we may be led in the direction of more radical change.

To insert vacuum energy into the Friedmann equation, we need the equation of state

$$w \equiv \frac{p}{\rho c^2}. \tag{3.26}$$

If this is constant, adiabatic expansion of the vacuum gives

$$\frac{8\pi G\rho}{3H_0^2} = \Omega_v a^{-3(w+1)}. \tag{3.27}$$

More generally, we can allow w to vary; in this case, we should regard $-3(w+1)$ as $d\ln\rho/d\ln a$, so

$$\frac{8\pi G\rho}{3H_0^2} = \Omega_v \exp\left(\int -3(w(a)+1)\, d\ln a\right). \tag{3.28}$$

In general, we therefore need

$$H^2(a) = H_0^2 \left[\Omega_v e^{\int -3(w(a)+1)\, d\ln a} + \Omega_m a^{-3} + \Omega_r a^{-4} - (\Omega - 1)a^{-2}\right]. \tag{3.29}$$

Some complete dynamical model is needed to calculate $w(a)$. Given the lack of a unique model, a common empirical parameterisation is

$$w(a) = w_0 + w_a(1-a). \tag{3.30}$$

It frequently is sufficient to stick with constant w; most experiments are sensitive to w at a particular redshift of order unity, and w at this redshift can be estimated with little dependence on whether we allow dw/dz to be non-zero.

If w is negative at all, this leads to models that become progressively more vacuum dominated as time goes by. When this process is complete, the scale factor should vary as a power of time. The case $w < -1$ is particularly interesting, sometimes known as phantom dark energy [8]. Here, the vacuum energy density will eventually diverge, which has two consequences: this singularity happens in a finite time rather than asymptotically; as it does so, vacuum repulsion will overcome the normal electromagnetic binding force of matter so that all objects will be torn apart in the big rip. Integrating the Friedmann equation forward, ignoring the current matter density, the time to this event is

$$t_{\rm rip} - t_0 \simeq \frac{2}{3} H_0^{-1} |1+w|^{-1} (1-\Omega_m)^{-1/2}. \tag{3.31}$$

Observable Effects of the Vacuum

The co-moving distance-redshift relation is one of the chief diagnostics of w. The general definition is

$$D \equiv R_0 r = \int_0^z \frac{c}{H(z)} \, dz. \tag{3.32}$$

Perturbing this about a fiducial $\Omega_m = 0.3$, $w = -1$ model shows a *sensitivity multiplier* of about 5 – i.e., a measurement of w to 10% requires D to 2%. Also, there is a near-perfect degeneracy with Ω_m, so this parameter must be known very well before the effect of varying w becomes detectable.

The other main diagnostic of w is its effect on the growth of density perturbations. These are also sensitive to the vacuum, as may be seen from the growth equation:

$$\ddot{\delta} + 2\frac{\dot{a}}{a}\dot{\delta} = 4\pi G \rho_0 \delta. \tag{3.33}$$

The vacuum energy manifests itself in the factor of H in the 'Hubble drag' term, $2(\dot{a}/a)\dot{\delta}$. For flat models with $w = -1$, the growing mode for density perturbations is approximately $g(a) \propto a\Omega(a)^{0.23}$ or (more accurately) $d \ln g / d \ln a = \Omega(a)^{0.55}$. For greater accuracy, the following expressions are good to a maximum error of 0.1% [41]. The cases of positive and negative Λ are somewhat distinct. For the positive case,

$$\delta(a) \simeq x(1 - x^{1.91})^{0.82} + 1.437 \left(1 - (1 - x^3)^{2/3}\right), \tag{3.34}$$

where x denotes $\Omega_v(a)^{1/3}$, and we choose the $a = 1$ point to correspond to equal density in matter and vacuum:

$$\Omega_v(a) = (1 + a^{-3})^{-1} \tag{3.35}$$

so that $\delta(a) \simeq a$ for small a. For the negative-Λ case, we need time as a coordinate since the scale factor is not monotonic:

$$a(t) = \left[\sin\left(\frac{3t}{2}\right)\right]^{2/3}, \tag{3.36}$$

where here we choose units such that $a = 1$ at the point of maximum expansion, and time is measured in units of $(8\pi G|\rho_v|/3)^{-1/2}$ so that Friedmann's equation is $(\dot{a}/a)^2 = a^{-3} - 1$ and $\Omega_v(a) = (1 - a^{-3})^{-1}$. Here, the approximation for the growth function is

$$\delta(t) = \frac{(3t/2)^{2/3}}{(1 + 0.37(t/t_{\text{coll}})^{2.18})(1 - (t/t_{\text{coll}})^2)}. \tag{3.37}$$

Again, the normalisation is that $\delta(a) \simeq a$ for small a. Note that the fluctuations diverge at the collapse time ($t_{\text{coll}} = 2\pi/3$) as $1/(t_{\text{coll}} - t)$; this corresponds to the decaying mode in the expanding phase.

In the real Universe, we are interested in the positive branch of the growth factor, and this is illustrated in Figure 3.3. The plot shows two alternative points of view: either a growing $\delta(t)$ that asymptotes to a constant or a potential $\Phi(t)$ that is initially independent the bandwidth needs to be very narrow, and it takes a long time to scan an interesting frequency range: the axion model probably will be fully explored and ruled out within the next decade – or it could be detected any day now.

Despite the lack of detection of axions, the greater level of investment in WIMP searches and their lack of success has tended to focus greater attention on the axion model. Although the original QCD axion is constrained to a rather specific mass, the term 'axion' now tends to be used for a wider range of scalar fields. The most exciting possibility is the case where these are ultralight, with masses below 10^{-20}eV. In this case, the Compton wavelength $\hbar/mc$ is large enough that wave-mechanical effects may be observed on astronomical scales (see, e.g., [38, 46]).

3.5.4 The Outlook for Dark Energy

One significant problem with this line of research is the lack of a clear target. Some models, such as the $w \simeq -0.8$ from simple power-law quintessence, have been ruled out, but there is no guaranteed minimum deviation from $w = -1$. Perhaps dark energy is exactly a cosmological constant, and we are condemned to a future of ever more challenging experiments yielding increasingly precise null results around $w = -1$; at what point would we abandon the search?

Trotta [54] gave a nice Bayesian answer to this question, which applies to the general issue of asking when a theory should be expanded to include a new phenomenon. Let the effect in question be characterised by a parameter a (w in our case), such that the 'new physics' corresponds to $a \neq 0$. Let the prior probability of a be $P(a)$ so that the Bayesian odds ratio between the two hypotheses A (new physics: $a \neq 0$ and in the range da) and B (no new physics: $a = 0$ exactly) is $L(a) \times P(a)\,da$, where L is the likelihood ratio between A and B, given some data. This neglects a prior ratio of beliefs in new physics vs no new physics, which is generally taken to be of order unity (although, as usual, a sufficiently firm prejudice against new physics is not capable of being overturned, however strong the experimental evidence). To get the overall odds ratios of the two models in the face of the data, we should integrate over a to get the evidence ratio discussed earlier:

$$E = \int L(a)\,P(a)\,da. \tag{3.63}$$

Consider a simple Gaussian example, where $L(a) = \exp[-\hat{a}^2/2\sigma^2]\,/\,\exp[(a-\hat{a})^2/2\sigma^2]$ in terms of a measured value $a = \hat{a}\pm\sigma$, and let the prior on a be uniform over a range Σ. If Σ is large compared to the measuring error, σ, this gives $E = \sqrt{2\pi}(\sigma/\Sigma)\exp[\lambda^2/2]$, where $\lambda = \hat{a}/\sigma$ is the 'number of sigmas' at which the data 'detect' a deviation from $a = 0$. This expression is the basis of Figure 3.6 and reveals two important facts: (1) if λ is large, this need not constitute a detection if $\sigma \ll \Sigma$; (2) if we have a null result ($\lambda \lesssim 1$), then we may strongly disfavour hypothesis A if $\sigma \ll \Sigma$ – i.e., demonstrate convincingly that, in fact, $a = 0$ exactly, so there is no point in trying to measure it with improved precision. These slightly paradoxical conclusions arise because we take our prior seriously: if the prior on w

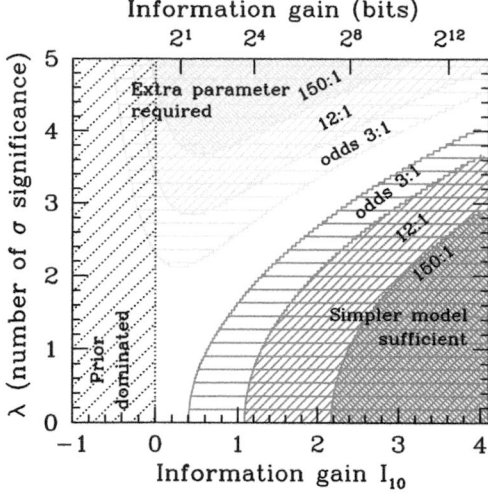

Figure 3.6 This plot, taken from [54], illustrates the interplay between the S/N of a measurement probing the existence of a new parameter (vertical axis) and the reciprocal fraction of the available parameter space ruled out by null measurements (horizontal axis). From a Bayesian point of view, improving the precision on parameter measurement sufficiently while still finding a null result can eventually yield compelling odds that the parameter will never be detected. In the context of dark energy, the initial prior range of interesting models was $|w + 1| \sim 1$; thus measuring, say, $w = -1 \pm 0.001$ would yield odds of 1,000:1 that the vacuum energy is exactly a cosmological constant.

ranged over -0.5 to -1.5, say, we are saying that the range -1.4 to -1.41 is just as likely as -1 to -1.01. But if we make a measurement with 0.01 precision, we have ruled out 99% of the possible options for a, so the original model starts to look like a poor bet.

With $\sigma/\Sigma \simeq 0.05$, dark energy is currently far from the situation envisaged in the preceding paragraph. But the next generation of experiments (Euclid; LSST) should push the measuring precision on w to below 1%, and a null result at that stage will start to be interesting from a Bayesian point of view. See, e.g., Laureijs [31].

Dark Energy vs Modified Gravity

A more radical approach to the problem of dark energy might be to wonder whether we are going entirely in the wrong direction. The acceleration of the cosmic expansion is surprising if our prejudice is that the Universe is dominated by the attractive gravitational effects of matter with normal equations of state. Introducing a new substance with $w < -1/3$ allows the acceleration to be explained and also makes new w-dependent predictions for the growth of density fluctuations. But both the expansion rate and the growth of perturbations are dictated by gravity, so is it possible that dark energy simply does not exist at all and that, in fact, Einstein's relativistic theory of gravity needs to be modified on cosmological scales?

This possibility has been explored energetically in recent cosmological research, and comprehensive reviews are given by Clifton *et al.* [12] and Joyce *et al.* [29]. There is clearly plenty of scope for building alternative models, since Einstein gravity is built on a simple scalar Lagrangian containing the Ricci scalar and the cosmological constant: $\mathcal{L} = R + 2\Lambda$. So it is only necessary to replace this with some other invariant, and we have a theory of modified gravity that is fully consistent with all the requirements of general covariance (thus, it is misleading to describe this area of research as 'testing general relativity'). The simplest such approach would be some non-linear function of R, and $f(R)$ gravity is one of the most widely explored theories of modified gravity. An alternative is to express the modification in the language of fifth forces and postulate an additional scalar field that contains explicit interaction terms with the gravitational and matter sectors (unlike the case of a simple quintessence field).

The behaviour of such models on non-linear scales is complicated, and much attention has been given to the linear regime, where we can concentrate on the behaviour of the metric perturbation potentials Ψ and Φ, which affect respectively the time and space parts of the metric. In Einstein's gravity, these potentials are both equal to the Newtonian grav-itational potential, which satisfies Poisson's equation: $\nabla^2\Phi/a^2 = 4\pi G\bar{\rho}\delta$. Empirically, modifications of gravity require us to explore a change with scale and with time of the 'slip' ($\eta \equiv \Psi/\Phi$) and the effective G on the right-hand side of Poisson's equation. The former aspect can only be probed via gravitational lensing, whereas the latter can be addressed on 10–100 Mpc scales via the growth of clustering. A common approach is to assume, as before, that the growth rate can be tied to the density parameter: $d \ln \delta/d \ln a = \Omega_m^\gamma(a)$. The parameter γ is close to 0.55 for standard relativistic gravity but can differ by around 0.1 from this value in many non-standard models. Clearly this parameterisation is incomplete, since it explicitly rejects the possibility of effects at early times ($\Omega_m(a) \to 1$ as $a \to 0$), but the recent onset of cosmological acceleration is used as a common justification for assuming that modifications of gravity only become significant at late times.

This reliance on the growth rate exposes a degeneracy, since we have seen that the growth rate is also sensitive to the equation of state of dark energy; thus, a deviation from ΛCDM growth could indicate modifications of gravity, or just that dark energy is not a simple Λ [47]. The way to break this degeneracy is by including geometrical probes that measure purely the expansion history (BAO and SNe) and, hence, measure the value of w (while being agnostic about whether this is genuinely the equation of state of a physical substance as opposed to being an effective value induced by the modification of gravity). We are then free to probe the two linear parameters that probe the effect of any modifications on the perturbations. A practical parameterisation of this is to express things in terms of factors that modify the strength of non-relativistic forces and relativistic forces (i.e., lensing) with respect to the values predicted in Einstein gravity:

$$\Psi = [1 + \mu(a,k)]\,\Psi_E$$
$$(\Psi + \Phi) = [1 + \Sigma(a,k)]\,(\Psi_E + \Phi_E), \tag{3.64}$$

where the relation to effective G and slip are $1+\mu = (G'/G)/\eta$; $1+\Sigma = (G'/2G)(1+1/\eta)$ (see, e.g., [14]). Current data are consistent with standard ΛCDM and exclude variations in slip or effective G of larger than of order 10% [48].

3.6 The Anthropic Vacuum

Whether or not one finds these approaches to dark energy compelling, there remains one big problem. All the models are constructed using Lagrangians with a particular zero level. All quintessence potentials have the field rolling down towards $V = 0$, and k-essence models lack a potential altogether. They are therefore subject to the classical dilemma of the cosmological constant: adding a pure constant to the Lagrangian has no affect on field dynamics but mimics a cosmological constant. With so many possible contributions to this vacuum energy from the zero-point energies of different fields (if nothing else), it seems contrived to force $V(\phi)$ to asymptote to zero without a reason.

To review why zero is a problematic value for the vacuum density, recall what we mean by the vacuum: $|0\rangle$, or zero occupation number for each wave mode inside a given box. But standard quantum mechanics assigns a zero-point energy of $\hbar\omega/2$ to each mode. Integrating $\hbar\omega/2c^2$ per mode over k-space (with a degeneracy of 2 for polarisation) gives a total density of

$$\rho_{\text{vac}} = \frac{\hbar}{2\pi^2 c^5} \int \omega^3 \, d\omega, \tag{3.65}$$

which diverges horribly. Is it possible that the upper limit of the integral should be finite? This would be the case if space were a lattice, which is perhaps conceivable on some unobservably small scale. However, even with a cut-off at the hardly microscopic level of $\lambda \sim 1$ mm, ρ_{vac} already exceeds the critical density of the Universe ($\sim 10^{-26}\text{kg m}^{-3}$). We can express things in terms of an energy scale E_v by writing the dimensional scaling

$$\rho_v = \frac{\hbar}{c} \left(\frac{E_v}{\hbar c}\right)^4, \tag{3.66}$$

or simply $\rho_v = E_v^4$ in natural units. if we adopt the values $\Omega_v = 0.7$ and $h = 0.7$ for the key cosmological parameters, then $E_v = 2.4$ meV is known to a tolerance of about 1%. What is a natural choice for E_v? A case can be made for E_v lying at the Planck scale, since quantum gravity effects must destroy the flat-space assumptions of quantum field theory. This would give a vacuum density 120 powers of 10 larger than observed. But this is over-dramatising the problem: one should focus on E_v rather than E_v^4. Also, the solution may lurk at much smaller energies. In unbroken supersymmetry, there would be an exact cancellation of the zero-point energy of bosonic and fermionic oscillators, and the scale of supersymmetry breaking could be as low as 10 TeV. So the vacuum problem is perhaps that the energy scale of the vacuum is 'only' 15 powers of 10 smaller than seems reasonable – a lot fewer than 120 powers of 10 but still enough to cause a problem.

The preceding argument is commonly given, but it should be taken with some caution. It is really the same argument as used to deduce black-body radiation, with a slightly different occupation number. Therefore, we would expect an equation of state $P = \rho c^2/3$: $w = +1/3$, so this is not at all a candidate calculation for the energy density of the vacuum. The problem is that the calculation is not relativistically invariant. Koksma and Prokopec [30] claim that a proper calculation changes the E_v^4 dependence to $M^4 \ln(E_v/M)$, where M is the particle rest mass associated with the field. Since we know of particles up to over $M = 100\,\text{GeV}$, this makes little practical change to the magnitude of the vacuum problem.

In any case, it should be clear that this prediction is hard to make fixed, partly because of our ignorance of the field content of the Universe, and because these zero-point contributions can be supplemented by classical contributions from $V(\phi)$ of any number of scalar fields. This problem has been sharpened by recent developments in string theory, known under the heading of the landscape. For the present purpose, this can be regarded as requiring the introduction of a large number of additional scalar fields, each with an associated potential. If we assume that a vacuum state is defined by these fields sitting at the minimum of their various potentials, then the effective cosmological constant can vary. It has been estimated that there are about 10^{500} distinct minima, which divides the natural vacuum density of E_p^4 into what is almost a continuous range from the point of view of observations – so we can have almost any effective value of Λ we like.

This leads us in the direction of anthropic arguments, which are able to limit Λ to some extent: if the Universe had become vacuum dominated at $z > 1,000$, gravitational instability would have been impossible – so that galaxies, stars, and observers would not have been possible [56]. Indeed, Weinberg made the astonishingly prescient prediction on this basis that a non-zero vacuum density would be detected at Ω_v of order unity, since there was no reason for it to be much smaller.

Many Universes

At first sight, this argument seems quite appealing, but it rapidly leads us into deep waters. How can we talk about changing Λ? It has the value that it has. We are implicitly invoking an ensemble picture in which there are many universes with differing properties. This is a big step (although exciting if this turns out to be the only way to explain the vacuum level we see). In fact, the idea of an ensemble emerges inevitably from the framework of inflationary cosmology, since the fluctuations in the scalar field can affect the progress of inflation itself. We have used this idea to look at the changes in when inflation ends – but fluctuations can affect the field at all stages of its evolution. They can be thought of as adding a random-walk element to the classical rolling of the scalar field down the trough defined by $V(\phi)$. In cases where ϕ is too close to the origin for inflation to persist for sufficiently long, it is possible for the quantum fluctuations to push ϕ further out – creating further inflation in a self-sustaining process. This is the concept of stochastic eternal inflation due to Linde *et al.* [34]. Sufficiently far from the origin, the random-walk effect of fluctuations becomes more marked and can overwhelm the classical downhill rolling. This means that some regions of space can inflate for an indefinite time, and a single inflating universe automatically breaks up into different bubbles with their own histories. Some random subset of these eventually random-walk close enough to the origin that the classical end of inflation can occur, thus creating a set of 'universes' each of which can potentially host observers.

With this as a starting point, the question now becomes whether we can arrange for the different members of this ensemble to have different values of Λ. This is easily achieved. Let there be some quintessence field with a very flat potential so that it is capable of simulating Λ effectively. Quantum fluctuations during inflation can also displace this field so that each member of the multiverse would have a different Λ.

The Distribution of Λ

We are now almost in a position to calculate a probability distribution for Λ, following [20]. First, we have to set some ground rules: what will vary and what will be held fixed? We should try to change as little as possible, so we assume that all universes have the same values for

(1) the baryon fraction $f_b = \rho_b/\rho_m$
(2) the entropy per particle $S = (T/2.725)^3/\Omega_m h^2$
(3) the horizon-scale inhomogeneity $\delta_{\mathrm{H}} \simeq 10^{-5}$

It is far from clear that these minimal assumptions are correct. For example, in the string theory landscape, there is no unique form for low-energy particle physics but, instead, a large number of possibilities in which numbers such as the fine-structure constant, neutrino masses, etc., are different. From the point of view of understanding Λ, we need there to be at least 10^{100} possible states so that at least some have Λ smaller than the natural m_{p}^4 density by a sufficient factor. The landscape hypothesis provides this variation in Λ but does not support the idea that particle physics is otherwise invariant. Still, it makes sense to start with the simplest forms of anthropic variation: if this can be ruled out, it might be taken as evidence in favour of the fuller landscape picture.

We then take a Bayesian viewpoint to the distribution of Λ, given the existence of observers:

$$P(\Lambda \mid \text{Observer}) \propto P_{\text{prior}}(\Lambda) P(\text{Observer} \mid \Lambda), \qquad (3.67)$$

where we need both the prior distribution of Λ between different members of the ensemble and how the chance of getting an observer is modified by Λ. The latter factor should be proportional to the number of stars, which is generally taken to be proportional to the fraction of the baryons that are incorporated into non-linear structures. We can estimate this using the Press-Schechter apparatus to get the collapse fraction into systems of a galaxy-scale mass. The exact definition of this is not very important since the CDM dimensionless power spectrum is very flat on small scales; any mass at all close to $10^{12} M_\odot$ gives similar answers.

The more difficult part is the prior distribution of Λ, and a common argument is to say that it has a uniform distribution – which seems reasonable enough if we are to allow it to have either sign but know that we will be interested in practice in a very small range near zero. This choice of prior is key to the results of the argument, and it is clear that different choices could completely transform the answer. For example, suppose we took a uniform prior in $\log(\Lambda)$ – or rather $\log(|\Lambda|)$, since Λ can be negative. This would multiply our flat prior by $1/|\Lambda|$, which gives a divergent spike at $\Lambda = 0$. This might have been attractive back in the days when $\Lambda = 0$ matched observations, but it actually is not an acceptable explanation. This particular prior says that there is something special about $\Lambda = 0$, whereas our whole problem is that we know of no physical argument why this should be so. As we saw earlier, the observed value of Λ is a mixture of the classical Λ and a physical vacuum density: $\Lambda_{\text{eff}} = \Lambda_{\text{bare}} + 8\pi G \rho_{\text{vac}}/c^2$. The natural value for the second term is very large, so somehow the classical term must cancel it to high precision, but there is no known reason for this to happen precisely (and, indeed, it does not). Therefore, we have to suppose that nothing violent happens to the prior as we cross from Λ being slightly positive to slightly negative – which is the basis of the uniform prior (see [57] for a detailed discussion).

We therefore have the startling proposition of the anthropic model: the effective vacuum density takes large ranges, and in almost all realisations, the values are comparable in magnitude to the natural scale m_{P}^4; such models are stupendously inimical to life. This is quantified by the simple model

$$dP(\rho_v) \propto f_c \, d\rho_v, \tag{3.68}$$

where f_c is the collapse fraction into galaxy-scale objects. For large values of Λ, growth ceases at high redshift, and f_c is exponentially suppressed. But things are less clear-cut if $\Lambda < 0$. Here, the Universe eventually recollapses, and the high density means that the collapse fraction always tends to unity. So why do we not observe $\Lambda < 0$? The answer is that we have to cut off the calculation at late stages of recollapse: once the Universe becomes too hot, star formation may be affected, and in any case, there is little time for life to form.

With this proviso, Figure 3.7 shows the posterior distribution of Λ conditional on the existence of observers in the multiverse. Provided we consider recollapse only to a maximum temperature of about 10 K, the observed figure is matched well by the anthropic prediction: with this cut-off, most observers will see a positive Λ, and something of order 10% of observers will see Λ as big as we do, or smaller.

So is the anthropic explanation the correct one? Many people find the hypothesis too radical: why postulate an infinity of universes in order to explain a detail of one of them? Certainly, if an alternative explanation for the 'why now' problem existed in the form of, e.g., a naturally successful quintessence model, one might tend to prefer that. But so far, there is no such alternative. The longer this situation persists, the more we will be forced to accept that the Universe we see can only be understood by making proper allowance for our role as observers.

Multiverse and Curvature

To some extent, curvature presents a parallel set of problems to the vacuum. There is a scale problem, in the sense that natural initial conditions might be thought to have a total

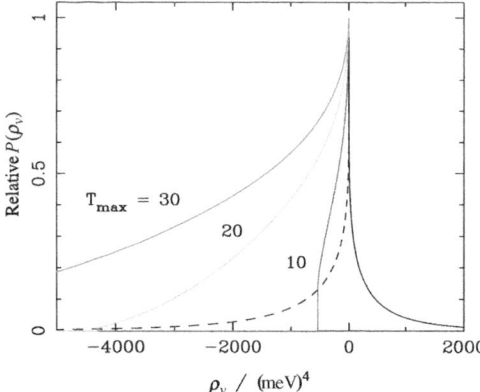

Figure 3.7 The collapse fraction as a function of the vacuum density, which is assumed to give the relative weighting of different models. The dashed line for negative density corresponds to the expanding phase only, whereas the solid lines for negative density include the recollapse phase, up to maximum temperatures of 10 K, 20 K, and 30 K.

$|\Omega - 1|$ of order unity, which would lead to a universe dominated by curvature long before today. There could also be a 'why now' problem if the present curvature was non-zero at the level of $|\Omega - 1| \sim 0.01$, which cannot currently be excluded. It is commonly assumed that inflation solves the curvature scale problem and also predicts that there is no 'why now' problem, but it is interesting to take an anthropic view of the problem. In particular, we might wonder why anthropic arguments were not applied to curvature decades ago, when many cosmologists were convinced that $\Lambda = 0$. At first sight, the issues are similar to Λ: curvature is negligible at high redshift, so we might consider a uniform prior in small early curvature values – as with Λ, appealing to the idea that zero curvature is not special. But in modern models where 'pocket' universes are formed by tunnelling, the result is an open universe, so priors on curvature might well have a discontinuity at zero (e.g., [23]). The idea of a uniform prior for curvature is therefore less well founded than it is for Λ.

In any case, the magnitude of curvature changes the situation. Decades ago, open models were seriously under consideration, and some would have argued for $\Omega_k \simeq 0.7$. This amount of curvature seriously suppresses structure formation, so an anthropic approach to explaining the density parameter in matter-only models would have yielded sensible answers. But today, we know that the Universe is flat to approximately $|\Omega_k| < 0.01$, and such a small value is suspiciously far below any anthropic upper limit. Therefore, even without the theory of inflation, we would not be attracted to anthropic explanations of the flatness problem – even though they would have worked well until perhaps the 1980s. Similarly, if we had no detection of Λ, a sufficiently strong upper limit would reject the anthropic approach, leading us to require a physical mechanism that forces $\Lambda = 0$. Anthropic reasoning is thus testable and could point to new physics. But this is not the situation we face: we have an actual detection of Λ rather than an ever-retreating upper limit, and no a priori

Λ-dominated Universe in which the gravitationally driven assembly of typical galaxies is largely suppressed by $z = 0$. This claim is to be tested by the detailed study of models for cosmological star formation in a variety of unusual contexts (see Chapter 6).

Galaxy formation requires an ability to follow the history of gas within dark matter halos, together with prescriptions for how cold gas turns into stars, followed by possible feedback of energy from the stars and from central black holes. All this must be calculated while following the hierarchical merging of dark matter halos. This can all be computed within an explicit N-body simulation, but there are disadvantages: finite resolution will mean that low-mass halos are not followed, and finite volume will limit the statistics of halos – especially at high masses. The alternative is to generate merger trees of halos via a rapid Monte Carlo algorithm: the semi-analytic approach. This is much faster than direct simulation and evades the problems of mass resolution and limited statistics – but at the price of losing the spatial relation between halos. Sudoh *et al.* [50] have applied such codes to estimate future star formation, as shown in Figure 3.8. They claim that little future star formation is permitted if Λ is raised; this is an interesting conclusion, but the calculations are only taken $\sim 10\,\mathrm{Gyr}$ beyond the present.

In principle, a more reliable alternative will be to run direct numerical simulations of 'counterfactual' universes. It is interesting to note that changing the cosmological constant at high redshift while keeping all other parameters fixed has the effect of transcribing the evolution onto another member of the ΛCDM family so that all the cosmological parameters will be different at the 'present' (i.e., when $T = 2.725\,\mathrm{K}$: we must keep this choice because the transfer function depends on the ratio of matter and radiation densities).

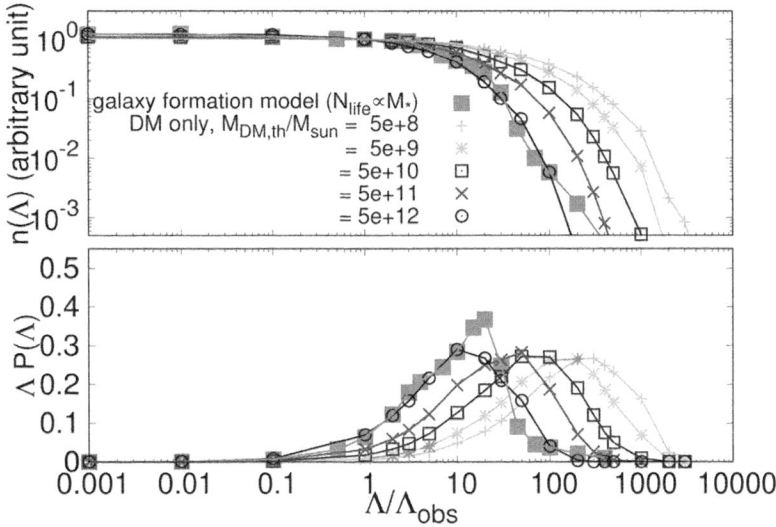

Figure 3.8 The dependence of the total stellar density on Λ, according to the semi-analytic calculations of Sudoh *et al.* [50].

Suppose we scale Λ by a factor α at some early time when the scale factor is a_i. The subsequent evolution of the Hubble parameter is $H^2 = H_i^2[(1 - \alpha\epsilon)(a_i/a)^3 + \alpha\epsilon)]$, where $\epsilon = \rho_v/\rho_{tot}$ at a_i. Note that we assume that spatial flatness is maintained, but provided ϵ is small enough, then we can neglect the $(1 - \alpha\epsilon)$ factor. For a reference standard model, $H^2 = H_{ref}^2(\Omega_{ref}a^{-3} + 1 - \Omega_{ref})$, which allows $H_i^2 a_i^3$ and $H_i^2\epsilon$ to be re-expressed in terms of the reference parameters so that the evolution of H is

$$H^2 = H_{ref}^2[\Omega_{ref}a^{-3} + \alpha(1 - \Omega_{ref})] = H_0^2[\Omega_m a^{-3} + (1 - \Omega_m)], \qquad (3.71)$$

where the last expression is ΛCDM in terms of new parameters. Thus, the new H_0 is $H_0 = H_{ref}[\Omega_{ref} + \alpha(1 - \Omega_{ref})]^{1/2}$, and the new density parameter must satisfy $(1 - 1/\Omega_m) = \alpha(1 - 1/\Omega_{ref})$. The new value of σ_8 is more complicated. If the co-moving length R is $8\,h_{ref}^{-1}$ Mpc, then $\sigma(R)$ is unchanged at high redshift, but $\sigma(R)$ is now altered at $z = 0$ because the change in Ω_m will alter the linear growth factor. Finally, because R is no longer $8\,h^{-1}$ Mpc in terms of the new h, we need to scale $\sigma(R)$ by a factor that depends on the shape of the power spectrum. The results of this exercise are shown in Figure 3.9.

The first simulation studies using this rescaling approach are now just starting to appear. Salcido *et al.* [45] used the EAGLE galaxy formation code to compute star formation ~ 10 Gyr into the future and also compare ΛCDM with an Einstein-de Sitter model in which Λ is set to zero at high redshift. The results are shown in Figure 3.10, and they are intriguing. The peak in the Lilly-Madau curve is still present independent of Λ (not so surprising, as Λ is rather subdominant at the redshift of the peak); star formation may decline into the future or revive, depending on the strength of feedback. These are early days for such studies, which need to integrate for longer and consider a wider range of

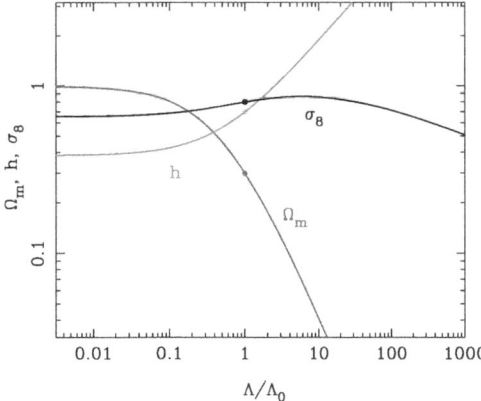

Figure 3.9 This plot shows how the principal ΛCDM parameters respond to a scaling of the cosmological constant while maintaining the high-redshift Universe otherwise unchanged and exactly flat. The 'present' is always defined as CMB temperature 2.725 K. Solid points show the default cosmology $(\Omega_m, h, \sigma_8) = (0.3, 0.7, 0.8)$, and how this scales as Λ is altered from the default value Λ_0 (which corresponds to $\Omega_v = 0.7$).

Acknowledgements

My research is currently supported by the European Research Council, under the COS-FORM grant, no. 670193.

References

[1] Abbott, T. M. C., *et al.* 2018. 'Dark Energy Survey Year 1 Results: Cosmological Constraints from Galaxy Clustering and Weak Lensing'. *Physical Review D*, **98**(4), 043526.

[2] Ade, P. A. R., *et al..* 2016. 'Planck 2015 Results. XIII. Cosmological Parameters'. *Astronomy and Astrophysics*, **594**(Sept.), A13.

[3] Armendariz-Picon, C., Mukhanov, V., and Steinhardt, P. J. 2001. 'Essentials of k-Essence'. *Physical Review D*, **63**(May), 103510.

[4] Barrow, J. D., and Tipler, F. J. 1986. *The Anthropic Cosmological Principle.* Oxford University Press.

[5] Bernstein, J. and Feinberg, G. 1989. *Cosmological Constants: Papers in Modern Cosmology: A Selection of the Library Science and Astronomy Book Clubs.* Columbia University Press.

[6] Bostrom, N. 2002. *Anthropic Bias: Observation Selection Effects in Science and Philosophy.* Routledge.

[7] Burbidge, E. M., *et al.* 1957. 'Synthesis of the Elements in Stars'. *Reviews of Modern Physics*, **29**, 547–650.

[8] Caldwell, R. R., Kamionkowski, M., and Weinberg, N. N. 2003. 'Phantom Energy: Dark Energy with w< −1 Causes a Cosmic Doomsday'. *Physical Review Letters*, **91**(7), 071301.

[9] Carr, B. 2009. *Universe or Multiverse?* Cambridge University Press.

[10] Carter, B. 1974. 'Large Number Coincidences and the Anthropic Principle in Cosmology', in Longair, M S., ed. *Confrontation of Cosmological Theories with Observational Data.* IAU Symposium, vol. 63, Pages 291–298.

[11] Carter, B. 1983. 'The Anthropic Principle and Its Implications for Biological Evolution'. *Philosophical Transactions of the Royal Society of London Series A*, **310**(Dec.), 347–363.

[12] Clifton, T., *et al.* 2012. 'Modified Gravity and Cosmology'. *Physics Reports*, **513**(Mar.), 1–189.

[13] Copeland, E. J., Sami, M., and Tsujikawa, S. 2006. 'Dynamics of Dark Energy'. *International Journal of Modern Physics D*, **15**(Jan.), 1753–1935.

[14] Daniel, S. F., *et al.* 2010. 'Testing General Relativity with Current Cosmological Data'. *Physical Review D*, **81**(12), 123508.

[15] Davé, R., *et al.* 2001. 'Baryons in the Warm-Hot Intergalactic Medium'. *Astrophysical Journal*, **552**(May), 473–483.

[16] Dawkins, R. 2004. *A Devil's Chaplain: Selected Essays.* Phoenix.

[17] de Graaff, A., *et al.* 2017. 'Missing Baryons in the Cosmic Web Revealed by the Sunyaev-Zel'dovich Effect'. *ArXiv e-prints*, Sept., arXiv:1709.10378.

[18] Dicke, R. H., 1961. 'Dirac's Cosmology and Mach's Principle'. *Nature*, **192**(Nov.), 440–441.

[19] Dirac, P. A. M. 1937. 'The Cosmological Constants'. *Nature*, **139**(Feb.), 323.

[20] Efstathiou, G. 1995. 'An Anthropic Argument for a Cosmological Constant'. *Monthly Notices of the Royal Astronomical Society*, **274**(June), L73–L76.

[21] Einstein, A. 2006. *Kosmologische Betrachtungen zur allgemeinen Relativittstheorie.* John Wiley & Sons, Ltd. Pages 119–130.

[22] Eke, V. R., *et al.* 2005. 'Where Are the Stars?'. *Monthly Notices of the Royal Astronomical Society*, **362**(Oct.), 1233–1246.

[23] Freivogel, B., *et al.* 2006. 'Observational Consequences of a Landscape'. *Journal of High Energy Physics*, **2006**(Mar.), 039.

[24] Garriga, J., and Vilenkin, A. 2006. 'Anthropic Prediction for Λ and the Q Catastrophe'. *Progress of Theoretical Physics Supplement*, **163**(Jan.), 245–257.

[25] Garriga, J., Livio, M., and Vilenkin, A. 2000. 'Cosmological Constant and the Time of Its Dominance'. *Physical Review D*, **61**(2), 023503.

[26] Gibbons, G. W., and Turok, N. 2008. 'Measure Problem in Cosmology'. *Physical Review D*, **77**(6), 063516.

[27] Guth, A. H. 2000. 'Inflation and Eternal Inflation'. *Physics Reports*, **333**(Aug.), 555–574.

[28] Hoyle, F., *et al.* 1953. 'A State in C^{12} Predicted from Astrophysical Evidence'. *Physical Review*, **92**, 1095.

[29] Joyce, A., *et al.* 2015. 'Beyond the Cosmological Standard Model'. *Physics Reports*, **568**(Mar.), 1–98.

[30] Koksma, J. F., and Prokopec, T. 2011. 'The Cosmological Constant and Lorentz Invariance of the Vacuum State'. *ArXiv e-prints*, May, arXiv:1105.6296.

[31] Laureijs, R., *et al.* 2011. 'Euclid Definition Study Report'. *ArXiv e-prints*, Oct., arXiv:1110.3193.

[32] Liddle, A. R., and Scherrer, R. J. 1999. 'Classification of Scalar Field Potentials with Cosmological Scaling Solutions'. *Physical Review D*, **59**(2), 023509.

[33] Linde, A. 2007. 'Sinks in the Landscape, Boltzmann Brains and the Cosmological Constant Problem'. *Journal of Cosmology and Astroparticle Physics*, **1**(Jan.), 022.

[34] Linde, A., Linde, D., and Mezhlumian, A. 1994. 'From the Big Bang Theory to the Theory of a Stationary Universe'. *Physical Review D*, **49**(Feb.), 1783–1826.

[35] Mac Low, M. M., 2013. 'From Gas to Stars over Cosmic Time'. *Science*, **340**(6140).

[36] Mackay, D. J. C. 2003. *Information Theory, Inference and Learning Algorithms.* Cambridge University Press.

[37] Malquarti, M., Copeland, E. J., and Liddle, A. R. 2003. 'k-Essence and the Coincidence Problem'. *Physical Review D*, **68**(July), 023512.

[38] Marsh, D. J. E. 2016. 'Axion Cosmology'. *Physics Reports*, **643**(July), 1–79.

[39] Olum, K. D. 2002. 'The Doomsday Argument and the Number of Possible Observers'. *The Philosophical Quarterly (1950-)*, **52**(207), 164–184.

[40] Page, D. N. 2008. 'Return of the Boltzmann Brains'. *Physical Review D*, **78**(6), 063536.

[41] Peacock, J. A. 2007. 'Testing Anthropic Predictions for Λ and the Cosmic Microwave Background Temperature'. *Monthly Notices of the Royal Astronomical Society*, **379**(Aug.), 1067–1074.

[42] Percival, W. J., *et al.* 2010. 'Baryon Acoustic Oscillations in the Sloan Digital Sky Survey Data Release 7 Galaxy Sample'. *Monthly Notices of the Royal Astronomical Society*, **401**(Feb.), 2148–2168.

[43] Ratra, B., and Peebles, P. J. E. 1988. 'Cosmological Consequences of a Rolling Homogeneous Scalar Field'. *Physical Review D*, **37**(June), 3406–3427.

where $\rho_T = \sum_{i=1}^{N} \rho_i$ and $p_T = \sum_{i=1}^{N} p_i$ denote the total energy density and pressure (assuming a vanishing cosmological constant or including its contribution in an extra fluid, as in (4.12) or (4.13)). Since the energy density of matter must be positive, we see that (4.11) requires a negative pressure – i.e., some exotic form of matter.

Even if the Einstein equations have been considerably simplified by the use of the cosmological principle, they remain difficult to solve analytically. However, it turns out that if the curvature term vanishes and if there is only one fluid with a constant equation of state, an exact solution to the Einstein equations is available. Of course, one can always solve these equations numerically, but exact solutions will be interesting when we discuss the puzzles of the hot Big Bang phase in the next sections. For this reason, we briefly present them. Since the equation of state is supposed to be constant, the conservation equation, which can be written as

$$\dot{\rho} + 3H(1+w)\rho = 0, \tag{4.12}$$

can be integrated exactly, and the solution reads

$$\rho(t) = \rho_f \left(\frac{a_f}{a}\right)^{3(1+w)}, \tag{4.13}$$

where ρ_f and a_f are the energy density and the scale factor expressed at a fiducial time t_f that can be chosen arbitrarily. Then, one inserts (4.13) in the Friedmann equation, namely

$$\left(\frac{1}{a}\frac{da}{dt}\right)^2 = \frac{\rho_f}{3M_{Pl}^2}\left(\frac{a_f}{a}\right)^{3(1+w)}, \tag{4.14}$$

whose solution can also be found and reads

$$\left(\frac{a}{a_f}\right)^{\frac{3(1+w)}{2}} = \frac{3(1+w)}{2}\frac{\rho_f^{1/2}}{\sqrt{3}M_{Pl}}t + C. \tag{4.15}$$

In this expression, C is an integration constant. Requiring that $a = a_f$ when $t = t_f$, one finds that $C = -3(1+w)\rho_f^{1/2}t_f/(2\sqrt{3}M_{Pl})+1$. Finally, noticing that $H_f = \rho_f^{1/2}/(\sqrt{3}M_{Pl})$, one arrives at

$$a(t) = a_f \left[\frac{3}{2}(1+w)H_f\left(t - t_f\right) + 1\right]^{\frac{2}{3(1+w)}}. \tag{4.16}$$

The corresponding Hubble parameter can be expressed as $H(t) = H_f/[3(1+w)H_f\left(t - t_f\right)/2+1]$. We notice that the scale factor vanishes when $t = t_{BB}$ with $t_{BB} = t_f - 2/[3(1+w)H_f]$. In some sense, 'time begins' at t_{BB}, and it would be meaningless to consider times such that $t < t_{BB}$. This is the famous Big Bang point where the classical analysis breaks down. This singularity is, of course, a serious problem for the hot Big Bang model. However, it is not considered to be a problem for inflation simply because inflation does not aim to address it. It could be solved if, prior to inflation, there is a bounce [19, 22] or if quantum gravitational effects take over and somehow regularise the singularity, as done, for instance, in quantum cosmology [54]. We see that the singularity problem can be treated separately and does not involve the inflationary scenario.

For future convenience, it is also interesting to rewrite the scale factor in terms of t_{BB}, and one obtains $a(t) = a_{\mathrm{f}} \left[\frac{3}{2}(1+w)H_{\mathrm{f}} \left(t - t_{\mathrm{BB}} \right) \right]^{\frac{2}{3(1+w)}}$ and $H(t) = 2/[3(1+w)\left(t - t_{\mathrm{BB}}\right)]$. If, in addition, one chooses $t_{\mathrm{BB}} = 0$ (which can always be done), then the scale factor takes the form (with this parameterisation, $H_{\mathrm{f}} = 2/\left[3(1+w)t_{\mathrm{f}}\right]$)

$$a(t) = a_{\mathrm{f}} \left(\frac{t}{t_{\mathrm{f}}} \right)^{\frac{2}{3(1+w)}}, \tag{4.17}$$

that is to say, a power-law function. For radiation, $w = 1/3$, the scale factor behaves as $a(t) \propto t^{1/2}$ and for pressureless matter, $w = 0$, one has $a(t) \propto t^{2/3}$. We also notice that the previous expressions are ill-defined if $w = -1$. This is just because, in that case, we have an exponential solution, namely $a(t) = a_{\mathrm{f}} \exp\left[H_{\mathrm{f}}\left(t - t_{\mathrm{f}}\right)\right]$, known as the de Sitter solution.

Putting aside the particular case $w = -1$, let us finally come back to the fact that, for $t = t_{\mathrm{BB}}$, the scale factor vanishes. This is clearly not an artefact of the coordinate system used, as is confirmed by a calculation of the scalar curvature

$$R = \frac{4(1-3w)}{3(1+w)^2} \frac{1}{\left(t - t_{\mathrm{BB}}\right)^2}, \tag{4.18}$$

which blows up when $t \to t_{\mathrm{BB}}$. This confirms the $t = t_{\mathrm{BB}}$ corresponds to a real singularity.[1]

Having introduced the theoretical tools needed in order to understand the hot Big Bang model, we now discuss the parameters that describe the model and how their values can be inferred from cosmological data.

4.2.2 The Real Universe

In order to describe our Universe, we need to know its energy budget, namely the contribution of the different forms of energy density present in the Universe. Our Universe is made of photons with energy density ρ_γ, neutrinos with energy density ρ_ν, baryons with energy density ρ_{b}, cold dark matter with energy density ρ_{c}, and dark energy with energy density ρ_Λ (here assumed to be a cosmological constant). Photons and neutrinos have an equation of state $1/3$, baryons and cold dark matter have a vanishing equation of state and, finally,

---

[1] Notice also that, for radiation, R is identically zero. Of course, this does not mean that there is no singularity in a radiation-dominated epoch. This can be shown by computing another invariant – for instance, $R_{\mu\nu}R^{\mu\nu}$, which reads

$$R_{\mu\nu}R^{\mu\nu} = R_{00}R^{00} + R_{ij}R^{ij} = 9\left(\frac{\ddot{a}}{a}\right)^2 + \left(\frac{\ddot{a}}{a} + 2\frac{\dot{a}^2}{a^2}\right)^2 g_{ij}g^{ij}$$
$$= 12\left(\frac{\ddot{a}}{a}\right)^2 + 12\frac{\ddot{a}}{a}\frac{\dot{a}^2}{a^2} + 12\left(\frac{\dot{a}}{a}\right)^4$$
$$= \frac{48(3w^2+1)}{27(1+w)^4} \frac{1}{\left(t - t_{\mathrm{BB}}\right)^4}. \tag{4.19}$$

Clearly, $R_{\mu\nu}R^{\mu\nu}$ blows up as $t \to t_{\mathrm{BB}}$ even if $w = 1/3$.

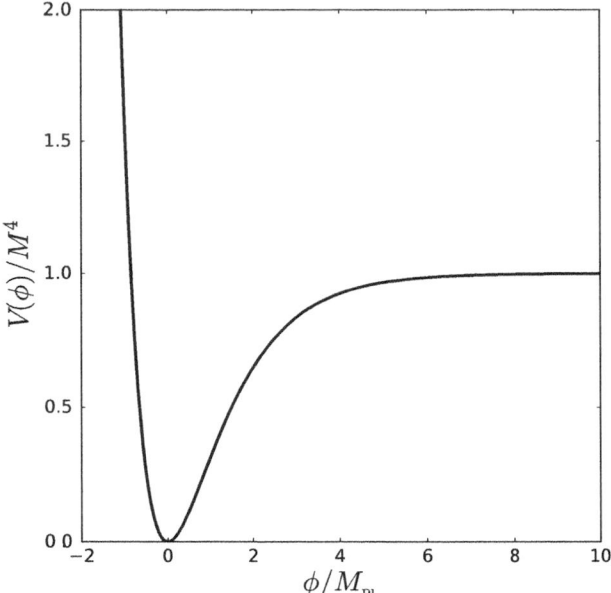

Figure 4.2 Example of a potential the Starobinsky potential (4.87) that can support inflation. Slow-roll inflation occurs along the plateau where the potential is almost flat, and the reheating phase takes place when the field oscillates around its minimum, here located at the origin.

$V(\phi)$. An example of a potential that supports inflation is given in Figure 4.2. From CMB data, one can constrain this function, and this will be discussed in the following. As already mentioned, the interaction term plays a crucial role in the process which ends inflation. Indeed, it controls how the inflaton field decays into particles describing ordinary matter. These decay products are then supposed to thermalise, and the radiation-dominated epoch starts at a temperature which is known as the reheating temperature $T_{\rm rh}$. This quantity is an important parameter of any inflationary model, and we will see that the CMB data can also say something about its value.

Following the preceding considerations, during inflation itself, the interaction term is neglected, and the evolution of the system is controlled by the Friedmann and Klein-Gordon equations, namely

$$H^2 = \frac{1}{3M_{\rm Pl}^2}\left[\frac{\dot{\phi}^2}{2} + V(\phi)\right], \tag{4.55}$$

$$\ddot{\phi} + 3H\dot{\phi} + V_\phi = 0, \tag{4.56}$$

where a subscript ϕ means a derivative with respect to the inflaton field. For an arbitrary potential, this system of equations cannot be solved analytically. This means that we have to use either numerical calculations or a perturbative method. In general, a perturbative method is based on the presence of a small parameter in the problem and on an expansion of the relevant quantities in terms of this small parameter. In the case of inflation, there exists such a small parameter which physically expresses the fact that the potential is flat. So it can be chosen as the curvature of the potential or, equivalently, as the kinetic to potential energy ratio or, given that inflation corresponds to an approximately constant Hubble parameter, as the derivative of H. Therefore, we introduce the Hubble flow functions ϵ_n defined by [67, 117]

$$\epsilon_{n+1} \equiv \frac{\mathrm{d}\ln|\epsilon_n|}{\mathrm{d}N}, \quad n \geq 0, \tag{4.57}$$

where $\epsilon_0 \equiv H_{\mathrm{ini}}/H$ starts the hierarchy and $N \equiv \ln(a/a_{\mathrm{ini}})$ is the number of e-folds. From the preceding expression, the first Hubble flow parameter can be written as

$$\epsilon_1 = -\frac{\dot{H}}{H^2} = 1 - \frac{\ddot{a}}{aH^2} = \frac{3\dot{\phi}^2}{2} \frac{1}{\dot{\phi}^2/2 + V(\phi)}, \tag{4.58}$$

and, therefore, inflation ($\ddot{a} > 0$) occurs if $\epsilon_1 < 1$. In terms of the Hubble flow parameters, the Friedmann and Klein-Gordon equations take the form

$$H^2 = \frac{V}{M_{\mathrm{Pl}}^2(3 - \epsilon_1)}, \tag{4.59}$$

$$\left(1 + \frac{\epsilon_2}{6 - 2\epsilon_1}\right)\frac{\mathrm{d}\phi}{\mathrm{d}N} = -M_{\mathrm{Pl}}^2 \frac{\mathrm{d}\ln V}{\mathrm{d}\phi}. \tag{4.60}$$

It is worth stressing the point that these expressions are exact. The condition $\epsilon_1 < 1$ during ~ 60 e-folds is sufficient to solve the fine-tuning problems of the Standard Model, as discussed earlier. But, if one wants to describe properly the CMB anisotropy (as in Eqs. (4.61)(4.63)), one needs $\epsilon_n \ll 1$, which is called the slow-roll regime. In this situation, the first three Hubble flow parameters can be approximated as [69]

$$\epsilon_1 \simeq \frac{M_{\mathrm{Pl}}^2}{2}\left(\frac{V_\phi}{V}\right)^2, \tag{4.61}$$

$$\epsilon_2 \simeq 2M_{\mathrm{Pl}}^2\left[\left(\frac{V_\phi}{V}\right)^2 - \frac{V_{\phi\phi}}{V}\right], \tag{4.62}$$

$$\epsilon_2\epsilon_3 \simeq 2M_{\mathrm{Pl}}^4\left[\frac{V_{\phi\phi\phi}V_\phi}{V^2} - 3\frac{V_{\phi\phi}}{V}\left(\frac{V_\phi}{V}\right)^2 + 2\left(\frac{V_\phi}{V}\right)^4\right]. \tag{4.63}$$

We see that the first Hubble flow parameter is also a measure of the steepness of the potential and of its first derivative. The second Hubble flow parameter is a measure of the second derivative of the potential, and so on. Therefore, if one can observationally constrain the values of the Hubble flow parameters, we can say something about the shape of the inflationary potential. The slow-roll approximation also allows us to simplify the equations of motion and to analytically integrate the inflaton trajectory. Indeed, in this

Let us now turn to a quantitative characterisation of the cosmological fluctuations. The amplitude of scalar perturbations is described by the curvature perturbations [18, 94] $\zeta(\eta,\boldsymbol{x}) \equiv \Phi + 2(\mathcal{H}^{-1}\Phi' + \Phi)/(3 + 3w)$, with $w = p/\rho$ the equation of state during inflation and Φ the Bardeen potential [17] (not to be confused with the scalar field ϕ). The Bardeen potential is the quantity that describes scalar perturbations as revealed by writing explicitly the perturbed metric in longitudinal gauge, $\mathrm{d}s^2 = a^2(\eta)[-(1-2\Phi)\mathrm{d}\eta^2 + (1 - 2\Phi)\delta_{ij}\mathrm{d}x^i\mathrm{d}x^j]$. Since we deal with a linear theory, we can go to Fourier space and follow the time evolution of the Fourier component $\zeta_{\boldsymbol{k}}(\eta)$. Then, the properties of the fluctuations are described by the power spectrum of scalar perturbations, which is given by

$$\mathcal{P}_\zeta(k) = \frac{k^3}{2\pi^2}|\zeta_{\boldsymbol{k}}|^2. \tag{4.69}$$

The power spectrum depends on the model of inflation – that is to say, for the simple class of models discussed here, on the potential $V(\phi)$. Unfortunately, there exists no exact analytic calculation of $\mathcal{P}_\zeta(k)$ for an arbitrary $V(\phi)$. Therefore, one must rely on either numerical calculations or perturbative methods. Here again, the slow-roll approximation can be used, and it leads to the following result [67]

$$\mathcal{P}_\zeta(k) = \mathcal{P}_{\zeta 0}(k_{\mathrm{p}}) \left[a_0^{(\mathrm{S})} + a_1^{(\mathrm{S})} \ln\left(\frac{k}{k_{\mathrm{p}}}\right) + \frac{a_2^{(\mathrm{S})}}{2} \ln^2\left(\frac{k}{k_{\mathrm{p}}}\right) + \cdots \right], \tag{4.70}$$

where k_{p} is a pivot scale, and the overall amplitude can be written as

$$\mathcal{P}_{\zeta 0} = \frac{H_*^2}{8\pi^2\epsilon_{1*}M_{\mathrm{Pl}}^2}. \tag{4.71}$$

In the preceding expression (and in the subsequent ones), a star means that the corresponding quantity has been evaluated at the time at which the pivot scale crossed out the Hubble radius during inflation, namely $k_{\mathrm{p}} \sim a_* H_*$. The amplitude of the spectrum depends on (the square of) the strength of the gravitational field during inflation, which is described by the expansion rate H_*. It is also inversely proportional to the first derivative of the potential through the presence of ϵ_{1*} in the denominator. The main property of $\mathcal{P}_{\zeta 0}$ is that it is does not depend on the wave number; in other words, it is scale independent. This result represents one of the main successes of inflation since a scale-invariant power spectrum was known for a long time to be in agreement with the observations. But there is even more. We see that the scale-invariant piece of the power spectrum receives scale-dependent logarithmic corrections, the amplitudes of which are controlled by the Hubble flow parameters and given by [29–31, 33, 46, 67, 78, 102, 117],

$$a_0^{(S)} = 1 - 2\left(C + 1\right)\epsilon_{1*} - C\epsilon_{2*} + \left(2C^2 + 2C + \frac{\pi^2}{2} - 5\right)\epsilon_{1*}^2$$

$$+ \left(C^2 - C + \frac{7\pi^2}{12} - 7\right)\epsilon_{1*}\epsilon_{2*} + \left(\frac{1}{2}C^2 + \frac{\pi^2}{8} - 1\right)\epsilon_{2*}^2$$

$$+ \left(-\frac{1}{2}C^2 + \frac{\pi^2}{24}\right)\epsilon_{2*}\epsilon_{3*} + \cdots, \tag{4.72}$$

$$a_1^{(S)} = -2\epsilon_{1*} - \epsilon_{2*} + 2(2C + 1)\epsilon_{1*}^2 + (2C - 1)\epsilon_{1*}\epsilon_{2*} + C\epsilon_{2*}^2 - C\epsilon_{2*}\epsilon_{3*} + \cdots, \tag{4.73}$$

$$a_2^{(S)} = 4\epsilon_{1*}^2 + 2\epsilon_{1*}\epsilon_{2*} + \epsilon_{2*}^2 - \epsilon_{2*}\epsilon_{3*} + \cdots, \tag{4.74}$$

$$a_3^{(S)} = \mathcal{O}(\epsilon_{n*}^3), \tag{4.75}$$

where $C \equiv \gamma_E + \ln 2 - 2 \approx -0.7296$, γ_E being the Euler constant. Since the coefficients $a_1^{(S)}, a_2^{(S)}$ etc.... are small (being proportional to the Hubble flow parameters), this means that the inflationary power spectrum is not exactly scale invariant but, in fact, almost scale invariant. This is the main prediction of inflation, and it was confirmed recently by the CMB Planck data. We stress that this is a prediction since it was made before it was measured. In terms of spectral index, being defined as the logarithmic derivative of $\ln \mathcal{P}_\zeta(k)$, one has

$$n_s = 1 - 2\epsilon_{1*} - \epsilon_{2*}, \tag{4.76}$$

where $n_s = 1$ corresponds to exact scale invariance. We see in this expression that the small deviations from exact scale invariance carry information about the shape of the inflationary potential since ϵ_1 and ϵ_2 respectively depend on the first and second derivative of $V(\phi)$. Therefore, an accurate measurement of the power spectrum can provide information about which version of inflation was realised in the early Universe.

We have also mentioned that gravitational waves are produced during inflation. The corresponding treatment is very similar to the one we have just described. In particular, the tensor power spectrum $\mathcal{P}_h$ can be written as

$$\mathcal{P}_h(k) = \mathcal{P}_{h0}(k_p)\left[a_0^{(T)} + a_1^{(T)}\ln\left(\frac{k}{k_p}\right) + \frac{a_2^{(T)}}{2}\ln^2\left(\frac{k}{k_p}\right) + \cdots\right], \tag{4.77}$$

with a scale-invariant overall amplitude that can be expressed as

$$\mathcal{P}_{h0} = \frac{2H_*^2}{\pi^2 M_{\text{Pl}}^2}. \tag{4.78}$$

This time, and contrary to scalar perturbations, the amplitude only depends on the Hubble parameter during inflation. This has a very important implication: if one can measure the amplitude of tensor power spectrum, then one immediately determines the expansion rate during inflation or, in other words, the energy scale of inflation. Unfortunately, the inflationary gravitational waves have not yet been detected. As for scalar perturbations,

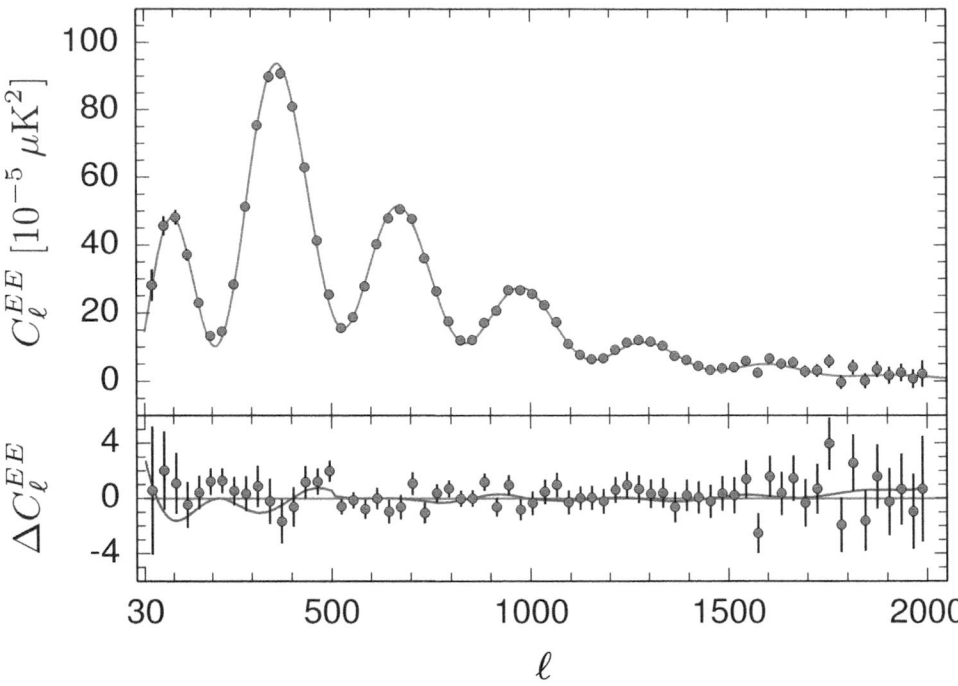

Figure 4.5 Same as in Figure 4.4 but for the E-mode power spectrum obtained from Planck 2015. Figure taken from Reference [3].

Planck satellite is probably the measurement of the scalar spectral index, which is found to be [4]

$$n_s = 0.9645 \pm 0.0049. \tag{4.85}$$

It is a crucial result since this is the first time that a deviation from $n_s = 1$ is measured at a statistically significant level (say, more than 5σ). It is clearly a strong point in favour of inflation. As was discussed previously, inflation also predicts the presence of a background of gravitational waves, and, unfortunately, we do not yet have a detection of those primordial gravity waves. This means that we only have an upper bound on the parameter r, namely

$$r \lesssim 0.07, \tag{4.86}$$

obtained by combining the Planck data and the BICEP/Keck data [2]. As already mentioned, the Planck data are also compatible with no non-Gaussianity [7] and no non-adiabatic perturbations [4], which is compatible with the simplest model of inflation.

One can also use the Planck data to constrain the shape of the inflationary potential. The performance of a model can be described by two numbers: the Bayesian evidence

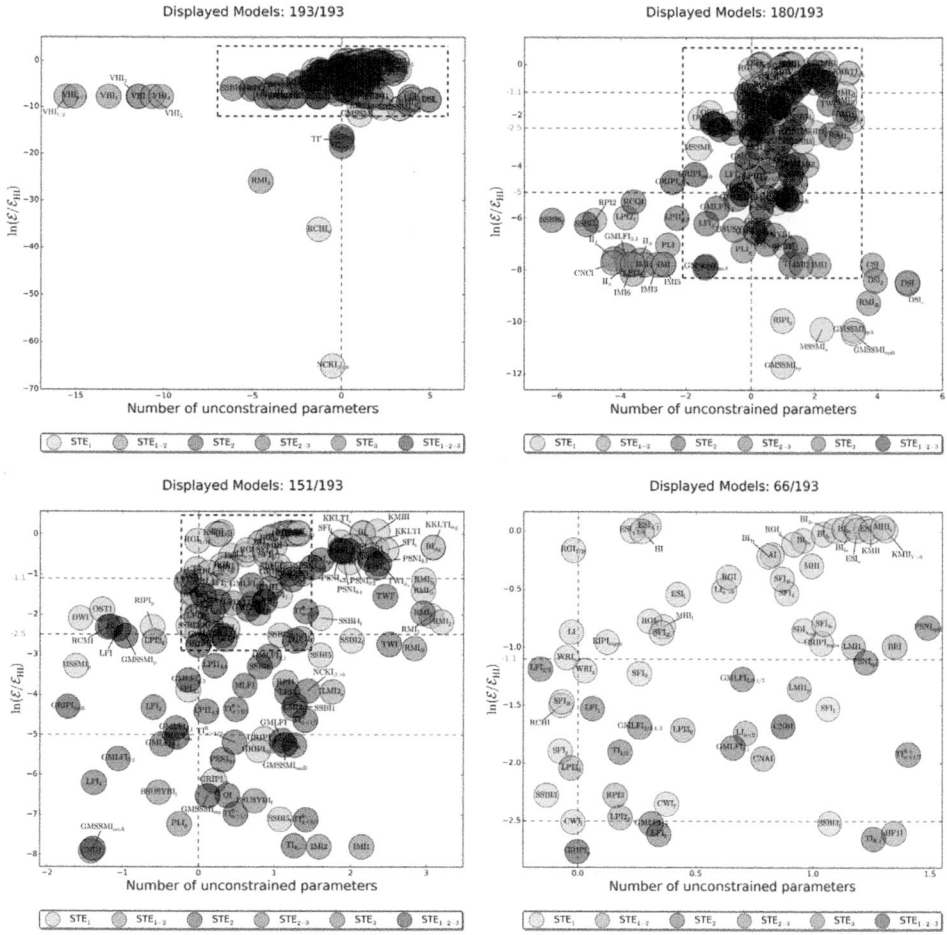

Figure 4.6 Inflationary models in the space $\left(N_{\mathrm{uc}}, \ln B_{\mathrm{REF}}^i\right)$. N_{nuc} represents the number of unconstrained parameters of a given model while B_{REF}^i is the evidence of a given model i to evidence of a reference model ratio. Each model is represented by a circle (the radius of which has no meaning) with its acronym, taken from Reference [103], written inside. The four panels corresponds to successive zooms towards the best region (indicated by the dashed rectangles). Figures taken from Reference [103].

[131, 132], which characterises the ability of the model to fit the data in a simple way, and the Bayesian complexity [63], which is related to the number of unconstrained parameters (given a data set). A good model is a model that has a large Bayesian evidence and no unconstrained parameters. In Figure 4.6, we have represented the Bayesian evidence and complexity for nearly 200 models of inflation, given the Planck data [76, 77, 84, 92, 97–99, 103]. Based on this analysis, it is found that potentials with a plateau are favoured by the

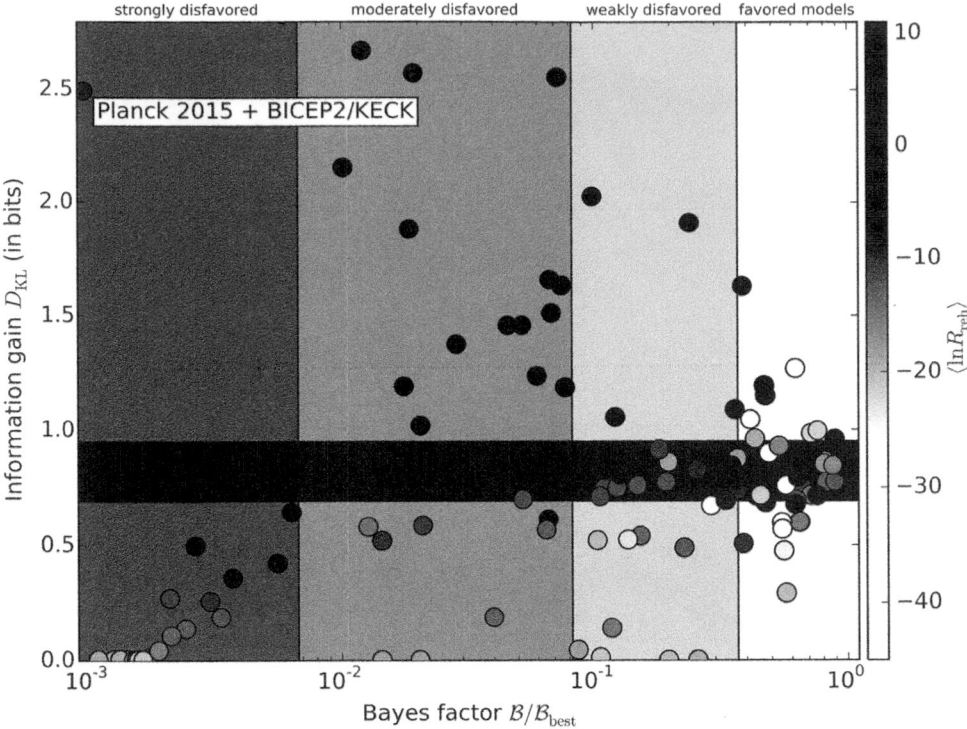

Figure 4.7 Observational constraints on reheating from the Planck and BICEP2/KECK data. The vertical axis a measure of how tight the constraint on the reheating parameter is while the horizontal axis represents the Bayesian evidence, namely the performance of a model. Each circle represents a model of inflation. The shadings give the best value of the reheating parameter.

data, the prototypical example being the Starobinsky model [121], for which the potential is given by

$$V(\phi) = M^4 \left(1 - e^{-\sqrt{2/3}\phi/M_{\mathrm{Pl}}}\right)^2 . \qquad (4.87)$$

Reheating can also be constrained by means of the Planck data [84, 92, 93, 100, 101]; see Figure 4.7. We have seen that the only piece of information about the end of inflation that can be extracted from CMB data is the posterior distribution of the reheating parameter. In order to quantify whether the constraint is tight or not, one then has to compare the posterior to the prior. In technical terms, this is given by the Kullback-Leibler divergence, D_{KL}, between the prior and the posterior. In Figure 4.7, we have represented D_{KL} as a function of the Bayesian evidence for the nearly 200 models of inflation, already studied in Figure 4.6. Each model is represented by a circle. The yellow band corresponds to the one-sigma deviation around the mean value, which is given by $\langle D_{\mathrm{KL}} \rangle = 0.82 \pm 0.13$. This corresponds to the information value of almost one bit, and, therefore, this confirms that reheating is constrained by CMB data. Of course, it is not straightforward to translate these constraints into constraints on the reheating temperature unless one specifies $\overline{w}_{\mathrm{reh}}$ explicitly, in which case the reheating parameter and the reheating temperature are in one-to-one correspondence.

4.5 Is Inflation Fine-Tuned? Choosing the Free Parameters of the Inflationary Potential

In this section, we turn to the question of whether inflation, which was invented in order to solve fine-tuning problems, is itself fine-tuned. Let us discuss the first aspect of the problem, namely how the parameters of the potential must be chosen and what their numerical values are in order for the model to correctly account for the data. Let us start with a particular model, namely Large Field Model (LFI, for Large Field Inflation) for which the potential is given by

$$V(\phi) = M^4 \left(\frac{\phi}{M_{\mathrm{Pl}}} \right)^p, \tag{4.88}$$

where M and p are two free parameters. Using Eq. (4.64), one can calculate the slow-roll trajectory, and one finds

$$\phi(N) = \sqrt{\phi_{\mathrm{ini}}^2 - 2pM_{\mathrm{Pl}}^2(N - N_{\mathrm{ini}})}. \tag{4.89}$$

In order to calculate the spectral index and the scalar-to-tensor ratio, one must calculate the Hubble flow parameters. Using the expressions of ϵ_1 and ϵ_2 in the slow-roll approximation, one obtains (see Eqs. (4.61) and (4.62))

$$\epsilon_1 = \frac{p^2 M_{\mathrm{Pl}}^2}{2\phi^2}, \quad \epsilon_2 = \frac{2pM_{\mathrm{Pl}}^2}{\phi^2}. \tag{4.90}$$

This immediately leads to the vacuum expectation value at which inflation ends since the condition $\epsilon_1 = 1$ implies $\phi_{\mathrm{end}}/M_{\mathrm{Pl}} = p/\sqrt{2}$. Then, we must evaluate the Hubble flow parameters at the time that was previously denoted with a star, namely the time at which the pivot scale crossed out the Hubble radius during inflation. Using the slow-roll trajectory, it is easy to show that $\phi_*^2/M_{\mathrm{Pl}}^2 = p^2/2 + 2p\Delta N_*$, where $\Delta N_* = N_{\mathrm{end}} - N_*$ with N_{end} the number of e-folds at the end of inflation and N_* the number of e-folds at Hubble radius exit. In terms of ΔN_*, the Hubble flow parameters read

$$\epsilon_{1*} = \frac{p}{4(\Delta N_* + p/4)}, \quad \epsilon_{2*} = \frac{1}{\Delta N_* + p/4}. \tag{4.91}$$

As a consequence, one has

$$n_{\mathrm{s}} - 1 = -\frac{p+2}{2\Delta N_* + p/2}, \quad r = \frac{4p}{\Delta N_* + p/4}. \tag{4.92}$$

The measurements of n_{s} and the constraints on r can therefore allow us to put constraints on the parameter p. But we also see that the spectral index and the tensor-to-scalar ratio do not depend on the other free parameter, namely M. This one is, in fact, fixed by the amplitude of the fluctuations (i.e., the 'COBE normalisation') – that is to say, by the fact that $\delta T/T \sim 10^{-5}$. Using Eq. (4.71) and the slow-roll approximation for the Friedmann equation, one obtains

$$\frac{M^4}{M_{\rm Pl}^4} = 12\pi^2 p^2 \left(\frac{\phi_*}{M_{\rm Pl}}\right)^{-p-2} \mathcal{P}_{\zeta 0} = 12\pi^2 p^2 \left(\frac{p^2}{2} + 2p\Delta N_*\right)^{-p/2-1} \mathcal{P}_{\zeta 0}. \qquad (4.93)$$

Knowing that the Planck 2015 data [2–4] indicate that

$$\ln\left(10^{10}\mathcal{P}_{\zeta 0}\right) = 3.094 \pm 0.0049, \qquad (4.94)$$

one finds that $M/M_{\rm Pl} \simeq 1.3 \times 10^{-3}$ for $p = 2$ and $M/M_{\rm Pl} \simeq 3 \times 10^{-4}$ for $p = 4$. In order to obtain these numbers, we have assumed $\Delta N_* = 55$ and a comment is in order at this stage. In principle, one should not assume a value for ΔN_* since it is determined once the reheating temperature and the mean equation of state parameter during reheating have been chosen [84, 92, 93, 100, 101]. It can be quite dangerous to choose a 'reasonable' value blindly because, sometimes, it could imply a reheating energy density higher than the energy density at the end of inflation, which is clearly meaningless. In fact, the dependence in ΔN_* of $n_{\rm s}$ and r is precisely the reason why one can use the CMB to put constraints on the reheating epoch, as explained in the previous sections. Indeed, ΔN_* cannot take arbitrary values; otherwise, the corresponding spectral index and tensor-to-scalar ratio would be incompatible with the data. But since ΔN_* depends on $T_{\rm reh}$ and $\overline{w}_{\rm reh}$, this means that those quantities cannot take arbitrary values as well, or, to put it differently, they are constrained by the CMB data. Nevertheless, one can show that, for large-field inflation, ΔN_* can vary in a quite small range around the value $\Delta N_* = 55$, and this is the reason why we choose this value. Considering another value would not affect our numerical estimate much and would change nothing in the present discussion.

The estimates of the mass-scale M derived earlier show that inflation in this model takes place around the GUT scale. But let us consider the case $p = 4$ and write the potential as $V(\phi) = \lambda\phi^4$, where λ is a dimensionless coupling constant. Clearly, $\lambda = M^4/M_{\rm Pl}^4$, which implies that $\lambda \sim 10^{-13}$. This very small value can be viewed as a fine-tuning, at least if one adopts the standard lore that absence of fine-tuning means that dimensionless quantities should be 'naturally' of order one. Let us now consider the case $p = 2$ and write the corresponding potential as $V(\phi) = m^2\phi^2/2$, where m is the mass of the inflaton field. In that case, one has $m = \sqrt{2}(M/M_{\rm Pl})^2 M_{\rm Pl}$, which leads to $m \sim 2 \times 10^{-6} M_{\rm Pl}$. Is this fine-tuning? In absence of a rigorous definition of fine-tuning, this is hard to tell. But one can notice that $m/H \sim \sqrt{6}(2 + 4\Delta N_*)^{-1/2} < 1$, which may be viewed as unnatural. Indeed, we expect the mass of the inflaton to be corrected by high-energy physics according to $m^2 \to m^2 + gM^2 \ln(\Lambda/\mu)$, where μ is the renormalisation scale, $M > \Lambda$ the mass of a heavy field, Λ the cut-off already discussed in Section 4.4.2, and g the coupling constant. The presence of these corrections implies $m/H \sim 1$, and keeping $m/H < 1$ may be problematic. This problem is also known as the η-problem of inflation [20]. But this at least illustrates the fact that the fine-tuning of the parameters (if any) can depend on the potential.

For this reason, it is worth studying the situation for the Starobinsky potential (4.87) since this is the favoured model.

The Starobinsky model can be derived from different assumptions. Historically, it was derived by considering R^2 corrections to the Einstein-Hilbert action. However, more recently, it was realised that it can also be viewed as a scenario in which the inflaton field is the Higgs field, this one being non-minimally coupled to gravity. In technical terms, the action of the model reads

$$ S = \frac{M_{\mathrm{Pl}}^2}{2} \int \mathrm{d}^4 x \sqrt{-g} \left[\left(1 + \xi h^2\right) R - g^{\mu\nu} \partial_\mu h \partial_\nu h - 2 M_{\mathrm{Pl}}^2 \frac{\lambda}{4} \left(h^2 - \frac{v^2}{M_{\mathrm{Pl}}^2} \right)^2 \right], \quad (4.95) $$

where v is the Higgs vacuum expectation value and λ the self-interacting coupling constant. The quantity ξ is a dimensionless constant which describes the non-minimal coupling. If one defines the field ϕ by $\mathrm{d}[\phi/(\sqrt{2}M_{\mathrm{Pl}})]/\mathrm{d}h = \sqrt{1 + \xi(1 + 6\xi)h^2}/[\sqrt{2}(1 + \xi h^2)]$ then this field has a standard Lagrangian with a potential which is exactly the potential of Eq. (4.87), the scale M being given by

$$ M^4 = \frac{M_{\mathrm{Pl}}^4 \lambda}{4\xi^2}. \quad (4.96) $$

Then the COBE normalisation, which constrains the value of M, leads to

$$ \xi \sim 46{,}000 \sqrt{\lambda}, \quad (4.97) $$

where $\lambda = m_{\mathrm{H}}^2/v^2$ with $v \simeq 175 \, \mathrm{GeV}$ and $m_{\mathrm{H}} \simeq 125 \, \mathrm{GeV}$. We see that $\xi \gg 1$, which can imply many issues as far as the consistency of the model is concerned.

The overall picture that emerges from this section is that it is difficult to say whether the parameters of the inflationary potential are necessarily fine-tuned if one wants to account for the data. It is clear that this question is model dependent. For some potentials, the fine-tuning is present, but for others (and, in particular, those that fit the data well), it is unclear whether this is the case. The situation of the Starobinsky model is particularly interesting. The coupling between gravity and the Higgs is not small, or it is not perturbative, which may lead to technical difficulties, but this strong coupling problem is not necessarily associated with a fine-tuning problem. Here, we are just missing an objective definition of what fine-tuning is.

In fact, one could argue that such a definition exists and is nothing but the Bayesian evidence considered in Section 4.4.4. Technically, the Bayesian evidence is the integral of the likelihood over prior space, but its meaning can easily be grasped intuitively. Let us consider a model depending on, say, one free parameter. If, for all values of the parameter in the prior range, one obtains a good fit, then the Bayesian evidence is 'good'. This is, for instance, the case of the model in Figure 4.8 (left panel). Different points correspond to different values of the reheating temperature but all points are within the 1σ Planck contour. On the contrary, if one needs to tune the value of the free parameter in order to have a good fit, then the Bayesian evidence will be 'bad'. This is the case for the model

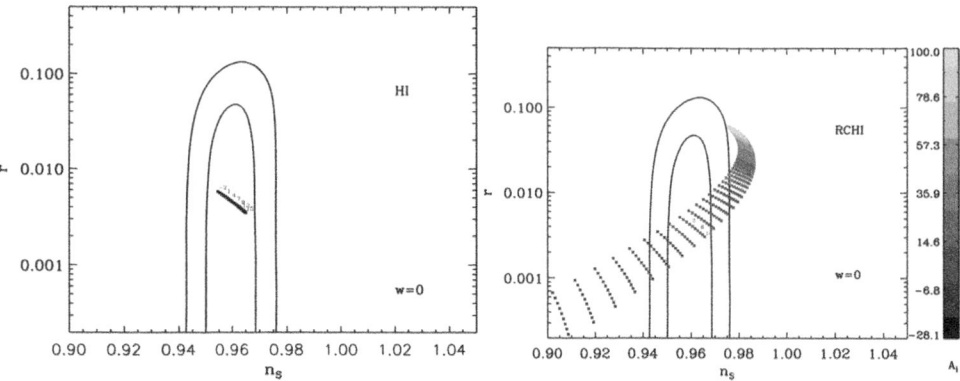

Figure 4.8 Predictions in the (n_S, r) space for two inflationary models, Higgs inflation (left panel) and Higgs inflation with quantum corrections (right panel); see Reference [99]. In both cases, a very good fit can be found, but in the case of Higgs inflation with quantum corrections, this requires a tuning of the free parameters characterising the model. As a consequence, the Bayesian evidence is smaller than that of Higgs inflation, and given the data, the model is seen as 'less good'.

in Figure 4.8 (right panel). In order to have a good compatibility with the data (i.e., points within the 1σ contour), one needs to tune the parameter A_I (which controls the amplitude of the quantum corrections), and the Bayesian evidence is 'bad'. In other words, the wasted parameter space is penalised. Obviously, the smaller the range of A_I leading to a good fit (compared to the prior), the smaller the evidence. We conclude that the evidence is a good, objective measure of fine-tuning. In this sense, the Starobinsky model is the best model because it is the less fine-tuned one.

4.6 Inflationary Initial Conditions

4.6.1 Homogeneous Initial Conditions

Let us now discuss another type of possible fine-tuning, namely the initial conditions (see also Reference [34] for a detailed discussion of this question). We have seen previously that one of the main motivations for inflation is to avoid the fine-tuning of the initial conditions that is needed in order for the Standard Model to work. If our solution to that issue were also fine-tuned, then one could wonder whether something has been gained or not. In fact, this problem has different facets. If we restrict ourselves to an homogeneous and isotropic solution, then the only question is how we should choose ϕ_{ini} and $\dot{\phi}_{\text{ini}}$. The slow-roll trajectory corresponds to $\dot{\phi}_{\text{ini}} \simeq -V_\phi(\phi_{\text{ini}})/[3H(\phi_{\text{ini}})]$, and, therefore, there could be the worry that we have to tune the initial velocity to this value. However, this is not the case because the slow-roll trajectory is an attractor as can be seen in Figure 4.9. It is true that, for some $V(\phi)$, the corresponding basin of attraction is very small. This is, for instance, the case for Small Field Inflation (SFI). However, on the contrary, it can be very large for other models, such as Large Field Inflation (LFI) (let us also notice that the existence of

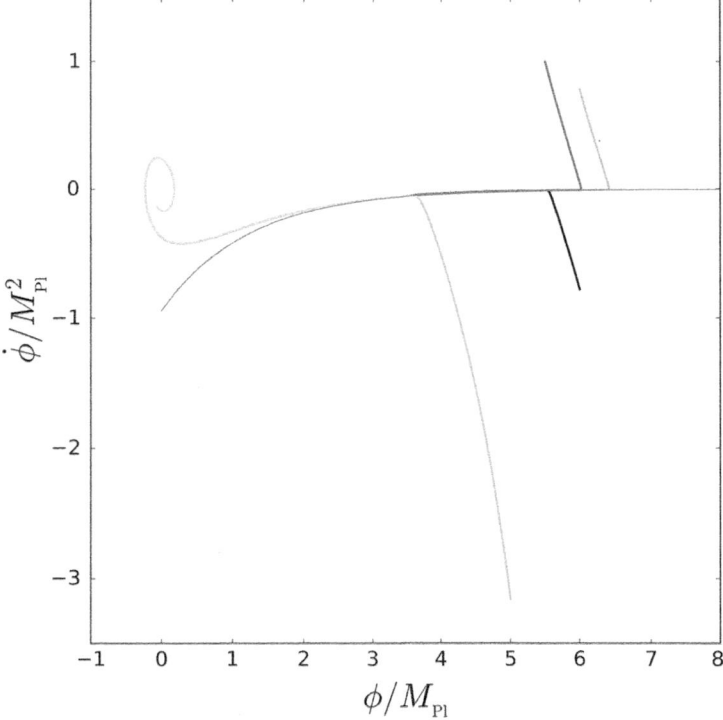

Figure 4.9 Phase space for the Starobinsky model. The central solid curve represents the slow roll trajectory while the grey lines correspond to exact trajectories with different initial conditions. It is evident from the plot that the slow-roll trajectory is an attractor in phase space.

an attractor is immune to stochastic effects; see Reference [47]). The interesting point is that it is also the case for the Starobinsky model and plateau potentials, namely the models favoured by the data. In this sense, in this restricted framework, there is no fine-tuning of the initial condition [56].

4.6.2 Anisotropic Initial Conditions

Obviously, however, the previous analysis is not entirely satisfactory. Indeed, we start from a homogeneous and isotropic situation while inflation is precisely supposed to explain why our Universe is homogeneous and isotropic. The analysis can be improved by considering that, initially, the Universe is not isotropic (but still homogeneous) [13, 127, 134]. For this purpose, let us consider the following metric (Bianchi I model):

$$\mathrm{d}s^2 = -\mathrm{d}t^2 + a_i^2(t)\big(\mathrm{d}x^i\big)^2; \tag{4.98}$$

that is to say, we now have one scale factor for each space direction. This metric can also be rewritten as $ds^2 = -dt^2 + a^2(t)\gamma_{ij}dx^i dx^j$, with

$$a(t) \equiv [a_1(t)a_2(t)a_3(t)]^{1/3}, \qquad (4.99)$$

and

$$\gamma_{ij} = \begin{pmatrix} e^{2\beta_1(t)} & 0 & 0 \\ 0 & e^{2\beta_2(t)} & 0 \\ 0 & 0 & e^{2\beta_3(t)} \end{pmatrix}, \qquad (4.100)$$

with $\sum_{i=1}^{i=3} \beta_i = 0$. As usual, one can also introduce the conformal time η in terms of which the metric can be expressed as $ds^2 = a^2(\eta)\left(-d\eta^2 + \gamma_{ij}dx^i dx^j\right)$. Then, the next step is to introduce the shear σ_{ij}, which is defined by (as usual, a prime denotes a derivative with respect to conformal time)

$$\sigma_{ij} = \frac{1}{2}\gamma'_{ij} = \begin{pmatrix} \beta'_1 e^{2\beta_1} & 0 & 0 \\ 0 & \beta'_2 e^{2\beta_2} & 0 \\ 0 & 0 & \beta'_3 e^{2\beta_3} \end{pmatrix}. \qquad (4.101)$$

Assuming that matter is described by a scalar field, it is then easy to write the Einstein equations. They read

$$3\frac{\mathcal{H}^2}{a^2} = \frac{\rho}{M_{\text{Pl}}^2} + \frac{\sigma^2}{2a^2} = \frac{1}{M_{\text{Pl}}^2}\left[\frac{\phi'^2}{2a^2} + V(\phi)\right] + \frac{\sigma^2}{2a^2}, \qquad (4.102)$$

$$-\frac{1}{a^2}\left(\mathcal{H}^2 + 2\mathcal{H}'\right) = \frac{p}{M_{\text{Pl}}^2} + \frac{\sigma^2}{2a^2} = \frac{1}{M_{\text{Pl}}^2}\left[\frac{\phi'^2}{2a^2} - V(\phi)\right] + \frac{\sigma^2}{2a^2}, \qquad (4.103)$$

$$\left(\sigma^i_j\right)' + 2\mathcal{H}\sigma^i_j = 0, \qquad (4.104)$$

where $\sigma^2 = \sigma_{ij}\sigma^{ij} = \sum_{i=1}^{i=3} \beta'^2_i$ and $\sigma^i_j = \gamma^{ik}\sigma_{kj}$; that is to say,

$$\sigma^i_j = \begin{pmatrix} \beta'_1 & 0 & 0 \\ 0 & \beta'_2 & 0 \\ 0 & 0 & \beta'_3 \end{pmatrix}. \qquad (4.105)$$

The solution for the shear can easily be found, namely $\sigma^i_j = S^i_j/a^2$, where S^i_j is a constant tensor. This implies that $\sigma^2 = S^2/a^4$, where $S^2 = S^i_j S^j_i$. As a consequence, one sees that the shear is, in fact, equivalent to a stiff fluid with an equation of state $w_\sigma \equiv p_\sigma/\rho_\sigma = 1$ and $\rho_\sigma = M_{\text{Pl}}^2 S^2/(2a^6)$. Therefore, if initially the shear dominates, $\rho_\sigma \gg \rho_\phi$, the Universe will expand as $a \propto t^{1/3}$, (see Eq. (4.17)) and the expansion will not be accelerated. However, since $\rho_\sigma \propto a^{-6}$ while ρ_ϕ is approximately constant, the scalar field will eventually take over, and inflation will start. We conclude that, even if the Universe is not initially isotropic, it will become so in the presence of a scalar field whose energy density is dominated by its potential. In this sense, it is legitimate to start from an isotropic situation, as was done previously. This is clearly not a fine-tuning but, rather, an attractor of the dynamical evolution.

4.6.3 Inhomogeneous Initial Conditions

Despite the fact that taking into account the shear represents an improvement, this still does not allow us to discuss the real issue. For that, we need a framework where the initial state of the Universe is neither isotropic nor homogeneous. Technically, this is clearly very complicated since we have to solve the Einstein equations in full generality. The only way to study these questions exactly is therefore numerical relativity. However, some schemes of approximation have also been developed, and we now discuss them. Of course, the perturbative approach described before (see Section 4.4.3) is one way of taking into account the inhomogeneities. However, by definition, these fluctuations must be small while we would like to see whether inflation 'homogenises' the Universe even if it is strongly inhomogeneous initially. Another method is the so-called effective-density approximation [44, 45]. The idea is to study an inhomogeneous scalar field on a (isotropic and homogeneous) FLRW background and to add to the Friedmann equation a term which describes the backreaction of the field gradient on the geometry [44, 45]. In practice, one writes

$$\phi(t, \boldsymbol{x}) = \phi_0(t) + \Re \left[\delta\phi(t) e^{i \boldsymbol{k} \cdot \boldsymbol{x}/a(t)} \right] \tag{4.106}$$

and assumes that the corresponding Klein-Gordon equation can be split into two equations, namely

$$\ddot{\phi}_0 + 3H\dot{\phi}_0 + V_\phi(\phi_0) = 0, \tag{4.107}$$

$$\ddot{\delta\phi} + 3H\dot{\delta\phi} + \frac{k^2}{a^2}\delta\phi = 0. \tag{4.108}$$

The Friedmann equation is then written as

$$H^2 = \frac{1}{3M_{\mathrm{Pl}}^2} \left[\frac{1}{2}\dot{\phi}_0^2 + V(\phi_0) + \frac{1}{2}\dot{\delta\phi}^2 + \frac{1}{2}\frac{k^2}{a^2}\delta\phi^2 \right] - \frac{\mathcal{K}}{a^2}. \tag{4.109}$$

The wave number k should be chosen such that the wavelength of the perturbations is much smaller than the Hubble radius, namely $2\pi k/a \ll H^{-1}$. In the opposite limit, the contribution of $\delta\phi$ should just be added to the background. The energy density of the inhomogeneities $\rho_{\delta\phi} = \rho_{\dot{\delta\phi}} + \rho_{\nabla}$, with $\rho_{\dot{\delta\phi}} = \dot{\delta\phi}^2/2$ and $\rho_{\nabla} = k^2\delta\phi^2/(2a^2)$ is supposed to dominate initially (i.e., the Universe is inhomogeneous initially), $\rho_{\delta\phi} \gg \rho_{\phi_0}$. The question is whether $\rho_{\delta\phi}$ can decrease (i.e., the Universe becomes homogeneous) such that, at some point, ρ_{ϕ_0} takes over and inflation starts.

Let us now discuss the initial conditions. We take $\dot{\phi}_0$ and ϕ_0 such that, in absence of inhomogeneities, slow-roll inflation starts. Initially, the Friedmann equations can be written as

$$\frac{3a^2H^2}{k^2} \simeq \frac{1}{2}\frac{\dot{\delta\phi}^2}{M_{\mathrm{Pl}}^2}\frac{a^2}{k^2} + \frac{1}{2}\frac{\delta\phi^2}{M_{\mathrm{Pl}}^2}, \tag{4.110}$$

since $\rho_{\delta\phi} \gg \rho_{\phi_0}$. For simplicity, we have taken $\mathcal{K} = 0$, but it is straightforward to include the case where curvature is not vanishing. We have already mentioned that the effective density approximation is valid only if the wavelength of $\delta\phi$ is smaller than the Hubble radius.

Historically, the first numerical solutions [42–45] were done under the assumption that space-time is spherically symmetric. This has the advantage to simplify the equations since they only depend on time and r, the radial coordinate. Of course, in that case, one still has to numerically solve partial differential equations. The metric considered in Reference [45] reads

$$ds^2 = -\left(N^2 - R^2\beta^2\right)dt^2 + 2R^2\beta d\chi\,dt + R^2\left(d\chi^2 + \sin^2\chi\,d\Omega^2\right), \tag{4.112}$$

where $0 \le \chi \le \pi$ so that the space-like sections are closed. The lapse and shift functions N and β depend on t and r as well as the 'scale factor' R. The matter content assumed in Reference [45] is a scalar field ϕ, which is the inflaton, and another scalar field ψ without potential and playing the role of radiation. Some important technical restrictions are also postulated on the initial data. First, it is assumed that the total energy density is constant. Given an initial inhomogeneous distribution for the inflaton $\phi(\chi)$, this is achieved by choosing the initial velocity of ψ to be such that the total energy density is constant. Second, the initial momentum is taken to vanish. Based on the previous calculations (see Eqs. (4.107), (4.108), and (4.109)), it is argued in Reference [45] that, at least for large-field models, this does not restrict the significance of the results. Third, the integration is performed for values of the inflaton self-coupling that are larger than the ones necessary to CMB-normalise the model. Different initial configurations for the inflaton field are considered. In particular, the Gaussian ansatz

$$\phi_{\mathrm{ini}}(\chi) = \phi_0 + \delta\phi\left[1 - \exp\left(-\frac{\sin^2\chi}{\Delta^2}\right)\right] \tag{4.113}$$

was studied in details. This initial profile depends on three parameters: ϕ_0, the value of the field at the origin $\chi = 0$; $\delta\phi$, which can be viewed as the value of the field on the other side of the Universe, $\phi(\pi/2) = \phi_0 + \delta\phi\left(1 - e^{-1/\Delta^2}\right)$; and Δ, which represents the width of the Gaussian.

Let us now describe the results obtained for large-field inflation. If $V(\chi = 0)$ and $V(\chi = \pi/2)$, or ϕ_0 and $\delta\phi$, are such that, in a homogeneous situation, inflation would start, then it also starts in the present case. If, on the contrary, $V(\chi = 0)$ is such that inflation would start in a homogeneous situation but not $V(\chi = \pi/2)$ (therefore, the gradients are important), then Reference [45] has shown that the outcome crucially depends on the width Δ. More precisely, the numerical simulations show that the crucial parameter is $R\Delta/H^{-1}$, which has to be large enough in order for inflation to start. Moreover, the larger the gradient, the shorter the duration of inflation. For small-field models, the sensitivity to the initial conditions is even greater.

A few years later, the analysis was improved in a significant way, and, in particular, the assumption of spherical symmetry was relaxed. Indeed, References [64–66] ran simulations of strongly inhomogeneous inflation with a three-dimensional numerical relativity code.

These simulations are such that the initial time slice has homogeneous total energy density, which means that $(\nabla\phi)^2/2 < 3M_{\mathrm{Pl}}^2 H^2$ implying that

$$\nabla\phi < \sqrt{6}\frac{M_{\mathrm{Pl}}}{H^{-1}}. \tag{4.114}$$

Thus, inhomogeneities that have wavelengths smaller than the Hubble radius must have a small amplitude or, to put it differently, large inhomogeneities must necessarily extend over many Hubble patches. The simulations were carried out for a quartic large field model with an initial configuration given by

$$\phi_{\mathrm{ini}}(t_{\mathrm{ini}}, \boldsymbol{x}) = \phi_0$$
$$+ \delta\phi \sum_{\ell,m,n=1}^{2} \frac{1}{\ell m n} \sin\left(\frac{2\pi \ell x}{L} + \theta_{x\ell}\right) \sin\left(\frac{2\pi \ell y}{L} + \theta_{ym}\right) \sin\left(\frac{2\pi \ell z}{L} + \theta_{zn}\right),$$
$$\tag{4.115}$$

where the θs are random phases. Two runs have been carried out in Reference [65], one with $L = H^{-1}$ and $\delta\phi = 0.0125m_{\mathrm{Pl}}$ and one with $L = 32H^{-1}$ and $\delta\phi = 0.4m_{\mathrm{Pl}}$. In both cases, one has $\phi_0 = 5m_{\mathrm{Pl}}$ and $H_0 = 0.1m_{\mathrm{Pl}}$. The simulations show that, in the first case, the inhomogeneities oscillate, and their amplitude is damped. At the end of the run, the inflaton field is homogeneous. But, in the second case, they do not oscillate (initially, they are larger than the Hubble radius) and are not damped.

In conclusion, it seems possible to start inflation with inhomogeneous initial conditions and to homogenise the Universe. However, admittedly, the numerical simulations that have been carried out so far all require some technical restrictions. The crucial question that emerges from the simulations is the size of the initial homogeneous patch. There is also a dependence on the model with large-field scenarios being the preferred class of scenarios. As a consequence, the Starobinsky model is (again) among the good models. Let us also notice that, even more recently, new simulations have been carried out; see References [35, 39]. These new works bring new insights into an issue that will probably be studied even more in the future.

A last comment is that we have good reasons to believe the quantum effects to play an important role at the beginning of inflation. For this reason, studying the initial conditions at only the classical level is maybe not sufficient, and even more elaborate investigations may be needed to settle this question.

4.6.4 Initial Conditions for the Perturbations

So far, we have discussed the question of the fine-tuning of the initial conditions related to the background. Obviously, there is the same question for the perturbations. We have seen that they are chosen such that the perturbations are initially placed in the vacuum state. However, if one traces back the scale of astrophysical interest today to the beginning of inflation, one notices that they correspond to physical lengths smaller than the Planck

Then, since inflation is an almost de Sitter expansion, what we have just described for a test field should also be true when the backreaction is taken into account, namely for the field the behaviour of which is controlled by Eq. (4.127).[2] Moreover, each patch is isolated from the others as can be seen by computing the event horizon in de Sitter space-time. Let us indeed consider a specific observer that we choose, for convenience, to be at the origin. Then, its future horizon (the part of the Universe with which the observer will be able to communicate in the future) is given by

$$d_{\rm E} = a_0 \int_{t_0}^{\infty} \frac{dt}{a(t)} = a_0 \int_{t_0}^{\infty} dt \frac{1}{a_0} e^{-H(t-t_0)} = \frac{1}{H}, \tag{4.132}$$

namely the size of the patch itself. In other words, each patch is causally disconnected from the others and this forever. These patches are sometimes referred to as 'pocket universes'. The number of these patches is growing with time. Indeed, in one e-fold, the 'size' of the Universe increases by a factor $e^3 \sim 20$ while the 'size' of a patch is constant (since the Hubble parameter is constant). As a consequence, each e-fold, one patch gives rise to about 20 new patches, all causally disconnected.

There is also some kind of ergodic argument at play. When, see for instance Figure 4.14, we have solved the Langevin equation, each realisation of the solution of this equation was supposed to represent a specific configuration of the field over the entire homogeneous and isotropic space-time. But one can also assume that one realisation corresponds to a specific value of the field in a given patch. And, as a consequence, different realisations correspond to different values of the field in different patches. So, in this interpretation, different realisations do not represent an ensemble of different field configurations over an homogeneous and isotropic space-time but, rather, the spatial distribution of $\phi_{\rm IR}$ in different patches.

The overall picture that emerges is that of an expanding space-time where the number of independent patches is increasing, the value of the field in each pocket universe being a stochastic quantity controlled by a Langevin equation. Since we have seen that, due to stochastic effects, the field can climb up its potential, there are patches where inflation will never stop. Obviously, the volume occupied by those patches, compared to the volume occupied by the patches where inflation stops, is growing, which means that patches where inflation is taking place occupy more and more regions of space-time. Globally, inflation will never stop, meaning that there are always regions of space-time undergoing inflation.

---

[2] For the potential $V(\phi) = m^2\phi^2/2$, this equation reads

$$\frac{d\phi_{\rm IR}}{dN} + \frac{2M_{\rm Pl}^2}{\phi_{\rm IR}} = \frac{m}{2\pi M_{\rm Pl}\sqrt{6}}\phi_{\rm IR}\,\xi. \tag{4.130}$$

It is of the Bernouilli type and, therefore, can be solved explicitly. The solution takes the form

$$\phi_{\rm IR}^2 = e^{\frac{m}{\pi M_{\rm Pl}\sqrt{6}}\int_{N_{\rm ini}}^{N}\xi\,dn}\left[\phi_{\rm ini}^2 - 4M_{\rm Pl}^2\int_{N_{\rm ini}}^{N} e^{-\frac{m}{\pi M_{\rm Pl}\sqrt{6}}\int_{N_{\rm ini}}^{n}\xi(\bar{n})d\bar{n}}\,dn\right]. \tag{4.131}$$

However, it is so complicated that it is not very useful. In particular, it seems very difficult to calculate the two-point correlation function of the field from this solution.

Of course, there will also be regions of space-time where inflation stops – those where, by chance, the stochastic fluctuations do not push the field upwards. This structure is referred to as 'eternal inflation'. The stochastic effects are said to produce a 'multiverse'. Notice that the word 'multiverse' is especially awkward in the present context since we do not produce many universes as in the many-world interpretation of quantum mechanics, for instance, but just a specific spatial configuration of our single Universe made of causally independent regions, the pocket universes.

Before discussing the reliability and the implications of eternal inflation, we would like to investigate the question of whether it is unavoidable or not.

4.7.3 Avoiding Self-Replication

Before discussing the robustness of eternal inflation, it is interesting to investigate whether this is an unavoidable consequence of inflation. As recently discussed in Reference [104], it turns out that this is not the case, and in this section, we closely follow this paper although we also present some new results. We have seen that the quantum-to-classical variation of the field is given by the amplitude of the scalar power spectrum; see Eq. (4.119). If there exists a field value for which this amplitude

$$\mathcal{P}_{\zeta 0}(\phi) = \frac{H^2(\phi)}{8\pi M_{\mathrm{Pl}}^2 \epsilon_1(\phi)}, \tag{4.133}$$

is of order one, then this means that the quantum fluctuations of the field are of order one and, if the considerations presented in the previous section are correct, the regime of eternal inflation starts. Usually, this happens in the regime where $\epsilon_1(\phi) \to 0$ since $\epsilon_1(\phi)$ stands at the denominator. But this also implies that, if the shape of the potential is such that there is a field range such that $\epsilon_1 \ll 1$ (in order to have inflation!), but otherwise, $\epsilon_1(\phi)$ is large, then there could be no regime where $\mathcal{P}_{\zeta 0} > 1$. One example was found by V. Mukhanov in Reference [104]. The corresponding potential is

$$V(\phi) = M^4 \left(1 - e^{-\phi/M_{\mathrm{Pl}}}\right)^2 \left(1 - \frac{\phi}{\phi_{\mathrm{m}}}\right)^{-\alpha} \tag{4.134}$$

and is represented in Figure 4.15. It looks like the Starobinsky model corrected by a term $(1 - \phi/\phi_{\mathrm{m}})^{-\alpha}$. The model depends on three parameters: M, ϕ_{m}, and α. As usual, M is fixed by the CMB normalisation.

The first two Hubble flow parameters are given by the following expressions

$$\epsilon_1 = \frac{1}{2} \left[2\frac{e^{-\phi/M_{\mathrm{Pl}}}}{1 - e^{-\phi/M_{\mathrm{Pl}}}} + \alpha \frac{M_{\mathrm{Pl}}}{\phi_{\mathrm{m}}} \left(1 - \frac{\phi}{\phi_{\mathrm{m}}}\right)^{-1} \right]^2, \tag{4.135}$$

$$\epsilon_2 = 4\epsilon_1 + 4e^{-\phi/M_{\mathrm{Pl}}} \left(1 - e^{-\phi/M_{\mathrm{Pl}}}\right)^{-1} - 4e^{-2\phi/M_{\mathrm{Pl}}} \left(1 - e^{-\phi/M_{\mathrm{Pl}}}\right)^{-2}$$
$$- 8\alpha \frac{M_{\mathrm{Pl}}}{\phi_{\mathrm{m}}} \left(1 - e^{-\phi/M_{\mathrm{Pl}}}\right)^{-1} \left(1 - \frac{\phi}{\phi_{\mathrm{m}}}\right)^{-1} - 2(\alpha + \alpha^2)\frac{M_{\mathrm{Pl}}^2}{\phi_{\mathrm{m}}^2} \left(1 - \frac{\phi}{\phi_{\mathrm{m}}}\right)^{-2}. \tag{4.136}$$

[4] Ade, P. A. R., *et al.* 2016b. 'Planck 2015 Results. XX. Constraints on Inflation'. *Astronomy and Astrophysics*, **594**, A20.

[5] Ade, P., *et al.* 2014b. 'Planck 2013 Results. XVI. Cosmological Parameters'. *Astronomy and Astrophysics*, **571**, A16.

[6] Ade, P., *et al.* 2014c. 'Planck 2013 Results. XXII. Constraints on Inflation'. *Astronomy and Astrophysics*, **571**, A22.

[7] Ade, P., *et al.* 2014d. 'Planck 2013 Results. XXIV. Constraints on Primordial Non-Gaussianity'. *Astronomy and Astrophysics*, **571**, A24.

[8] Aghanim, N., *et al.* 2018. 'Planck 2018 Results. VI. Cosmological Parameters'. *Astronomy and Astrophysics*, DOI: 10.1051/0004-6361/201833910.

[9] Akrami, Y., *et al.* 2018a. 'Planck 2018 Results. I. Overview and the Cosmological Legacy of Planck'. *Astronomy and Astrophysics*, DOI:10.1051/0004-6361/201833880.

[10] Akrami, Y., *et al.* 2018b. 'Planck 2018 Results. X. Constraints on Inflation'. *Astronomy and Astrophysics*, DOI:10.1051/0004-6361/201833887.

[11] Albrecht, A., Brandenberger, R. H., and Matzner, R. 1985. 'Numerical Analysis of Inflation'. *Physical Review D*, **32**, 1280.

[12] Amin, M. A., *et al.* 2014. 'Nonperturbative Dynamics of Reheating after Inflation: A Review'. *International Journal of Modern Physics D*, **24**, 1530003.

[13] Anninos, P., *et al.* 1991. 'How Does Inflation Isotropize the Universe?' *Physical Review D*, **43**, 3821–3832.

[14] Armendariz-Picon, C., Mukhanov, V. F., and Steinhardt, P. J. 2001. 'Essentials of k Essence'. *Physical Review D*, **63**, 103510.

[15] Assadullahi, H., *et al.* 2016. 'Multiple Fields in Stochastic Inflation'. *Journal of Cosmology and Astroparticle Physics*, **1606**(06), 043.

[16] Astier, P., *et al.* 2006. 'The Supernova Legacy Survey: Measurement of ωM, $\omega \lambda$ and W from the First Year Data Set'. *Astronomy and Astrophysics*, **447**, 31–48.

[17] Bardeen, J. M. 1980. 'Gauge Invariant Cosmological Perturbations'. *Physical Review D*, **22**, 1882–1905.

[18] Bardeen, J. M., Steinhardt, P. J., and Turner, M. S. 1983. 'Spontaneous Creation of Almost Scale-Free Density Perturbations in an Inflationary Universe'. *Physical Review D*, **28**, 679.

[19] Battefeld, D., and Peter, P. 2015. 'A Critical Review of Classical Bouncing Cosmologies'. *Physics Reports*, **571**, 1–66.

[20] Baumann, D., and McAllister, L. 2015. *Inflation and String Theory*. Cambridge University Press.

[21] Bennett, C. L., *et al.* 1996. 'Four year COBE DMR Cosmic Microwave Background Observations: Maps and Basic Results'. *The Astrophysical Journal Letters*, **464**, L1–L4.

[22] Brandenberger, R., and Peter, P. 2017. 'Bouncing Cosmologies: Progress and Problems'. *Foundations of Physics*, **47**(6), 797–850.

[23] Brandenberger, R. H., and Kung, J. H. 1990. 'Chaotic Inflation as an Attractor in Initial Condition Space'. *Physical Review D*, **42**, 1008–1015.

[24] Brandenberger, R. H., and Martin, J. 2001. 'The Robustness of Inflation to Changes in SuperPlanck Scale Physics'. *Modern Physics Letters A*, **16**, 999–1006.

[25] Brandenberger, R. H., and Martin, J. 2002. 'On Signatures of Short Distance Physics in the Cosmic Microwave Background'. *International Journal of Modern Physics A*, **17**, 3663–3680.

[26] Brandenberger, R. H., and Martin, J. 2005. 'Back-Reaction and the Trans-Planckian Problem of Inflation Revisited'. *Physical Review D*, **71**, 023504.

[27] Brandenberger, R. H., and Martin, J. 2013. 'Trans-Planckian Issues for Inflationary Cosmology'. *Classical and Quantum Gravity*, **30**, 113001.

[28] Calzetta, E., and Sakellariadou, M. 1992. 'Inflation in Inhomogeneous Cosmology'. *Physical Review D*, **45**, 2802–2805.

[29] Casadio, R., *et al.* 2005a. 'Higher Order Slow-Roll Predictions for Inflation'. *Physics Letters B*, **625**, 1–6.

[30] Casadio, R., *et al.* 2005b. 'Improved WKB Analysis of Cosmological Perturbations'. *Physical Review D*, **71**, 043517.

[31] Casadio, R., *et al.* 2005c. 'Improved WKB Analysis of Slow-Roll Inflation'. *Physical Review D*, **72**, 103516.

[32] Chialva, D., and Mazumdar, A. 2015. 'Cosmological Implications of Quantum Corrections and Higher-Derivative Extension'. *Modern Physics Letters A*, **30**(03n04), 1540008.

[33] Choe, J., Gong, J. O., and Stewart, E. D. 2004. 'Second Order General Slow-Roll Power Spectrum'. *Journal of Cosmology and Astroparticle Physics*, **0407**, 012.

[34] Chowdhury, D., *et al.* 2019. 'Inflation after Planck: Judgment Day'. *arXiv e-prints*, Feb, arXiv:1902.03951.

[35] Clough, K., *et al.* 2016. 'Robustness of Inflation to Inhomogeneous Initial Conditions'. arXiv:1608.04408.

[36] Coule, D. H., and Martin, J. 2000. 'Quantum Cosmology and Open Universes'. *Physical Review D*, **61**, 063501.

[37] Cyburt, R. H., *et al.* 2016. 'Big Bang Nucleosynthesis: Present Status'. *Reviews of Modern Physics*, **88**(1), 015004.

[38] Durrer, R., Gangui, A., and Sakellariadou, M. 1996. 'Doppler Peaks: A Fingerprint of Topological Defects'. *Physical Review Letters*, **76**, 579–582.

[39] East, W. E., *et al.* 2016. 'Beginning Inflation in an Inhomogeneous Universe'. *Journal of Cosmology and Astroparticle Physics*, **1609**(09), 010.

[40] Fixsen, D. J. 2009. 'The Temperature of the Cosmic Microwave Background'. *The Astrophysical Journal*, **707**(Dec.), 916–920.

[41] Gangui, A., Durrer, R., and Sakellariadou, M. 1996. 'Global Textures and the Doppler Peaks'. *Astronomical Society of the Pacific Conference Series*, **94**, 335–340.

[42] Goldwirth, D. S. 1991. 'On Inhomogeneous Initial Conditions for Inflation'. *Physical Review D*, **43**, 3204–3213.

[43] Goldwirth, D. S., and Piran, T. 1989. 'Spherical Inhomogeneous Cosmologies and Inflation. 1. Numerical Methods'. *Physical Review D*, **40**, 3263.

[44] Goldwirth, D. S., and Piran, T. 1990. 'Inhomogeneity and the Onset of Inflation'. *Physical Review Letters*, **64**, 2852–2855.

[45] Goldwirth, D. S., and Piran, T. 1992. 'Initial Conditions for Inflation'. *Physics Reports*, **214**, 223–291.

[46] Gong, J. O., and Stewart, E. D. 2001. 'The Density Perturbation Power Spectrum to Second Order Corrections in the Slow Roll Expansion'. *Physics Letters B*, **510**, 1–9.

[47] Grain, J., and Vennin, V. 2017. 'Stochastic Inflation in Phase Space: Is Slow Roll a Stochastic Attractor?' *Journal of Cosmology and Astroparticle Physics*, **1705**(05), 045.

[48] Gross, D. J., 2005. 'Where Do We Stand in Fundamental String Theory'. *Physica Scripta*, **T117**, 102–105.

[49] Gunn, J. E., and Peterson, B. A. 1965. 'On the Density of Neutral Hydrogen in Intergalactic Space'. *The Astrophysical Journal*, **142**(Nov.), 1633–1641.

[50] Guth, A., Vachaspati, T., and Winitzki, S. 'Energy Conditions in Inflation'.

[51] Guth, A. H. 1981. 'The Inflationary Universe: A Possible Solution to the Horizon and Flatness Problems'. *Physical Review D*, **23**, 347–356.

[52] Guth, A. H. 2000. 'Inflation and Eternal Inflation'. *Physics Reports*, **333**, 555–574.

[53] Guth, A. H. 2007. 'Eternal Inflation and Its Implications'. *Journal of Physics A*, **40**, 6811–6826.

[54] Hartle, J. B., and Hawking, S. W. 1983. 'Wave Function of the Universe'. *Physical Review D*, **28**, 2960–2975.

[55] Hassan, S. F., and Sloth, M. S. 2003. 'TransPlanckian Effects in Inflationary Cosmology and the Modified Uncertainty Principle'. *Nuclear Physics B*, **674**, 434–458.

[56] Ijjas, A., Steinhardt, P. J., and Loeb, A. 2013. 'Inflationary Paradigm in Trouble after Planck2013'. *Physics Letters B*, **723**, 261–266.

[57] Ijjas, A., Steinhardt, P. J., and Loeb, A. 2014. 'Inflationary Schism'. *Physics Letters B*, **736**, 142–146.

[58] Kofman, L., Linde, A. D., and Starobinsky, A. A. 1997. 'Towards the Theory of Reheating after Inflation'. *Physical Review D*, **56**, 3258–3295.

[59] Kolb, E. W., and Turner, M. S. 1990. 'The Early Universe'. *Frontiers in Physics*, **69**, 1–547.

[60] Kosowsky, A. 1996. 'Cosmic Microwave Background Polarization'. *Annals of Physics*, **246**, 49–85.

[61] Kowalski, M., *et al.* 2008. 'Improved Cosmological Constraints from New, Old and Combined Supernova Datasets'. *The Astrophysical Journal*, **686**, 749–778.

[62] Kung, J. H., and Brandenberger, R. H. 1989. 'The Initial Condition Dependence of Inflationary Universe Models'. *Physical Review D*, **40**, 2532.

[63] Kunz, M., Trotta, R., and Parkinson, D. 2006. 'Measuring the Effective Complexity of Cosmological Models'. *Physical Review D*, **74**, 023503.

[64] Kurki-Suonio, H., *et al.* 1987. 'Inflation from Inhomogeneous Initial Data in a One-Dimensional Back Reacting Cosmology'. *Physical Review D*, **35**, 435–448.

[65] Kurki-Suonio, H., Laguna, P., and Matzner, R. A. 1993. 'Inhomogeneous Inflation: Numerical Evolution'. *Physical Review D*, **48**, 3611–3624.

[66] Laguna, P., Kurki-Suonio, H., and Matzner, R. A. 1991. 'Inhomogeneous Inflation: The Initial Value Problem'. *Physical Review D*, **44**, 3077–3086.

[67] Leach, S. M., *et al.* 2002. 'Cosmological Parameter Estimation and the Inflationary Cosmology'. *Physical Review D*, **66**, 023515.

[68] Lemoine, M., *et al.* 2002. 'The Stress Energy Tensor for TransPlanckian Cosmology'. *Physical Review D*, **65**, 023510.

[69] Liddle, A. R., Parsons, P., and Barrow, J. D. 1994. 'Formalizing the Slow Roll Approximation in Inflation'. *Physical Review D*, **50**, 7222–7232.

[70] Linde, A. 2017. 'A Brief History of the Multiverse'. *Reports on Progress in Physics*, **80**(2), 022001.

[71] Linde, A. D. 1982a. 'A New Inflationary Universe Scenario: A Possible Solution of the Horizon, Flatness, Homogeneity, Isotropy and Primordial Monopole Problems'. *Physics Letters B*, **108**, 389–393.

[72] Linde, A. D. 1982b. 'Nonsingular Regenerating Inflationary Universe'. Print-82-0554, Cambridge University preprint, available at https://web.stanford.edu/~alinde/1982.pdf

[73] Linde, A. D. 1986. 'Eternally Existing Selfreproducing Chaotic Inflationary Universe'. *Physics Letters B*, **175**, 395–400.

[74] Linde, A. D., and Mezhlumian, A. 1995. 'Inflation with $\Omega \neq 1$'. *Physical Review D*, **52**, 6789–6804.

[75] Linde, A. D., Linde, D. A., and Mezhlumian, A. 1994. 'From the Big Bang Theory to the Theory of a Stationary Universe'. *Physical Review D*, **49**, 1783–1826.

[76] Lorenz, L., Martin, J., and Ringeval, C. 2008a. 'Brane Inflation and the WMAP Data: A Bayesian Analysis'. *Journal of Cosmology and Astroparticle Physics*, **0804**, 001.

[77] Lorenz, L., Martin, J., and Ringeval, C. 2008b. 'Constraints on Kinetically Modified Inflation from WMAP5'. *Physical Review D*, **78**, 063543.

[78] Lorenz, L., Martin, J., and Ringeval, C. 2008c. 'K-Inflationary Power Spectra in the Uniform Approximation'. *Physical Review D*, **78**, 083513.

[79] Lorenz, L., Martin, J., and Yokoyama, J. 2010. 'Geometrically Consistent Approach to Stochastic DBI Inflation'. *Physical Review D*, **82**, 023515.

[80] Maartens, R. 2011. 'Is the Universe Homogeneous?' *Philosophical Transactions of the Royal Society*, **369**, 5115–5137.

[81] Martin, J. 2004. 'Inflation and Precision Cosmology'. *Brazilian Journal of Physics*, **34**, 1307–1321.

[82] Martin, J. 2005. 'Inflationary Cosmological Perturbations of Quantum-Mechanical Origin'. *Lecture Notes in Physics*, **669**, 199–244.

[83] Martin, J. 2008. 'Inflationary Perturbations: The Cosmological Schwinger Effect'. *Lecture Notes in Physics*, **738**, 193–241.

[84] Martin, J. 2016. 'The Observational Status of Cosmic Inflation after Planck'. *Astrophysics and Space Science Proceedings*, **45**, 41–134.

[85] Martin, J., and Brandenberger, R. 2003. 'On the Dependence of the Spectra of Fluctuations in Inflationary Cosmology on TransPlanckian Physics'. *Physical Review D*, **68**, 063513.

[86] Martin, J., and Brandenberger, R. H. 2001. 'The TransPlanckian Problem of Inflationary Cosmology'. *Physical Review D*, **63**, 123501.

[87] Martin, J. and Musso, M. 2006a. 'On the Reliability of the Langevin Perturbative Solution in Stochastic Inflation'. *Physical Review D*, **73**, 043517.

[88] Martin, J., and Musso, M. 2006b. 'Solving Stochastic Inflation for Arbitrary Potentials'. *Physical Review D*, **73**, 043516.

[89] Martin, J., and Ringeval, C. 2004a. 'Addendum to "Superimposed oscillations in the WMAP data?"'. *Physical Review D*, **69**, 127303.

[90] Martin, J., and Ringeval, C. 2004b. 'Superimposed Oscillations in the WMAP Data?' *Physical Review D*, **69**, 083515.

[91] Martin, J., and Ringeval, C. 2005. 'Exploring the Superimposed Oscillations Parameter Space'. *Journal of Cosmology and Astroparticle Physics*, **0501**, 007.

[92] Martin, J., and Ringeval, C. 2006. 'Inflation after WMAP3: Confronting the Slow-Roll and Exact Power Spectra to CMB Data'. *Journal of Cosmology and Astroparticle Physics*, **0608**, 009.

[93] Martin, J., and Ringeval, C. 2010. 'First CMB Constraints on the Inflationary Reheating Temperature'. *Physical Review D*, **82**, 023511.

[94] Martin, J., and Schwarz, D. J. 1998. 'The Influence of Cosmological Transitions on the Evolution of Density Perturbations'. *Physical Review D*, **57**, 3302–3316.

[95] Martin, J., and Vennin, V. 2012. 'Stochastic Effects in Hybrid Inflation'. *Physical Review D*, **85**, 043525.

[96] Martin, J., and Vennin, V. 2016. 'Quantum Discord of Cosmic Inflation: Can We Show that CMB Anisotropies Are of Quantum-Mechanical Origin?' *Physical Review D*, **93**(2), 023505.

[97] Martin, J., Ringeval, C., and Trotta, R. 2011. 'Hunting Down the Best Model of Inflation with Bayesian Evidence'. *Physical Review D*, **83**, 063524.

[98] Martin, J., *et al.* 2014a. 'Compatibility of Planck and BICEP2 in the Light of Inflation'. *Physical Review D*, **90**(6), 063501.

[99] Martin, J., Ringeval, C., and Vennin, V. 2014b. 'Encyclopædia Inflationaris'. *Physics of the Dark Universe*, **5–6**, 75–235.

[100] Martin, J., Ringeval, C., and Vennin, V. 2015. 'Observing Inflationary Reheating'. *Physical Review Letters*, **114**(8), 081303.

[101] Martin, J., Ringeval, C., and Vennin, V. 2016. 'Information Gain on Reheating: The One Bit Milestone'. *Physical Review D*, **93**(10), 103532.

[102] Martin, J., Ringeval, C., and Vennin, V. 2013. 'K-Inflationary Power Spectra at Second Order'. *Journal of Cosmology and Astroparticle Physics*, **1306**, 021.

[103] Martin, J., *et al.* 2014c. 'The Best Inflationary Models after Planck'. *Journal of Cosmology and Astroparticle Physics*, **1403**, 039.

[104] Mukhanov, V. 2015. 'Inflation without Selfreproduction'. *Fortschritte der Physik*, **63**, 36–41.

[105] Mukhanov, V. F., and Chibisov, G. 1981. 'Quantum Fluctuation and Nonsingular Universe'. *Journal of Experimental and Theoretical Physics Letters*, **33**, 532–535.

[106] Mukhanov, V. F., and Chibisov, G. 1982. 'The Vacuum Energy and Large Scale Structure of the Universe'. *Journal of Experimental and Theoretical Physics*, **56**, 258–265.

[107] Mukhanov, V. F., Feldman, H., and Brandenberger, R. H. 1992. 'Theory of Cosmological Perturbations. Part 1. Classical Perturbations. Part 2. Quantum Theory of Perturbations. Part 3. Extensions'. *Physics Reports*, **215**, 203–333.

[108] Peebles, P. J. E. 1968. 'Recombination of the Primeval Plasma'. *The Astrophysical Journal*, **153**(July), 1.

[109] Peebles, P. J. E. 1999. *Summary: Inflation and Traditions of Research. Proceedings of the Pritzker Symposium on the Status of Inflationary Cosmology*, preprint available at arXiv:astro-ph/9905390.

[110] Penrose, R. 1989. 'Difficulties with Inflationary Cosmology'. *Annals of the New York Academy of Sciences*, **571**, 249–264.

[111] Perez, R. S., and Pinto-Neto, N. 2011. 'Spherically Symmetric Inflation'. *Gravitation and Cosmology*, **17**, 136–140.

[112] Perlmutter, S., *et al.* 1997. Measurements of the Cosmological Parameters Omega and Lambda from the First 7 Supernovae at z>=0.35'. *The Astrophysical Journal*, **483**, 565.

[113] Perlmutter, S., *et al.* 1998. 'Discovery of a Supernova Explosion at Half the Age of the Universe and Its Cosmological Implications'. *Nature*, **391**, 51–54.

[114] Peter, P. and Uzan, J. P. 2009. *Primordial Cosmology*. Oxford Graduate Texts. Oxford University Press.

[115] Polarski, D. and Starobinsky, A. A. 1996. 'Semiclassicality and Decoherence of Cosmological Perturbations'. *Classical and Quantum Gravity*, **13**, 377–392.

[116] Ringeval, C., Sakellariadou, M., and Bouchet, F. 2007. 'Cosmological Evolution of Cosmic String Loops'. *Journal of Cosmology and Astroparticle Physics*, **0702**, 023.

[117] Schwarz, D. J., Terrero-Escalante, C. A., and Garcia, A. A. 2001. 'Higher Order Corrections to Primordial Spectra from Cosmological Inflation'. *Physics Letters B*, **517**, 243–249.

[118] Silverstein, E. 2019. 'The Dangerous Irrelevance of String Theory'. In R. Dardashti, R. Dawid, R., and K. Thébault (eds.), Why Trust a Theory?: Epistemology of Fundamental Physics (pp. 365–376). Cambridge University Press, DOI:10.1017/9781108671224.025.

[119] Smoot, G. F., *et al.* 1992. 'Structure in the COBE Differential Microwave Radiometer First Year Maps'. *The Astrophysical Journal Letters*, **396**, L1–L5.

[120] Starobinsky, A. A. 1979. 'Spectrum of Relict Gravitational Radiation and the Early State of the Universe'. *Journal of Experimental and Theoretical Physics Letters*, **30**, 682–685.

[121] Starobinsky, A. A. 1980. 'A New Type of Isotropic Cosmological Models without Singularity'. *Physics Letters B*, **91**, 99–102.

[122] Starobinsky, A. A. 1982. 'Dynamics of Phase Transition in the New Inflationary Universe Scenario and Generation of Perturbations'. *Physics Letters B*, **117**, 175–178.

[123] Starobinsky, A. A. 1986. 'Stochastic De Sitter (Inflationary) Stage in the Early Universe'. *Lecture Notes in Physics*, **246**, 107–126.

[124] Starobinsky, A. A. 1992. 'Spectrum of Adiabatic Perturbations in the Universe When There Are Singularities in the Inflation Potential'. *JETP Letters*, **55**, 489–494.

[125] Starobinsky, A. A. 2005. 'Inflaton Field Potential Producing the Exactly Flat Spectrum of Adiabatic Perturbations'. *JETP Letters*, **82**, 169–173.

[126] Starobinsky, A. A., and Yokoyama, J. 1994. 'Equilibrium State of a Selfinteracting Scalar Field in the De Sitter Background'. *Physical Review D*, **50**, 6357–6368.

[127] Steigman, G., and Turner, M. S. 1983. 'Inflation in a Shear or Curvature Dominated Universe'. *Physics Letters B*, **128**, 295–298.

[128] Stein-Schabes, J. A. 1987. 'Inflation in Spherically Symmetric Inhomogeneous Models'. *Physical Review D*, **35**, 2345.

[129] Steinhardt, P. J. 1982. 'Natural Inflation'. In Gibbons, G,. Hawking, S., and Siklos, S. (eds.) The Very Early Universe. Cambridge University Press. Page 251.

[130] Traschen, J. H., and Brandenberger, R. H. 1990. 'Particle Production during Out-of-Equilibrium Phase Transitions'. *Physical Review D*, **42**, 2491–2504.

[131] Trotta, R. 2008. 'Bayes in the Sky: Bayesian Inference and Model Selection in Cosmology'. *Contemporary Physics*, **49**, 71–104.

[132] Trotta, R. 2017. *Bayesian Methods in Cosmology*. arXiv:1701.01467.

[133] Turner, M. S. 1983. 'Coherent Scalar Field Oscillations in an Expanding Universe'. *Physical Review D*, **28**, 1243.

[134] Turner, M. S., and Widrow, L. M. 1986. 'Homogeneous Cosmological Models and New Inflation'. *Physical Review Letters*, **57**, 2237–2240.

[135] Vachaspati, T. 2003. *Eternal Inflation and Energy Conditions in De Sitter Space-Time.*

[136] Vennin, V., and Starobinsky, A. A. 2015. 'Correlation Functions in Stochastic Inflation'. *European Physical Journal C*, **75**, 413.

[137] Wands, D. 2008. 'Multiple Field Inflation'. *Lecture Notes in Physics*, **738**, 275–304.

[138] Weinberg, S. 1972. *Gravitation and Cosmology: Principles and Applications of the General Theory of Relativity*. Wiley.

[139] Winitzki, S. 2001. *Null Energy Condition Violations in Eternal Inflation.*

[140] Ávila, S., Martin, J., and Steer, D. 2014. 'Superimposed Oscillations in Brane Inflation'. *Journal of Cosmology and Astroparticle Physics*, **1408**, 032.

5

Is the Universal Matter-Antimatter Asymmetry Fine-Tuned?

GARY STEIGMAN AND ROBERT J. SCHERRER*

Abstract

The asymmetry between matter and antimatter (baryons and anti-baryons or nucleons and anti-nucleons, along with their accompanying electrons and positrons) is key to the existence and nature of our Universe. A measure of the matter-antimatter asymmetry of the Universe is provided by the present value of the universal ratio of baryons (baryons minus anti-baryons) to photons (or the ratio of baryons to entropy). The baryon asymmetry parameter is an important physical and cosmological parameter. But how fine-tuned is it? A 'natural' value for this parameter is zero, corresponding to equal amounts of matter and antimatter. Such a universe would look nothing like ours and would be unlikely to host stars, planets, or life. Another, also possibly natural, choice for this dimensionless parameter would be of order unity, corresponding to nearly equal amounts (by number) of matter (and essentially no antimatter) and photons in every co-moving volume. However, observations suggest that in the Universe we inhabit, the value of this parameter is non-zero but smaller than this natural value by some nine to ten orders of magnitude. In this contribution we review the evidence, observational as well as theoretical, that our Universe does *not* contain equal amounts of matter and antimatter. An overview is provided of some of the theoretical proposals for extending the Standard Models of particle physics and cosmology in order to generate such an asymmetry during the early evolution of the Universe.

Any change in the magnitude of the baryon asymmetry parameter necessarily leads to a universe with physical characteristics different from those in our own. Small changes in this parameter will barely affect cosmic evolution, while large changes might alter the formation of stars and planets and affect the development of life. The degree of fine-tuning in the baryon asymmetry parameter is determined by the width of the range over which it can be varied and still allow for the existence of life. Our results suggest that the baryon asymmetry parameter can be varied over a very wide range without impacting the prospects for life; this result is *not* suggestive of fine-tuning.

---

* Following the untimely death of Gary Steigman, the second author was brought in to complete this chapter. He has endeavoured to adhere as closely as possible to the original format and spirit of the manuscript constructed by the first author.

We note that, according to those extensions of the Standard Models of particle physics and cosmology that allow for a non-zero baryon number, the Universe began with zero baryon number at a time (temperature) when baryon number was conserved. As the Universe expanded and cooled, baryon number conservation was broken at some high temperature- (mass-/energy-) scale, and a non-zero baryon number was created. However, even though baryon non-conservation is strongly suppressed at late times (low temperatures), baryon number is not conserved, so matter (protons, the lightest baryons) might eventually decay, with the baryon number reverting back to zero. Ashes to ashes, dust to dust, the Universe began with zero baryon number and may well end that way.

5.1 Introduction and Overview

The asymmetry between matter and antimatter (baryons and anti-baryons or nucleons and anti-nucleons, along with their accompanying electrons and positrons) is key to the existence and nature of our Universe. Any causal Lorentz-invariant quantum theory allows for particles to come in particle-antiparticle pairs. The discovery of the antiproton [6] in 1955 quickly stimulated serious consideration of the antimatter content of the Universe [3, 18] and led to constraints on the amount of antimatter based on the astrophysical effects of interacting matter and antimatter [4]. At the time, and for many years after, the prevailing view in the physics community was that baryon number (the quantum number that distinguishes baryons and anti-baryons) was absolutely conserved, and this assumption led to two differing points of view. Either the Universe is and always has been symmetric between matter and antimatter or the Universe is and always has been asymmetric, with an excess of matter over antimatter that has remained unchanged from the beginning of the expanding Universe (the Big Bang). Those who believed the Universe to be symmetric between matter and antimatter were undeterred by the fact that that the only antimatter seen up to that time (not counting positrons) was the handful of antiprotons created in collisions at high-energy accelerators. Those who believed the Universe to be asymmetric had to come to grips with the dilemma of creating such a universe if the laws of physics dictated that particles are always created (and destroyed) in pairs and that baryon number is absolutely conserved.

Most ignored this dilemma. Andrei Sakharov [35] did not. To set the stage for Sakharov's seminal work, it is useful to recall the 1965 discovery of the cosmic microwave background (CMB) radiation [9, 30], which transformed the study of cosmology from philosophy and mathematics to physics and astronomy. It quickly became clear that the discovery of the radiation content of the Universe, along with its observed expansion, ensured that very early in its evolution, when the temperature and densities (both number and energy densities) were very high, collisions among particles would be very rapid and energetic and, at sufficiently high temperatures, particle-antiparticle pairs would be produced (and would annihilate). Sakharov explored the requirements necessary for such high-energy collisions in the early Universe to create a matter-antimatter asymmetry if none existed initially. Sakharov's recipe for cooking a universal baryon asymmetry has three ingredients. One obvious condition is that baryon number cannot be absolutely conserved; baryon number (B) conservation must be violated. Although the Standard Model of particle physics at the time did not allow for violation of baryon number conservation, the later development of grand unified theories (GUTs) did. Sakharov also noted that the discrete symmetries of parity (P) and charge conjugation (C), replacing particles with antiparticle, or of CP, would need to be broken as well. Current models, in agreement with accelerator data, do allow for P and CP violation. Sakharov's third ingredient is not from particle physics but from cosmology, relying on the expansion of the Universe. The third ingredient in the recipe requires that thermodynamic equilibrium not be maintained when the B, P, and CP violating collisions occur in the early Universe. Although at the time of Sakharov's work there was no evidence that conservation of B and CP were violated, it was already known that parity is not conserved in the weak interactions and that the expansion of the Universe could possibly provide the required departure from thermodynamic equilibrium. Sakharov set the stage for consideration of a universe with unequal amounts of matter and antimatter. We will revisit Sakharov's three conditions for baryogenesis in Section 5.4.

In the hot, dense thermal soup of the very early Universe, matter and antimatter (baryons and anti-baryons) are as abundant as all the other particles whose mass is less than the temperature. As the Universe expands and cools, particle-antiparticle pairs annihilate, leaving behind only the lightest particles, along with any particle-antiparticle pairs that evaded annihilation in the early Universe or, perhaps, in an asymmetric universe, an initial matter excess that escaped annihilation. In the late Universe, when the temperature (in energy units) is far below the masses of the unstable particles of the Standard Model (SM) of particle physics, only photons and the lightest stable (or very long lived) SM particles remain: nucleons (and possibly anti-nucleons), electrons (and possibly positrons), and the three SM neutrinos. In cosmology, it is conventional to refer to all ordinary matter consisting of nucleons and electrons (nuclei, atoms, and molecules), as 'baryons' (B)[1] to distinguish it from dark matter (DM). Electrons are not baryons, but their (very small) contribution to the present-day matter density is included in this definition of the baryon density. The photons and neutrinos are often referred to as 'radiation'. The matter-antimatter asymmetry is the difference between the numbers of baryons and anti-baryons. Since this is an extensive quantity, scaling with the size of the volume considered, it is useful to introduce the ratio (by number) of baryons to photons to quantify the size of any matter-antimatter asymmetry. The ratio of the baryon (minus the anti-baryon) and photon number densities, $\eta_B = n_B/n_\gamma$, provides a measure of the matter-antimatter asymmetry of the Universe. However, as the Universe expands and cools, the heavier, unstable SM particles annihilate and decay, increasing the number of photons N_γ in a co-moving volume V, where $N_\gamma = n_\gamma V$, while the baryon number in the same co-moving volume is unchanged (at least during those epochs when baryons are conserved). Instead, it is the entropy, $S = sV$, in the co-moving volume, not the number of photons, that is conserved as the Universe expands adiabatically. The entropy

---

[1] Throughout this article, the terms baryons, nucleons, ordinary matter, and normal matter are used interchangeably.

and the number of photons in a co-moving volume are related by $S = 1.8g_s N_\gamma$, where the total entropy is related to the entropy in photons alone by $S \equiv (g_s/2)S_\gamma$, and $S_\gamma = 4/3$ $(\rho_\gamma/T)V = 4/3(\langle E_\gamma \rangle/T)N_\gamma$ and $\langle E_\gamma \rangle = 2.7\,T$. The quantity $g_s = g_s(T)$ counts the number of degrees of freedom contributing to the entropy at temperature T. For the SM of particle physics with three families of quarks and leptons at temperatures above the mass of the heaviest SM particle (the top quark), $g_s \approx 427/4$. For temperatures below the electron mass, after the three flavours of weakly interacting neutrinos have decoupled and the photons have been heated relative to the neutrinos by the annihilation of the $e^\pm$ pairs, $g_s \to g_{s0} \approx 43/11$. As a result, as the Universe cools from above the top quark mass to below the electron mass, the number of photons in a co-moving volume increases by a factor of ≈ 27, and the baryon to photon ratio is diluted by this same factor. In an adiabatically expanding Universe (as ours is assumed to be) the entropy in a co-moving volume is conserved, along with the net number of baryons minus anti-baryons (during those epochs when baryon number non-conservation is strongly suppressed). Therefore, the ratio of baryon number to entropy, $N_B/S = n_B/s$, provides a measure of the baryon asymmetry whose value is unchanged as the Universe expands and cools. Evaluated in the late Universe, after $e^\pm$ annihilation is complete, $s/n_\gamma \to (s/n_\gamma)_0 = 1.8\,g_{s0} \approx 7.0$, so that $n_B/s \approx \eta_B/7.0$. Consistent with most of the published literature, η_B is evaluated here in the late Universe, so $\eta_B \equiv \eta_{B0} \equiv (n_B/n_\gamma)_0$. In the discussion here η_B and n_B/s will both be referred to as the 'baryon asymmetry parameter'.

In a matter-antimatter *symmetric* Universe, the baryon asymmetry parameter $\eta_B = 0$. For a quantity that could, in principle, have any value between $-\infty$ and $+\infty$[2], zero might seem to be a 'natural' choice.

When is a physical parameter, such as the baryon asymmetry parameter, considered to be fine-tuned? The criteria for answering this question, along with a discussion of the degeneracies with other physical parameters, are discussed in Section 5.2. In Section 5.3, the overwhelming observational and theoretical evidence that our Universe is *not* matter-antimatter symmetric is reviewed, excluding the natural choice of $\eta_B = 0$. Faced with the necessity that a universe hosting stars, planets, and life requires $\eta_B \neq 0$, Section 5.4 provides an overview of the multitude of particle physics (and cosmology) models proposed to generate a non-zero baryon asymmetry during the early evolution of the Universe. These models are capable of generating a baryon asymmetry that is much smaller or much larger than that observed in our Universe, suggesting that there might be universes with almost any non-zero values of η_B. In an asymmetric universe, the quantitative value of the baryon asymmetry parameter plays an important role in primordial nucleosynthesis (Big Bang nucleosynthesis: BBN), regulating the abundances of the nuclides produced in the early Universe, before any stellar processing. BBN is reviewed for a large range of η_B in Section 5.5. The degeneracy of the baryon asymmetry parameter with other cosmological parameters is discussed in Section 5.6, and a variety of alternate cosmological models allowing

---

[2] In a universe with more 'matter' than 'antimatter', $\eta_B > 0$. For the opposite case, where $\eta_B < 0$, the definitions of matter and antimatter could be interchanged. Therefore, without loss of generality, it is assumed here that $\eta_B \geq 0$.

for a range of η_B values are presented in Section 5.7. The criterion used here to judge the viability of alternate cosmological models is whether their universes are capable of hosting stars, planets, and life. Our results and conclusions are summarised in Section 5.8.

5.2 Definition of Fine-Tuning of the Baryon Asymmetry Parameter

How fine-tuned is the baryon asymmetry parameter? Here we will adopt a definition of fine-tuning based on the capability of the Universe to harbour life. Clearly, small changes in the asymmetry parameter will have little effect on cosmic evolution. However, large changes in this parameter will have major effects, notably altering the production of elements in the early universe and changing the process of structure formation through the growth of primordial density perturbations. We will see that the former, even in extreme cases, is unlikely to have any effect on the development of life in the Universe, while the latter can have profound effects. In particular, if the process of galaxy and star formation is too inefficient, then there will be no planetary systems to harbour life. One must be cautious, of course, in defining the limits on environments that can support life; our argument will be based on life as we observe it, which exists on planets orbiting stars. It is always possible that more extreme environments might harbour life in ways that we have not considered; for example, Avi Loeb has pointed out that the cosmic microwave background can provide an energy source for life when the Universe was only 10 million years old and the temperature of the CMB was between the freezing and boiling points of water [27] (see Chapter 12). While we will not consider such extreme possibilities here, caution is always advised when defining the conditions needed for the existence of life.

The extent to which the value of η_B is fine-tuned will depend on how widely it can be varied while still allowing for the existence of life. The issue of the fine-tuning of η_B is not, of course, a true-false question: the best we can do is to determine an allowed range for η_B. The width of this range can then suggest the plausibility (or lack thereof) of the need for special initial conditions or special values for the underlying fundamental parameters that determine η_B. But the question, 'Is the baryon asymmetry parameter fine-tuned?' does not have a yes or no answer.

In considering the variation of one or more physical parameters, a choice must be made: do we consider the variation of the baryon asymmetry parameter alone, or do we allow other parameters to vary at the same time? In the latter case, changes in the value of one parameter may be compensated, at least in part, by changes in other parameters.

As an example, consider the way in which the relation between the baryon to entropy ratio and the baryon to photon ratio depends on the number of neutrino flavours, as well as on the neutrino decoupling temperature – which depends, in turn, on the strength of the weak interactions. In an alternate universe where there are N_ν flavours of neutrinos, instead of the SM value of $N_\nu = 3$, $g_{s0} = 43/11 + 7(N_\nu - 3)/11 = 43/11(1 + 7(N_\nu - 3)/43)$ and $g_{\rho 0} = 3.36 + 0.454(N_\nu - 3) = 3.36(1 + 0.135(N_\nu - 3))$. For these results, it has been assumed that when $N_\nu \neq 3$, the usual weak interactions are unchanged and all neutrinos decouple when $T_{dec} \gg m_e$ (but $T_{dec} \ll m_\mu$) so that $(T_\nu/T_\gamma)_0 \approx 4/11$. With these caveats, for $N_v \neq 3$, the late time entropy per photon is $(s/n_\gamma)_0 \approx 7.0(1 + 7(N_v - 3)/43)$, and the relation between η_B and n_B/s is changed,

$$\eta_B = (n_B/n_\gamma)_0 \approx 7.0(1 + 7(N_v - 3)/43)\,(n_B/s)\,. \tag{5.1}$$

For example, in an alternate universe with only one neutrino flavour ($N_v = 1$), $\eta_B \approx 4.7(n_B/s)$, while in one with eight flavours of neutrinos, ($N_v = 8$),[3] $\eta_B \approx 12.8(n_B/s)$.

In general, allowing multiple parameters to vary simultaneously will weaken the constraints provided when only one of them is varied, which is likely to be an issue with many of the other essays in this volume. For example, consider the atomic energy scale,

$$\epsilon \equiv \mu_H c^2 \alpha^2 = \left(\frac{m_e m_p}{m_e + m_p}\right) c^2 \left(\frac{e^2}{\hbar c}\right)^2, \tag{5.2}$$

where μ_H is the reduced mass of the proton-electron system, and the fine-structure constant is $\alpha = e^2/\hbar c \approx 1/137$ (when measured at low energies). For $m_e c^2 \approx 0.51\,\text{MeV}$ and $m_p \approx 0.94\,\text{GeV}$, $\epsilon \approx m_e c^2 \alpha^2 \approx 27\,\text{eV}$. Since ϵ is not a dimensionless parameter, perhaps it is the dimensionless parameter $\epsilon/\mu_H c^2 = \alpha^2 \approx 5.3 \times 10^{-5}$ that is fundamental. Suppose that α and $\mu_H c^2$ are allowed to change, while the atomic energy scale, $\mu_H c^2 \alpha^2$, is kept unchanged. For example, m_e and m_p might change while $m_e/m_p \ll 1$ might be (nearly) unchanged. Atomic energy levels will be largely unchanged while nuclear energies will be changed. How much freedom is there to change α along with other fundamental parameters (e.g., m_e, m_p, m_e/m_p), while leaving most of 'ordinary' atomic and nuclear physics unchanged? This issue of 'degeneracy' among physical parameters will rear its head in the subsequent discussion of the fine-tuning of the baryon asymmetry of the Universe. When exploring model universes with different values of η_B, we will keep all other parameters (e.g., α, m_e/m_p, N_v, etc.) fixed. However, we need to remain aware that the results presented here can be considerably altered if multiple parameters are simultaneously varied.

5.3 The Case against a Symmetric Universe

Over the years, experiments at ever higher energies have confirmed that particles are created (and annihilated) in pairs and that in all collisions studied so far, baryon (and lepton) number is conserved. Perhaps only at the very highest energies, inaccessible to the current terrestrial accelerators, or in searches for proton decay, will non-conservation of baryon (and lepton) number be revealed. However, it is not unreasonable to ask how our present Universe would differ if baryon number were absolutely conserved. A complementary approach is to ask what astrophysical observations can tell us about the amount of antimatter (if any) in gas, stars, galaxies, and clusters of galaxies in the current Universe (e.g., [4]). These two approaches are explored here. The discussion here is based on several earlier

---

[3] For $N_v \leq 8$, QCD is asymptotically free, allowing for quark confinement and bound nuclei [20].

papers by the first author (e.g., [41–43, 45, 46]); the reader is urged to see those papers for details and for many further references.

5.3.1 The Observational Evidence against a Symmetric Universe

To paraphrase remarks by the first author in a 1976 review of the status of antimatter in the Universe [42], it is quite easy to determine if an unknown sample is made of matter or antimatter. The most rudimentary detector will suffice. Simply place your sample in the detector and wait. If the detector disappears (annihilates), your sample contained antimatter. Indeed, if you had handled your sample, you would have already known the answer. Astrophysical sources have been repeating this experiment over cosmological times. The first lunar and Venus probes confirmed that the Moon and Venus are made of matter, not antimatter. Indeed, the solar wind, sweeping past the planets of the solar system revealed by the absence of annihilation gamma rays that the Sun and the planets and other solar system bodies are all made of what we have come to define as matter. Were any of the planets made of antimatter, they would be the strongest gamma ray sources in the sky (if they had not already annihilated away). As may be inferred from the discussion in Section 5.3.2, if there were any antimatter in the material that collapsed to form the planets and other solid body objects in the solar system (the pre–solar system gas cloud) , it would have annihilated long before the solar system formed. The same is true for the stars in our galaxy. On theoretical grounds, is is highly unlikely that in a universe some 14 Gyr old, there are any non-negligible amounts of antimatter surviving in our galaxy.

In a typical nucleon – anti-nucleon annihilation, $\sim$ 5–6 pions are produced. The pions decay to muons, neutrinos, and photons, and the muons decay to electrons ($e^{\pm}$ pairs) and neutrinos. The $e^{\pm}$ pairs may annihilate in flight or, being tied to local magnetic fields, they may lose energy by Compton emission and annihilate nearly at rest (producing a characteristic 511 keV line) [4]. Photons from matter-antimatter annihilations provide the most sensitive, albeit indirect, probe of the presence of antimatter, mixed with ordinary matter, on galactic and extragalactic scales. In the galaxy, gas (clouds of atomic or molecular gas) and stars are inevitably mixed. If either contained significant amounts of antimatter, the result would be annihilation, along with the corresponding production of gamma rays. The lifetime against annihilation of an antiparticle (e.g., an antiproton) in the gas in the interstellar medium (ISM) of the galaxy is very short, $t_{ann} \approx 300\,\mathrm{yr}$ [42]. It is therefore not surprising that observations of galactic gamma ray emission set very strong constraints on the antimatter fraction in the ISM, $f_{ISM} \lesssim 10^{-15}$ [42]. There can be no significant amounts of antimatter in the gas in the galaxy.

What about anti-stars? When gas collapses to form stars, the annihilation rate grows as the number density while the collapse rate increases only as the square root of the density. As a result, unless there were no normal matter in the gas that might collapse to form an anti-star, the anti-star would never form. Setting this aside, let us suppose that anti-stars had somehow formed in the galaxy. As the gas in the ISM flowed past these anti-stars, there would be annihilation, resulting in gamma rays. Using by now outdated

(40-year-old!) gamma ray data, the first author [42] determined that the absence of gamma rays indicated that the nearest anti-star in the galaxy is at least 30 pc away. This result sets an upper limit on the total number of anti-stars, N, that could be in the galaxy: $N < 10^7$, a small fraction of all the stars in the galaxy. Although more recent gamma ray data can refine these bounds, the old data were already sufficiently strong to argue against any significant amounts of antimatter in the galaxy.

Galactic cosmic rays, coming to us from outside of the solar system, provide a valuable direct probe of antimatter in the galaxy. Whatever the sources of the galactic cosmic rays, the discovery of anti-nuclei in the cosmic rays would provide direct evidence (a 'smoking gun') for the presence of antimatter in the galaxy (for more details but obsolete data, see the discussion in [42]). The antiproton would be the lightest anti-nucleus, but in high-energy collisions between cosmic rays and interstellar gas, some 'secondary' antiprotons will be produced. Indeed, antiprotons have been observed in the cosmic rays, but their numbers are consistent with a secondary origin. However, production of more complex anti-nuclei in high-energy cosmic rays – interstellar gas collisions (secondary anti-nuclei) is strongly suppressed, and to date, no anti-deuterons [16] or anti-alpha [2] particles have been detected in the cosmic rays. For example, the 1999 AMS upper bound [2] to the cosmic ray anti-helium-to-helium ratio is $< 10^{-6}$, providing a strong supplement to the gamma ray data suggesting our galaxy has no significant amounts of antimatter. The absence of primary antimatter in the cosmic rays is evidence that the sources of the galactic cosmic rays contain little, if any, antimatter (indeed, if there were some antimatter mixed with a predominant amount of ordinary matter in the cosmic ray sources, they likely would have annihilated over the lifetimes of the sources).

What of external galaxies or extragalactic high-luminosity sources such as AGNs or QSOs? If annihilations deposit their energy locally, then the gamma ray flux and the luminosity of an annihilation-powered source are connected [41]. If Φ_γ is the photon flux from annihilations (photons $cm^{-2} s^{-1}$) and Φ_E is the energy flux from the same source (ergs $cm^{-2} s^{-1}$), then $\Phi_\gamma \gtrsim 10^4 \Phi_E$ [41]. Although annihilation was proposed as a panacea for the energy budgets of QSOs and other high-luminosity sources [4], the detailed emission mechanisms required enormous magnetic fields, compounding the problems of an already stretched energy budget. Steigman and Strittmatter [46] explored whether observations of the annihilation neutrino flux could constrain models of annihilation-driven infrared emission in Seyfert galaxies. For individual sources, it was estimated [42] that the neutrino flux would be at least five orders of magnitude smaller than was observed at the time. The difficulty of detecting the relatively low energy ($\lesssim 500\,MeV$) neutrinos, combined with improved models for the energy sources in QSOs, Seyferts, etc., have made annihilation neutrinos an unlikely probe.

Moving further away, outside our own galaxy, the strongest constraints, on the largest scales, come from observations of X-ray emitting clusters of galaxies [42, 43, 45]. Most of the baryons in clusters of galaxies are in the hot intracluster gas. The same collisions between particles in the intracluster gas responsible for producing the observed X-ray emission would result in annihilation gamma rays if some fraction of the gas consisted of antiparticles. The virtue of using X-ray emitting clusters of galaxies is that there is a direct proportionality between the X-ray emission from thermal bremsstrahlung and gamma ray emission from annihilation. This approach leads to bounds on the antimatter fraction (the fraction of antimatter mixed with ordinary matter) on the largest scales in the universe ($M \sim 10^{14} - 10^{15} M_{\odot}$, $R \sim$ few Mpc) [45]. Using data from 55 X-ray emitting clusters of galaxies [12] in combination with the upper bounds to the gamma ray fluxes [33], it was found that the antimatter fraction from that sample is limited to $f < 10^{-6}$ [45]. However, even stronger bounds exist for some individual clusters. For the Perseus cluster, $f < 8 \times 10^{-9}$, and for the Virgo cluster, $f < 5 \times 10^{-9}$. Perhaps the most interesting upper bound on antimatter on the largest scales comes from colliding clusters. Analysis of the Bullet Cluster gives $f < 3 \times 10^{-6}$ on the scale $M \sim 3 \times 10^{15} h^{-1} M_{\odot}$, where h is the Hubble parameter in units of 100 km sec^{-1} Mpc^{-1} [45].

5.3.2 The Problem of a Symmetric Universe

Very shortly after the discovery of the CMB [9, 30], Ya. B. Zeldovich [54][4] and H. Y. Chiu [7] independently considered the fate of matter and antimatter emerging from the early stages of the evolution of a hot universe. The result, whose derivation is outlined here, is easily summarised. At high temperatures, above the quark – hadron transition, there are many quark-antiquark pairs, and in a symmetric universe, there are equal numbers of quarks and antiquarks. As the Universe expands and cools, the quarks (and gluons) are confined into nucleons (neutrons and protons), which – because the strong interaction is strong – are in thermal equilibrium with the cosmic plasma (e.g., photons, neutrinos, and the light leptons and bosons). In this regime, the nucleon mass exceeds the temperature so that annihilation of nucleon–anti-nucleon pairs proceeds on a timescale short compared to the expansion rate of the Universe. But, since $m \gg T$, creation of new nucleon–anti-nucleon pairs from collisions in the background plasma is strongly (exponentially) suppressed so that up to spin-statistics factors of order unity, the ratio of nucleons (and anti-nucleons) to photons is $n_N/n_\gamma = n_{\bar{N}}/n_\gamma = n_{eq}/n_\gamma \propto (m/T)^{3/2} e^{-(m/T)} \ll 1$. Even though the abundances of nucleons and anti-nucleons (e.g., relative to photons) are very small, the strong interaction is strong, ensuring that $n_N \approx n_{eq}$ is maintained down to very low temperatures, $T \ll m_N$. However, eventually, the abundance of the nucleon–anti-nucleon pairs becomes so small that they no longer can find each other to annihilate (and the creation of new pairs is exponentially suppressed), and the abundance of nucleons (and anti-nucleons) 'freezes out' at a 'relic' abundance $(n_N/n_\gamma)_0$. The evolution of the nucleon-anti-nucleon abundances follows an evolution equation, described next, that accounts for creation, annihilation, and the expansion of the Universe. The solution, presented in the following discussion, shows that the relic abundance of the nucleon–anti-nucleon pairs in a symmetric universe is some nine orders of magnitude smaller than the nucleon abundance observed in our Universe, providing an important nail in the coffin of the symmetric Universe.

---

[4] It is interesting that Zeldovich's article was written prior to the discovery of the CMB. As a result, in his review, Zeldovich considered both hot and cold universes.

As first derived from an argument of detailed balance by Zeldovich [54] and later rediscovered and supported by many textbook derivations based on the Boltzmann equation, the evolution of the abundance of a particle (and its antiparticle) produced and annihilated in pairs, is described by the standard evolution equation (SEE); see, e.g., [7, 19, 25, 37, 41–43, 47] and references therein. For equal numbers of particles and antiparticles (no asymmetry, zero chemical potential), the SEE may be written as

$$\frac{1}{V}\left(\frac{dN}{dt}\right) = \frac{dn}{dt} + 3Hn = \langle \sigma v \rangle (n_{eq}^2 - n^2), \tag{5.3}$$

where $N = nV$ is the number of particles (and antiparticles) in a co-moving volume V. As the Universe expands and the cosmic scale factor, a, increases, the co-moving volume grows as $V \propto a^3$. In Eq. (5.3), the number density of particles and antiparticles is n, the total annihilation cross section is $\langle \sigma v \rangle$, and $H = a^{-1}(da/dt)$ is the Hubble parameter. The SEE is a form of the Ricatti equation, for which there are no known closed-form solutions except in special cases. Although the SEE may be integrated numerically, here the approximate analytic approach first outlined by Zeldovich [54] and employed extensively in [19, 25, 37, 42, 43, 47] and elsewhere is followed.

For the approximate analytic solution to the SEE, it is convenient to write $n = (1 + \Delta)$ n_{eq} where, in the non-relativistic (NR) regime $(T < m)$, $n_{eq} = (g T^3/(2\pi)^{3/2})x^{3/2}$ $e^{-x} f(x)$, where $x \equiv m/T$ and $g = 2$ is the number of spin states of the proton (neutron) and of the antiproton (anti-neutron). Here, $f(x)$ is an asymptotic series in x for which $f(x) \to 1$ as $x \to \infty$. For the range of x of interest in tracking the evolution of nucleon–anti-nucleon pairs, $f(x) \approx 1$ is a very good approximation. Therefore, the evolution of the equilibrium number density (as a function of x) in the NR regime is very well described by $n_{eq} \propto T^3 x^{3/2} e^{-x} \propto x^{-3/2} e^{-x}$. Note that since the photon number density varies as T^3, $n_{eq}/n_\gamma \propto x^{3/2} e^{-x}$ in the NR regime. Instead of following the time evolution of the thermal relic abundance, it is more convenient to track its evolution as a function of x. Neglecting small logarithmic corrections involving derivatives related to the entropy and photon densities, the derivatives with respect to time and x (or T) are related by

$$dt \approx \frac{1}{H}\left(\frac{dx}{x}\right) \approx -\frac{1}{H}\left(\frac{dT}{T}\right), \tag{5.4}$$

where $H = H(T)$ is the Hubble parameter evaluated at temperature T. Now we define the quantity $g_\rho(T)$ (in analogy to g_s) by $g_\rho/2 \equiv \rho/\rho_\gamma$, where ρ is the *total* mass/energy density and ρ_γ is the energy density in photons alone. During those epochs in the evolution of the Universe when the energy density is dominated by the contribution from relativistic particles (radiation dominated: RD), $H \propto \rho_R^{1/2} \propto g_\rho^{1/2}\rho_\gamma \propto g_\rho^{1/2}T^2$. In terms of Δ and x, the SEE may be rewritten as

$$\frac{d(\ln(1 + \Delta)N_{eq})}{d(\ln x)} = -\left(\frac{\Gamma_{eq}}{H}\right)y, \tag{5.5}$$

where $\Gamma_{eq} \equiv \langle \sigma v \rangle n_{eq}$ and $y \equiv \Delta(2 + \Delta)/(1 + \Delta)$.

For $x \sim O(1)$ $(m \approx T)$, Δ is very small and $n = n_{eq}$ is a very good approximation. As the Universe expands and cools, x increases and Δ grows exponentially (while n_{eq} decreases exponentially), and the departure from equilibrium grows. Define x_* to be the value of x for which $\Delta(x_*) \equiv \Delta_* \sim O(1)$, so the true abundance, n_*, exceeds the equilibrium density, n_{eq*}, by factor $1 + \Delta_* > 1$. (A more precise definition of x_* is given after Eq. (5.13)). For $x \gtrsim x_*$, $\Delta \gtrsim \Delta_*$ and $n/n_{eq} > 1$ increases. In this regime, where $n \gtrsim n_{eq}$, the SEE simplifies,

$$dN/dt = \langle \sigma v \rangle (n_{eq}^2 - n^2) V \approx -\langle \sigma v \rangle n^2 V = -\langle \sigma v \rangle N^2 / V. \tag{5.6}$$

This equation can be integrated directly from $t = t_*$ (when $T = T_*$ and $x = x_*$) to $t = t_0$ (when $T = T_0 \ll T_*$ and $x \gg x_*$). Replacing the evolution with time (or with x) by the evolution with temperature,

$$\frac{dN}{N^2} \approx \frac{\langle \sigma v \rangle}{VH} \frac{dT}{T}, \tag{5.7}$$

where the Hubble parameter varies as $H \approx H_*(T/T_*)^2$ and the co-moving volume increases with decreasing temperature as $V \approx V_*(T_*/T)^3$. Integrating from $T = T_*$ to $T = T_0 \ll T_*$ results in

$$N_0/N_* = [1 + (\Gamma/H)_*]^{-1}, \tag{5.8}$$

where $\Gamma_* = n_* \langle \sigma v \rangle$. For nucleon–anti-nucleon annihilation, $(\Gamma/H)_* \gg 1$, so $N_0/N_* \approx (\Gamma/H)_*^{-1} \ll 1$. When $T = T_*$ $(x = x_*)$, the number of particles (neutrons or protons) in the co-moving volume, N_*, may be compared to the number of photons in the same volume, $N_{\gamma *}$,

$$\left(\frac{N}{N_\gamma}\right)_* = \left(\frac{n}{n_\gamma}\right)_* = \left(\frac{H}{n_\gamma \langle \sigma v \rangle}\right)_* \left(\frac{\Gamma}{H}\right)_*. \tag{5.9}$$

In terms of x_*,

$$\left(\frac{H}{n_\gamma \langle \sigma v \rangle}\right)_* = \frac{6.5 \times 10^{-36} g_{\rho *}^{1/2} x_*}{m \langle \sigma v \rangle}, \tag{5.10}$$

where m is in GeV and $\langle \sigma v \rangle$ is in $cm^3 s^{-1}$. As the Universe expands and cools from $T = T_*$ to $T = T_0$, the surviving nucleon (and anti-nucleon) abundance(s) decreases (decrease) to an asymptotic ('frozen out') value (ratio to photons) given by,

$$\left(\frac{N}{N_\gamma}\right)_0 = \left(\frac{N}{N_\gamma}\right)_* \left(\frac{N_0}{N_*}\right) \left(\frac{N_{\gamma *}}{N_{\gamma 0}}\right), \tag{5.11}$$

where – from entropy conservation – $N_{\gamma *}/N_0 = g_{s0}/g_{s*}$. Note that $(N/N_\gamma)_0$ is the frozen out ratio of neutrons or protons to photons (long after annihilation has ceased[5]) and is identical to the ratio of anti-neutrons or antiprotons to photons. Even though $(N/N_\gamma)_0 \neq 0$,

---

[5] Annihilations never really cease. They simply become so rare that they are unable to continue to reduce the relic abundance.

the baryon asymmetry parameter in a symmetric universe is $\eta_B = 0$. Combining the preceding equations,

$$\left(\frac{N}{N_\gamma}\right)_0 \approx \frac{2.5 \times 10^{-35}}{m \langle \sigma v \rangle} \left(\frac{g_{\rho*}^{1/2}}{g_{s*}}\right) x_*, \tag{5.12}$$

where $g_{s0} = 43/11$ – corresponding to $N_\nu = 3$ – has been adopted. For neutrons or protons (in the approximation here, they are assumed to have the same mass, $m \approx 0.94\,\text{GeV}$), the total (s-wave) annihilation cross section[6] is $\langle \sigma v \rangle \approx 1.5 \times 10^{-15}\,\text{cm}^3\,\text{s}^{-1}$ so that

$$\left(\frac{N}{N_\gamma}\right)_0 \approx 1.8 \times 10^{-20} \left(\frac{g_{\rho*}^{1/2}}{g_{s*}}\right) x_*. \tag{5.13}$$

To find x_* and $T_* = m/x_*$, in order to evaluate $g_{\rho*} = g_\rho(T_*)$ and $g_{s*} = g_s(T_*)$,[7] we impose the condition defining x_* – that is, when $x = x_*$, $\Delta(x) = \Delta(x_*) \equiv \Delta_*$. Although $\Delta_* \sim O(1)$, a specific choice needs to be made for Δ_* in order to find the corresponding value of x_* (and it needs to be checked and confirmed that the final result is insensitive to this specific choice). Here, $\Delta_* = 0.618$ (related to the 'Golden Mean') is adopted, so $y_* = \Delta_*(2 + \Delta_*)/(1 + \Delta_*) = 1$.

It may be verified that $d(\ln(1+\Delta))/d(\ln x) \ll d(\ln N_{eq})/d(\ln x)$, so Eq. (5.5) reduces to

$$-\left(\frac{\Gamma_{eq}}{H}\right) y \approx \frac{d(\ln N_{eq})}{d(\ln x)}, \tag{5.14}$$

where $N_{eq} = n_{eq} V \propto V T^3 x^{3/2} e^{-x}$. Generally, $V T^3 \propto (aT)^3 \approx$ constant, so the logarithmic derivative of $V T^3$, depending on $d(\ln g_s)/d(\ln dT)$, may be neglected, further simplifying Eq. (5.5) to an algebraic equation,

$$d(\ln N_{eq})/d(\ln x) \approx -(x - 3/2) \approx -(\Gamma_{eq}/H)\, y. \tag{5.15}$$

For $x = x_*$, $\Delta(x_*) = \Delta_* = 0.618$ and $y = y_* = 1$. As a result,

$$x_* - 3/2 = (\Gamma_{eq}/H)_* = n_{eq*} \langle \sigma v \rangle / H_* = A_* g_{\rho*}^{-1/2} x_*^{1/2} e^{-x_*}, \tag{5.16}$$

where $A_* \equiv 4 \times 10^{34} g\, m \langle \sigma v \rangle$; g is the number of neutron or proton spin states, and, as before, the mass m is in GeV and $\langle \sigma v \rangle$ is in cm³/s. For $g = 2$, $m = 0.94$, and $\langle \sigma v \rangle = 1.5 \times 10^{-15}$, $A_* = 1.1 \times 10^{20}$. The transcendental equation for x_*, Eq. (5.16), may be solved iteratively. The solution is $x_* \approx 43.1$, corresponding to $T_* = m/x_* \approx 21.8\,\text{MeV}$, for which $g_{\rho*} \approx 11.5$ and $g_{s*} \approx 11.4$ [24]. Substituting these values into Eq. (5.13) results in the frozen-out ratios of the surviving numbers of neutrons, protons, anti-neutrons, and antiprotons to photons,

$$\left(\frac{n_n}{n_\gamma}\right)_0 = \left(\frac{n_p}{n_\gamma}\right)_0 = \left(\frac{n_{\bar n}}{n_\gamma}\right)_0 = \left(\frac{n_{\bar p}}{n_\gamma}\right)_0 \approx 2.3 \times 10^{-19}. \tag{5.17}$$

---

[6] Even though $T_* \ll m$, the nucleons are moving sufficiently rapidly that Coulomb (Sommerfeld) enhancement of the proton-antiproton annihilation cross section, relative to the neutron–anti-neutron annihilation cross section, is unimportant.

[7] For $g_\rho(T)$ and $g_s(T)$, the results of Laine and Schroeder [24] are used here.

The corresponding nucleon (neutron plus proton) and anti-nucleon–to–photon ratios are $(n_N/n_\gamma)_0 = (n_{\bar{N}}/n_\gamma)_0 \approx 4.6 \times 10^{-19}$.

Of course, for this symmetric universe, $\eta_B \equiv (n_N/n_\gamma)_0 - (n_{\bar{N}}/n_\gamma)_0 = 0$. The present mass density of matter (nucleon plus anti-nucleon) is $\rho_B = m(n_N + n_{\bar{N}}) \approx 3.5 \times 10^{-16} \, \text{GeV cm}^{-3}$, or $\Omega_B h^2 \approx 3.3 \times 10^{-11}$. In contrast, for our observed *asymmetric* universe, where annihilation of any relic anti-nucleons is very efficient, $(n_N/n_\gamma)_0 \approx 6.1 \times 10^{-10} \gg (n_{\bar{N}}/n_\gamma)_0 \approx 0$ and $\Omega_B h^2 \approx 0.022$. In a symmetric Universe, the abundance of nucleons surviving annihilation in the early universe is smaller than the abundance of nucleons in our asymmetric Universe by some nine orders of magnitude.

Notice that when $T = T_*$, the ratio of the annihilation rate to the expansion rate is very large, $(\Gamma/H)_* \approx (1 + \Delta_*)(x_* - 3/2) \approx 67 \gg 1$. Neither annihilations nor the relic abundances freeze out when $T = T_*$. For $T < T_*$, annihilations continue to reduce the abundances of nucleons, and anti-nucleons and the ratio of the annihilation rate to the expansion rate, Γ/H, continues to decrease. Eventually, for $T \equiv T_f \approx T_*/2$, $(\Gamma/H)_f = 1$, and the relic abundances freeze out (although, depending on T_f, the number of photons in the co-moving volume may continue to increase until $T \lesssim m_e$, further reducing the relic baryon to photon ratio). For $T < T_f$, $n_N = n_{\bar{N}} = n_{Nf} (T/T_f)^3$. For temperatures even slightly below T_f, $(\Gamma/H) \approx H_f/H = (T_f/T)^2 < 1$. Thereafter, the annihilation rate scales as $n\langle\sigma v\rangle \propto T^3$ (for s-wave annihilation) while the expansion rate of the Universe scales as $H \propto T^2$ (during RD epochs in the evolution), so after freeze-out ($T \ll T_f$), $\Gamma/H \approx T/T_f \ll 1$. During matter-dominated (MD) epochs in the evolution of the universe, $H \propto T^{3/2}$, so $\Gamma/H \propto T^{3/2}$, and it is still the case that $\Gamma/H \ll 1$.

By the same argument, nuclear reactions in this universe are extremely suppressed by the very low nucleon density. There can be no primordial nucleosynthesis in a symmetric universe. After freeze-out, as the universe expands and cools, neutrons decay, and the universe is left with protons (and antiprotons) and electrons (and positrons). Note that as the protons and electrons (and antiprotons and positrons) cool and become non-relativistic, the long-range Coulomb interaction enhances, through Sommerfeld enhancement [40], the annihilation cross section, $\langle\sigma v\rangle \rightarrow 2\pi(\alpha c/v)\langle\sigma v\rangle \propto T^{-1/2}$. Even so, the ratio of the annihilation rate to the expansion rate still decreases (as $T^{1/2}$ during RD epochs and as T during MD epochs). Recombination cannot occur in such a low-baryon-density universe. In the absence of non-baryonic dark matter, it is unlikely that any collapsed structures (e.g., stars or galaxies) could form in such a low-density, ionised universe; for a more detailed discussion, see Chapter 6. The history (and future) of a symmetric universe is very bleak. The story barely changes if a symmetric universe contains non-baryonic dark matter. If, for example, the presence of DM in a symmetric universe allows collapsed DM structures to form, the relic matter and antimatter would fall into the DM potential wells, increasing their number densities, leading to renewed annihilation, further reducing their already very small abundances. A matter-antimatter symmetric universe simply bears no resemblance to our Universe.

Even in an asymmetric universe, during the very early evolution of the universe, when the temperature is very high, the *equilibrium* abundance of nucleons and anti-nucleons may be much larger than the relic abundance of nucleons in our Universe, $\eta_B = (n_N/n_\gamma)_0 \approx 6 \times 10^{-10}$. These pairs will annihilate until, at some temperature, T, $(n_N/n_\gamma)(g_s(T)/g_{s0}) \approx 6 \times 10^{-10}$. For lower temperatures, the anti-nucleons continue to be annihilated, but the nucleons, due to the asymmetry, are frozen out. For nucleons (protons plus neutrons), $g = 4$, and their equilibrium abundance relative to photons is $n_N/n_\gamma = 0.26 \, g \, x^{3/2} e^{-x} \approx x^{3/2} e^{-x}$, where $x = m_B/T$, and prior to BBN, the average mass per baryon is $m_B \approx 939$ MeV [44] so that $x \approx 939/T$, with T in MeV. Here, we have assumed that $f(x) \approx 1$. To find T, we need to solve $(939/T)^{3/2} \exp(-939/T) \approx 1.5 \times 10^{-10} g_s(T)$. Using [24] for $g_s(T)$, the solution is $T \approx 38 \, \text{MeV}$ ($x \approx 25$). To avoid the annihilation catastrophe in a symmetric universe, the baryon asymmetry must have been created when $T > 38 \, \text{MeV}$ (or when $T \gg 38 \, \text{MeV}$). Recall that $T_* \approx 22 \, \text{MeV}$, so $T > T_*$, as expected. In the extensions of the Standard Models of particle physics and cosmology that allow for a baryon asymmetry at low temperatures, the energy-/temperature-/mass-scales are orders of magnitude larger than this conservative estimate.

5.4 Particle Physics Models for Generating the Universal Matter-Antimatter Asymmetry

It is clear from the preceding two sections that a universe containing equal abundances of baryons and anti-baryons is *not* the Universe we actually observe. At some point in its evolution, the Universe must have developed an asymmetry between matter and antimatter. How did this asymmetry come about?

One possibility is that the Universe actually began in an asymmetric state, with more baryons than anti-baryons. This is, however, a very unsatisfying explanation. Furthermore, if the Universe underwent a period of inflation (i.e., very rapid expansion followed by reheating), then any pre-existing net baryon number would have been erased. A more natural explanation is that the Universe began in an initially symmetric state, with equal numbers of baryons and anti-baryons, and that it evolved later to produce a net baryon asymmetry.

As we noted in the introduction, Sakharov introduced three conditions necessary to produce a net baryon asymmetry in a universe that began with zero net baryon number. These Sakharov conditions form the basis of nearly all modern theories of baryogenesis, so we will review them in more detail here. These conditions are as follows:

1. *Baryon number violation.* This is the most obvious component needed for baryogenesis. If the Universe began with zero net baryon number, and baryon number were conserved, then it would still have zero net baryon number today.
2. *C and CP violation.* The operator C changes particles into antiparticles and vice versa, while CP also flips all three coordinate axes. A universe that is baryon–anti-baryon symmetric is unchanged when C or CP is applied, while the same is not true for a universe with a net baryon excess. Hence, the production of a baryon asymmetry requires C and CP violation.

3. *A departure from thermodynamic equilibrium.* If baryon and C/CP were violated while thermal equilibrium conditions prevailed, then the chemical potentials for baryons would be driven to zero, and the only possible difference between particle and antiparticle abundances would arise if there were a mass difference between them. But CPT invariance implies that the masses of particles and antiparticles are the same. Hence, Sakharov conditions 1 and 2 allow for a net baryon number to be created only when the particles of interest are out of thermal equilibrium.

While we know the general conditions necessary to generate a baryon asymmetry from an initially symmetric state, we are far from having a single accepted theory of baryogenesis. Here we will outline some of the ideas that have been proposed over the years. For some of the earliest work in this field, see References [10, 13, 28, 50, 53]. For reviews of this topic, see References [11, 34].

Perhaps the simplest class of models (and one of the earliest to be investigated) involves the decay of massive particles. Consider a particle-antiparticle, X and $\bar{X}$, that has dropped out of thermal equilibrium in the early Universe, in the sense defined in Section 5.3.2. Suppose the X can decay into two different channels, with baryon numbers B_1 and B_2, respectively, while $\bar{X}$ decays into the corresponding 'anti'-channels, with baryon numbers $-B_1$ and $-B_2$, respectively. Invariance under CPT guarantees that the *total* decay rate for an antiparticle must be equal to the decay rate for the corresponding particle. However, it says nothing about individual branching ratios. So it is possible, for instance, for the branching ratio of X into the channel with baryon number B_1 (which we will take to be r) to be different from the branching ratio of $\bar{X}$ into the channel with baryon number $-B_1$, which we will call $\bar{r}$. The possibility of such a difference is the key idea underlying this mechanism for baryogenesis. Note that $r \neq \bar{r}$ is only possible if C and CP are violated.

With the previously defined branching ratios and baryon numbers, the net baryon number produced from each pair of X and $\bar{X}$ decays is

$$B = B_1 r + B_2(1 - r) - B_1 \bar{r} - B_2(1 - \bar{r}),$$
$$= (B_1 - B_2)(r - \bar{r}). \tag{5.18}$$

Eq. (5.18) illustrates the necessity of the three Sakharov conditions. If C and CP were not violated, we would have $r = \bar{r}$, and the right-hand side of Eq. (5.18) would be zero. Similarly, the possibility that X can decay into two different channels with different baryon numbers is only possible if B is not conserved; otherwise, we would have $B_1 = B_2$ and again the right-hand side of Eq. (5.18) would be zero. Finally, we assumed out-of-equilibrium conditions in setting up this scenario; i.e., when they decay, X and $\bar{X}$ are not in equilibrium with the thermal background, either through annihilations with each other or through inverse decays. If this were not the case, the particles produced in the X and $\bar{X}$ decays would simply assume thermal equilibrium abundances, which would yield equal baryon and anti-baryon densities.

The scenario we have sketched out here is a toy model; for more detailed models see, e.g., Reference [22]. Models of this sort were first advanced in connection with physics at the GUT (grand-unified) scale, $T \sim 10^{15} - 10^{16}$ GeV. However, these ideas run into trouble if inflation is assumed to occur in the early Universe. The reason is that, as we have noted, inflation wipes out any pre-existing baryon asymmetry, so baryogenesis must occur after inflation, and currently favoured models of inflation do not reheat the Universe to a temperature as high as the GUT scale.

Another possibility for baryogenesis is the Affleck-Dine mechanism [1]. This model is motivated by supersymmetry, in which all of the particles of the Standard Model have corresponding super-partners with opposite spin statistics (fermions are paired with bosonic super-particles and bosons with fermionic super-partners). The Affleck-Dine mechanism invokes a scalar field that can carry a net baryon number. The field is initially frozen at early times but begins oscillating when the Hubble parameter drops below its mass. During these oscillations, the scalar field acquires a net baryon number, which is transferred at later times into Standard Model particles.

Electroweak baryogenesis [23] is based on the idea that the Universe underwent an electroweak phase transition at a temperature $T \sim 100$ GeV, when the Higgs field dropped into its vacuum state, giving masses to the quarks, leptons, and gauge bosons. If the electroweak phase transition is first order, it can temporarily drive the Universe out of thermal equilibrium as bubbles of the low-temperature vacuum nucleate, expand, and collide, ultimately occupying all of space. The production of baryons occurs in this out-of-equilibrium state near the walls of these expanding bubbles. Electroweak baryogenesis does require physics beyond the Standard Model, as the measured Higgs boson mass implies that the phase transition would not be first order in the Standard Model. This new physics would couple to the Higgs boson, altering its production and decay. Thus, the viability of these models can be tested in the laboratory.

Another proposal goes under the heading of leptogenesis [17]. These models are based on a result by 't Hooft [48], who showed that even in the Standard Model, baryon number is violated by non-perturbative electroweak processes. These processes conserve B−L but not B (the baryon number) and L (the lepton number) separately. Furthermore, while the rates for such processes are very low at low temperatures, they can be much higher in the early Universe. Leptogenesis, then, is the production of a net lepton asymmetry in the early Universe – e.g., through massive particle decay as discussed before. Then non-perturbative electroweak effects transfer some of the net lepton number into a net baryon number.

This is by no means an exhaustive list of models for baryogenesis, which remains very much an open and active field of research. At this point, we are confident of the ingredients required in any successful model (the Sakharov conditions), and we have a very accurate measure of the desired outcome (the observed baryon asymmetry), but the determination of the correct model for baryogenesis remains an ongoing effort.

5.5 The Baryon Asymmetry Parameter and Primordial Nucleosynthesis

In the Standard Model of particle physics and cosmology, the baryon asymmetry parameter plays a key role in BBN, regulating the rates of the nuclear reactions synthesizing (and destroying) the nuclides heavier than hydrogen. BBN in the Standard Model (SBBN) and in extensions of the SM when various nuclear physics and other parameters are allowed to vary is described in Uzan's contribution to this volume (Chapter 7). Here we are mainly concerned with the BBN predicted primordial (prestellar) abundances of the light nuclides, along with the CNO abundances.

5.5.1 Standard BBN

The SBBN-predicted abundances [29], the ratios by number compared to hydrogen, are shown as a function of η_B in Figure 5.1 for a factor of 1,000 range in η_B, for the SM case of $N_\nu = 3$. Agreement between the predicted and the observationally inferred deuterium abundance and the Planck observations of the CMB power spectrum imply a value of $\eta_B \sim 6 \times 10^{-10}$. This value is shown by the dashed vertical line in Figure 5.1. This value of η_B also provides good agreement with the primordial ^{4}He abundance derived from observations. However, it predicts a primordial ^{7}Li abundance roughly three times larger than the

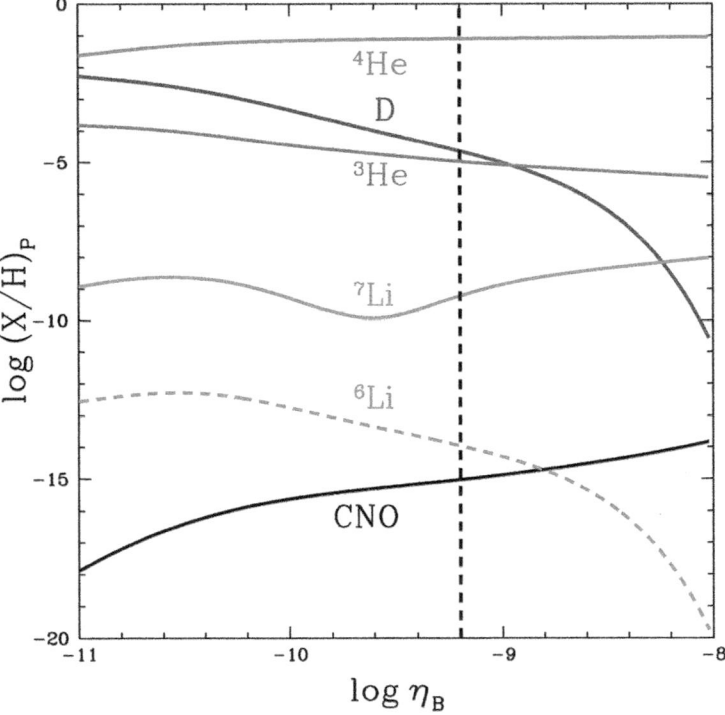

Figure 5.1 The primordial abundances predicted by SBBN [29] for a large range of the present value of the baryon to photon ratio $\eta_B = (n_B/n_\gamma)_0$. For all abundances (including ^{4}He), the ratio to hydrogen by number is shown. The dashed vertical line indicates the current SM value of $\eta_B \approx 6 \times 10^{-10}$.

observationally inferred abundance; this primordial lithium problem remains unresolved at present (see Reference [14] for a recent review).

As seen in Figure 5.1, over this large range in the baryon asymmetry parameter, the abundance trends are quite simple: as η_B increases, the ^{4}He abundance increases monotonically, but very slowly ($\sim$ logarithmically); the abundances of D, ^{3}He, and ^{6}Li are all monotonically decreasing, while the abundances of the CNO nuclides increase. In contrast, the evolution of the abundance of ^{7}Li is non-monotonic. Starting from very small values of η_B, as η_B increases, the ^{7}Li abundance first increases (until $\eta_B \sim 3 \times 10^{-11}$), then decreases (until $\eta_B \sim 3 \times 10^{-10}$), and finally increases again (eventually, for even larger values of η_B, the ^{7}Li abundance will decrease, being replaced by CNO and heavier nuclides). As η_B increases from 10^{-11} to 10^{-8}, ^{4}He/H increases by a factor of ~ 4, from ^{4}He/H ~ 0.024 ($Y_P \sim 0.09$) to ^{4}He/H ~ 0.093 ($Y_P \sim 0.27$), and the deuterium abundance decreases dramatically, from D/H $\sim 5 \times 10^{-3}$ to D/H $\sim 3 \times 10^{-11}$. Over the same range in η_B, the ^{3}He abundance decreases more slowly, from $\sim 10^{-4}$ to $\sim 3 \times 10^{-6}$, and the ^{7}Li abundance ranges from $\gtrsim 10^{-10}$ to $\lesssim 10^{-8}$, while the abundance of the CNO nuclides increases from $\sim 10^{-18}$ to $\sim 10^{-14}$.

For the value of the baryon asymmetry parameter inferred for the observed Universe, $\eta_B \sim 6 \times 10^{-10}$, ^{4}He/H ~ 0.082 ($Y_P \sim 0.25$), D/H $\sim 2.5 \times 10^{-5}$, ^{3}He/H $\sim 1.1 \times 10^{-5}$, ^{7}Li/H $\sim 5.4 \times 10^{-10}$, and the abundances of all the other primordial nuclides are $\lesssim 10^{-14}$. For a very wide range in the baryon asymmetry parameter, the gas that will become the first stars in the Universe consists mainly of hydrogen and helium (^{4}He), with only trace amounts of any other, heavier nuclides. Note, however, that for $\eta_B \gg 10^{-8}$, the primordial abundances of the CNO and heavier nuclides may become non-negligible (see Section 5.5.2). In the absence of significant CNO (or D) abundances, it is the hydrogen and helium content of the primordial gas that will most influence the formation, structure, and evolution of the first stars.

5.5.2 BBN for a Larger Range of Baryon Asymmetries

In the seminal BBN paper of Wagoner, Fowler, and Hoyle (WFH) [52], and in several follow up papers by Wagoner [51], Schramm and Wagoner [39], and Schramm [38], a much larger range in the baryon asymmetry parameter was explored than is shown here in Figure 5.1. In the WFH paper, a range of some 8.5 orders of magnitude was considered, $-12 \lesssim \log \eta_B \lesssim -3.5$, while in the other cited papers, the range is five orders of magnitude, $-11 \lesssim \log \eta_B \lesssim -6$. Although the quantitative BBN yields in those papers, based on what are now outdated nuclear and weak interaction rates (especially the much revised neutron lifetime), should be taken with a large grain of salt, the trends of the yields with η_B revealed in those papers are likely robust.

For example, over the entire range explored, the helium mass fraction increases (and the hydrogen mass fraction decreases) monotonically with η_B. Over the same range in η_B, the D and ^{3}He mass fractions decrease monotonically, with the deuterium abundance falling much more rapidly than the ^{3}He abundance. The evolution of the ^{7}Li mass fraction, X_7, is more interesting. At the lowest baryon asymmetries, X_7 increases from being negligible at $\log \eta_B \sim -12$ to a local maximum, a hint of which may be seen in Figure 5.1, when $\log \eta_B \sim -10.5$. Then, as η_B continues to increase, X_7 decreases to a local minimum at $\log \eta_B \sim -9.5$, as may be seen in Figure 5.1. For $\eta_B \gtrsim 3 \times 10^{-10}$, X_7 increases to another local maximum when $\log \eta_B \sim -6$, after which X_7 decreases monotonically for all larger values of η_B. For $\log \eta_B \lesssim -8$, the abundances of the CNO and heavier nuclides are negligible. As η_B continues to increase, so, too, do the CNO abundances, surpassing the ^{3}He and ^{7}Li abundances for $\log \eta_B \gtrsim -6$. However, almost as soon as the CNO nuclides become large enough to be of possible interest, they decrease as η_B continues to increase, being replaced by even heavier nuclides. The trend seen at the very highest values of the baryon asymmetry parameter in the WFH paper suggests that at sufficiently high values of η_B, the iron peak elements might be produced during primordial nucleosynthesis. It is interesting to speculate if even larger baryon to photon ratios might lead to the r-process elements.

As discussed in Section 5.2, in determining if the baryon asymmetry parameter is fined tuned, we are asking if stars, planets, and life could exist in alternate universes with different values of η_B. In this case, we need to check if the primordial abundances in alternate universes allow for the cooling and collapse of primordial gas clouds to form the first stars and if, in the course of evolution of those stars, the elements required for life can be synthesised.

5.6 Relation between the Baryon Asymmetry Parameter and the Observable Cosmological Parameters

In our present-day Universe, the parameter η_B is not a directly observable quantity. Instead, we measure quantities such as the baryon density or the CMB temperature, from which η_B can be inferred. In this section, we examine the relation between η_B and the observable cosmological quantities.

In a matter-antimatter asymmetric universe such as ours, the baryon asymmetry parameter is related to the contribution of baryons (normal matter) to the total mass density. As a result, the magnitude of the baryon asymmetry plays a role in the evolution of the Universe and in the growth and evolution of structure in it. For the discussion here, it is assumed that the Universe is, on average, homogeneous and is expanding isotropically so that its evolution is described by the 'Friedman equation',

$$\left(\frac{H}{H_0}\right)^2 = \Omega_R \left(\frac{a_0}{a}\right)^4 + \Omega_B \left(\frac{a_0}{a}\right)^3 + \Omega_{DM} \left(\frac{a_0}{a}\right)^3 + \Omega_k \left(\frac{a_0}{a}\right)^2 + \Omega_\Lambda. \qquad (5.19)$$

In Eq. (5.19), the subscript 0 indicates the present ($t = t_0$) value of the parameters, and $H = a^{-1}(da/dt)$ is the Hubble parameter, quantifying the expansion rate of the Universe, where $a = a(t)$ is the cosmic scale factor. The subscripts, R, B, DM, k, and Λ stand, respectively, for the contributions to the total mass/energy density from 'radiation' – i.e., massless particles or particles whose total energy (rest mass plus kinetic) far exceeds the rest mass energy), baryons ('normal' or 'ordinary' matter), dark matter (non-baryonic matter),[8] curvature, and a cosmological constant. For simplicity, we assume here that the observed accelerated expansion of the Universe is driven by a cosmological constant rather than a time-varying dark-energy component. Since the mass densities of baryonic and dark matter evolve the same way (e.g., $\rho \propto a^{-3}$), it is convenient to introduce a parameter describing the 'matter density', the total mass density in non-relativistic particles, $\Omega_M \equiv \Omega_B + \Omega_{DM}$. At the present epoch ($t = t_0$), a 'critical density' of the Universe may be identified, $\rho_{crit\,0} \equiv 3H_0^2/8\pi G = 1.05 \times 10^{-5} h^2 \, \mathrm{GeV \, cm^{-3}}$, where $H_0 \equiv 100\,h \, \mathrm{km \, s^{-1} \, Mpc^{-1}}$ and G is Newton's gravitational constant. Here and elsewhere, we will often set $c = 1$ and express masses in energy units. The parameters Ω_i that appear in Eq. (5.19) are the ratios of the various contributions to the present energy densities, normalised to the present critical density: $\Omega_i \equiv (\rho_i/\rho_{crit})_0$.

Consider the relation between the baryon asymmetry parameter (η_B), the baryon mass density parameter (Ω_B), and the Hubble constant (H_0). The present mass/energy density in ordinary (baryonic) matter is $\rho_{B0} = m_B n_{B0} = \Omega_B \, \rho_{crit\,0} \approx 1.05 \times 10^{-5} \, \Omega_B h^2 \, \mathrm{GeV \, cm^{-3}}$. For an average mass per baryon of $m_B \approx 0.938 \, \mathrm{MeV}$ [44] [9], the present baryon number density is $n_{B0} \approx 1.12 \times 10^{-5} \, \Omega_B h^2 \, \mathrm{cm^{-3}}$. If the present temperature of the CMB photons is T_0 (in degrees Kelvin), then the present photon number density is $n_{\gamma 0} \approx 20.3 \, T_0^3 \, \mathrm{cm^{-3}}$, and the present baryon-to-photon ratio is $\eta_B \approx 5.54 \times 10^{-7} \, (\Omega_B h^2/T_0^3)$ so that $\eta_B \approx 5.54 \times 10^{-7} \, (\Omega_B h^2/T_0^3)$ and $n_B/s \approx 7.87 \times 10^{-8} \, (\Omega_B h^2/T_0^3)$ (for three flavours of SM neutrinos). Note that the connection between the baryon asymmetry parameter (η_B or n_B/s) and the present mass density in ordinary matter ($\propto \Omega_B h^2$) depends on the present ($t = t_0$) value of the photon temperature (T_0). If the baryon asymmetry parameter were to change by some factor, the combination $\Omega_B h^2/T_0^3$ would change by the same factor, resulting in changes to the other universal observables (e.g., Ω_B, h, T_0), separately or in combination. The baryon asymmetry parameter is degenerate with these other cosmological parameters. In particular, changes in Ω_B alone would change the expansion history of the Universe, as may be seen from the Friedman equation. The interconnections (degeneracies) among the cosmological observables complicate any discussion of the effect on the history and evolution of the Universe resulting from changes to any one of them (e.g., the baryon asymmetry parameter).

If the Friedman equation, Eq. (5.19), is evaluated at present ($t = t_0$), when $H = H_0$, there is one condition on the five parameters,

$$1 = \Omega_R + \Omega_B + \Omega_{DM} + \Omega_k + \Omega_\Lambda = \Omega_R + \Omega_M + \Omega_k + \Omega_\Lambda, \tag{5.20}$$

leaving four free parameters. For our observed Universe, $\Omega_k \ll 1$ and $\Omega_R \ll 1$ so that $\Omega_B + \Omega_{DM} + \Omega_\Lambda \approx 1$. There are still three parameters and only one constraint, leaving

---

[8] For agreement with observations of structure formation and its growth as the Universe evolves, it is assumed that the DM is 'cold', in the sense that for those epochs when deviations from homogeneity occur, the DM particles are moving slowly ($v \ll c$).

[9] In the post-BBN universe, when the baryons are mainly protons and alpha particles (hydrogen and helium), the average mass per baryon depends on the helium abundance (mass fraction, Y_P). For $Y_P \approx 0.25$, $m_B \approx 938.112 + 6.683$ ($Y_P - 0.250$) MeV [44].

two free parameters. By writing $\Omega_M = \Omega_B + \Omega_{DM}$, it might appear that there are only two parameters and one constraint, $\Omega_M + \Omega_\Lambda \approx 1$. However, the ratio Ω_B / Ω_{DM} remains a free parameter, so there are still three parameters with one constraint among them.

In the next section, we will consider how the evolution of the Universe changes when η_B differs from its observed value. While our intention is to keep all of the other cosmological parameters constant, there remains an ambiguity in the way we treat them. Note that η_B is a dimensionless ratio of two quantities, the baryon and photon number densities. When we alter this quantity, we can consider two different possibilities: (1) changing n_B relative to the other cosmological parameters while leaving n_γ unchanged relative to these parameters or (2) keeping n_B fixed while changing n_γ relative to the other cosmological parameters. While each of these possibilities produces a change in η_B, they differ in their treatment of the way that n_B and n_γ change relative to the other cosmological quantities of interest. (Of course, these are only the two simplest possibilities; one could consider allowing the ratios of *both* n_B and n_γ relative to the other cosmological parameters to change, but by different amounts, thus changing η_B as well).

Which of the two approaches spelled out in the previous paragraph is the correct one? Absent a particular model for a different universe with a different value of η_B, it is impossible to say. However, the first possibility seems to be the more natural one. If we assume that baryogenesis is independent of the processes that led to dark matter or dark energy, then tweaking the model for baryogenesis will alter n_B by the same factor relative to all of the other cosmological parameters of interest. This is the case we will consider in detail.

Let F be the ratio of the value of η_B in some hypothetical universe relative to its value in our Universe; our goal will be to understand what constraints, if any, can be placed on F. We will use a tilde to denote physical quantities in a hypothetical universe in which η_B has changed, and quantities without a tilde will denote the corresponding values of these quantities in our Universe, so

$$\widetilde{\eta}_B = F\eta_B. \tag{5.21}$$

In case (1) discussed earlier, the ratios ρ_B/ρ_{DM}, ρ_B/ρ_Λ, and n_B/n_ν change in proportion to the change in η_B, while n_γ/ρ_{DM}, n_γ/ρ_Λ, and n_ν/n_γ remain the same. Thus, we have

$$\widetilde{\rho}_B/\widetilde{\rho}_{DM} = F\rho_B/\rho_{DM}, \tag{5.22}$$

$$\widetilde{\rho}_B/\widetilde{\rho}_\Lambda = F\rho_B/\rho_\Lambda, \tag{5.23}$$

$$\widetilde{n}_B/\widetilde{n}_\nu = Fn_B/n_\nu. \tag{5.24}$$

Of course, there are other possibilities that we will not explore here. In an alternate universe with a late production of entropy, n_B would remain unchanged while the ratio of n_γ to all of the other cosmological parameters would be altered. Alternately, if baryogenesis were linked to the process that produced dark matter (as it is in some models), one might consider the possibility of changing η_B while leaving ρ_B/ρ_{DM} fixed. Nonetheless, we feel that the model spelled out in Eqs. (5.22)–(5.24) is the most natural way in which to modify η_B, and this is the case we will now attempt to constrain.

5.7 Alternate Universes with Different Baryon Asymmetry Parameters

Changing η_B alters the evolution of the Universe in two ways: it changes BBN, and it alters the processes that give rise to structure formation and ultimately yield stars and planets. We will consider both effects in turn.

First, consider our Universe at present. Our Universe is very well described by a ΛCDM cosmological model with $\Omega_k \approx 0$, $\Omega_R \ll 1$, and $\Omega_B < \Omega_{DM} < \Omega_\Lambda$ ($\Omega_B + \Omega_{DM} + \Omega_\Lambda \approx 1$). For our observed Universe, a good approximation to the 2015 Planck CMB observations [31] is $\Omega_\Lambda \approx 0.7$, $\Omega_M \approx 0.3$, $\Omega_B \approx 0.05$, $\Omega_{DM} \approx 0.25$. For a ΛCDM cosmology with $\Omega_\Lambda \approx 0.7$, $H_0 t_0 \approx 0.96$, so for $H_0 \approx 68 \, \mathrm{km \, s^{-1} \, Mpc^{-1}}$, $t_0 \approx 13.8 \, \mathrm{Gyr}$. For the present CMB temperature, the Fixsen *et al.* [15] result may be approximated by $T_0 \approx 2.7 \, \mathrm{K}$, corresponding to a CMB photon number density $n_{\gamma 0} \approx 400 \, \mathrm{cm^{-3}}$ (compared to the more accurate results, $T_0 = 2.7255 \, \mathrm{K}$ and $n_{\gamma 0} \approx 411 \, \mathrm{cm^{-3}}$).

5.7.1 Effect on BBN

What happens to BBN when we allow for extreme variations in η_B? As noted earlier, the most important effect of increasing η_B is to increase the primordial ^{4}He mass fraction at the expense of hydrogen. One might imagine that a universe in which stellar evolution begins with almost pure ^{4}He might be less hospitable to life. For example, Hall *et al.* [21] pointed out that in such a universe, halo cooling takes longer, stellar lifetimes are reduced, and there is less hydrogen to support organic chemistry. (The calculations in Reference [21] are focused on variations in the weak scale rather than the magnitude of the baryon asymmetry). However, even extreme increases in η_B do not produce primordial ^{4}He mass fractions close to 100%. For example, a value of η_B as large as 10^{-3} (more than six orders of magnitude larger than the observed value) yields a ^{4}He mass fraction of only 0.4 [36].

Large values of η_B also open up the possibility of producing heavier elements in BBN. Consider first the CNO elements. In standard BBN, these are produced in very small amounts, with abundances relative to hydrogen of CNO/H $\sim 10^{-15} - 10^{-14}$ [8]. However, the abundances of these elements are an increasing function of η_B, peaking at CNO/H $\sim 10^{-8}$ for $\eta_B \sim 10^{-5}$ and decreasing for larger values of η_B [52]. Even a small primordial abundance of CNO/H could affect the evolution of the first generation of stars, as noted in Reference [5]; this evolution begins to change when CNO/H increases above 10^{-11}. Nonetheless, it seems unlikely that such a change would affect the ability of the Universe to harbour life. For $\eta_B > 10^{-5}$, the abundance of the CNO elements begins to decrease as the nuclei are converted into even heavier elements [36, 52]. However, even extreme increases in the value of η_B result in only trace amounts of such heavy elements. In terms of models that can support life, it does not appear that BBN provides a useful upper bound on η_B, and it certainly does not provide a bound competitive with arguments from structure/galaxy/star formation.

Now consider BBN in the limit of very low values for η_B. In this limit, the ^{4}He abundance becomes negligible while ^{2}H increases, reaching a peak abundance of order

D/H $\sim 10^{-2}$ when $\eta_B \sim 2 \times 10^{-12}$. For smaller values of η_B, even the deuterium abundance decreases as η_B is reduced, yielding – in the limit $\eta_B \to 0$ – a primordial universe consisting essentially of pure hydrogen. The reduction in primordial helium for small values of η_B is likely to reduce the cooling of galaxies that results from the collisional excitation of ionised helium, but this is unlikely to have a major impact [21]. On the other hand, a significantly larger abundance of deuterium would lead to enhanced molecular cooling through an increase in the HD abundance [26]. While interesting, this is also unlikely to affect the prospects for a life-bearing universe.

Our conclusion, then, is that BBN provides essentially *no* constraints on universes with different values of η_B. The formation of stars and planets and the development of life is nearly completely insensitive to variations in the primordial element abundances, at least within the ranges of η_B that we have considered here.

5.7.2 *Effect on Large-Scale Structure: The Linear Regime*

In the Standard Model for structure formation, small initial fluctuations in the density are imprinted on the matter and radiation by inflation or some other process early in the evolution of the Universe. When the Universe is radiation dominated, these fluctuations cannot grow inside of the horizon; subhorizon fluctuations begin to grow once matter dominates the radiation. If $\delta\rho/\rho$ represents the magnitude of the fluctuation in the matter density relative to the mean matter density, then after matter domination begins, $\delta\rho/\rho$ grows proportional to the scale factor a,

$$\delta\rho/\rho \propto a. \tag{5.25}$$

Eq. (5.25) applies only as long as $\delta\rho/\rho \ll 1$; in this case, the density fluctuations are said to be in the *linear* regime. Once $\delta\rho/\rho > 1$, the Universe enters the non-linear regime, and the analytic solution given by Eq. (5.25) no longer applies. Numerical simulations are necessary to evolve the density field further forward in time. In the non-linear regime, the fluctuations in the matter density grow much more rapidly, and the dark matter ultimately collapses into halos.

This process applies in a straightforward way only to dark matter, which is collisionless. The baryons evolve in a more complicated way. At high temperatures ($T \gg 10^3$ K), the matter is ionised, and the cross section for scattering off of photons is very high. Thus, the baryons are frozen to the radiation background, and baryonic density perturbations cannot grow. As the temperature drops, the electrons become bound to the protons and to the primordial helium nuclei in a process known as recombination.[10] At this point, the density perturbations in the baryons can begin to grow along with the dark matter perturbations. A further complication is that in the non-linear regime, the baryonic matter, unlike the dark matter, is not pressureless and can also radiate away energy in the form of photons. Thus, at late times, the baryons evolve very differently than the dark matter. The end result is

---

[10] Note that this term is a bit misleading, as the electrons and atomic nuclei were never 'combined' to begin with.

that the baryons ultimately bind into fairly compact disks or ellipsoids (galaxies), fragment into stars, and form planets, while the dark matter remains in the form of diffuse halos surrounding the galaxies.

In considering the effect of changing η_B, we must therefore consider the change in two key parameters: the redshift of equal matter and radiation and the redshift at which recombination occurs. However, redshifts are defined relative to the present day, so they are not particularly useful in determining whether a modified universe can support life because we are not restricting life to form at redshift zero as it does in our Universe. Instead, we should examine the temperature of equal matter and radiation and the temperature of recombination. In our Universe, the temperature of equal matter and radiation, T_{eq} is given in terms of the present-day temperature, T_0, by $T_{eq} = T_0(\rho_M/\rho_\gamma)_0$. For the parameter values given at the beginning of this section, we obtain $T_{eq} = 9,000$ K. How does this change when η_B is altered? To determine this, note that the redshift of equal matter and radiation is given by this ratio of present-day densities:

$$1 + z_{eq} = \left(\frac{\rho_{DM} + \rho_B}{\rho_\gamma + \rho_\nu}\right)_0. \tag{5.26}$$

Here we are ignoring the fact that the neutrinos can become non-relativistic at very late times. Then we have

$$1 + \tilde{z}_{eq} = \left(\frac{\rho_{DM} + F\rho_B}{\rho_\gamma + \rho_\nu}\right)_0. \tag{5.27}$$

Using the values for the preceding cosmological parameters, we can trace out the effect of F on T_{eq}. We have $\rho_{DM}/\rho_B \approx 5$. Thus, z_{eq} changes little for $F \lesssim 5$ while, for $F \gtrsim 5$, we have $(1 + \tilde{z}_{eq}) = (F/5)(1 + z_{eq})$. Then we have

$$\tilde{T}_{eq} \approx T_{eq} \quad (F \lesssim 5), \tag{5.28}$$

$$\tilde{T}_{eq} \approx \frac{F}{5} T_{eq} \quad (F \gtrsim 5). \tag{5.29}$$

Now consider the effect of altering η_B on the recombination temperature T_{rec}. While recombination is a gradual process and does not occur suddenly at a single temperature, for the purposes of this study, it will be sufficient to take $T_{rec} \approx 3,000$ K. The process of recombination depends primarily on the ratio of the photon temperature to the binding energy of hydrogen, but there is also a residual dependence on η_B. This dependence comes about because η_B^{-1} determines the number of photons per hydrogen atom; an increase in this number makes it easier for photons to ionise the hydrogen, delaying recombination, while the reverse is true if the number of photons per hydrogen atom decreases. However, the temperature at which a given ionisation fraction is reached varies roughly logarithmically with η_B. This is a much smaller effect than the change in T_{eq} with η_B, so we will ignore it in what follows and take the recombination temperature to be roughly insensitive to changes in η_B.

Now we can investigate the effect of changing η_B on large-scale structure in the linear regime. We will not consider any possible changes in the magnitude of the primordial

[39] Schramm, D. N., and Wagoner, R. V. 1977. 'Element Production in the Early Universe'. *Annual Review of Nuclear and Particle Science*, **27**, 37–74.

[40] Sommerfeld, A. 1931. 'Über die Beugung und Bremsung der Elektronen'. *Annalen der Physik*, **403**, 257–330.

[41] Steigman, G. 1969. 'Antimatter, Galactic Nuclei and Theories of the Universe: Antimatter and Cosmology'. *Nature*, **224**(Nov.), 477–481.

[42] Steigman, G. 1976. 'Observational Tests of Antimatter Cosmologies'. *Annual Review of Astronomy and Astrophysics*, **14**, 339–372.

[43] Steigman, G. 1979. 'Cosmology Confronts Particle Physics'. *Annual Review of Nuclear and Particle Science*, **29**, 313–338.

[44] Steigman, G. 2006. 'The Cosmological Evolution of the Average Mass per Baryon'. *Journal of Cosmology and Astroparticle Physics*, **10**(Oct.), 016.

[45] Steigman, G. 2008. 'When Clusters Collide: Constraints on Antimatter on the Largest Scales'. *Journal of Cosmology and Astroparticle Physics*, **10**(Oct.), 001.

[46] Steigman, G. and Strittmatter, P. A. 1971. 'Neutrino Limits on Antimatter Sources of Energy in Seyfert Galaxies'. *Astronomy and Astrophysics*, **11**(Mar.), 279.

[47] Steigman, G., Dasgupta, B., and Beacom, J. F. 2012. 'Precise Relic WIMP Abundance and Its Impact on Searches for Dark Matter Annihilation'. *Physical Review D*, **86**(2), 023506.

[48] 't Hooft, G. 1976. 'Symmetry Breaking through Bell-Jackiw Anomalies'. *Physical Review Letters*, **37**(July), 8–11.

[49] Tegmark, M., *et al.* 2006. 'Dimensionless Constants, Cosmology, and Other Dark Matters'. *Physical Review D*, **73**(2), 023505.

[50] Toussaint, D., *et al.* 1979. 'Matter-Antimatter Accounting, Thermodynamics, and Black-Hole Radiation'. *Physical Review D*, **19**(Feb.), 1036–1045.

[51] Wagoner, R. V. 1973. 'Big-Bang Nucleosynthesis Revisited'. *Astrophysical Journal*, **179**(Jan.), 343–360.

[52] Wagoner, R. V., Fowler, W. A., and Hoyle, F. 1967. 'On the Synthesis of Elements at Very High Temperatures'. *Astrophysical Journal*, **148**(Apr.), 3.

[53] Yoshimura, M. 1979. 'Unified Gauge Theories and the Baryon Number of the Universe'. *Physical Review Letters*, **42**(Mar.), 746.

[54] Zeldovich, Y. 1965. 'Survey of Modern Cosmology'. *Advances in Astronomy and Astrophysics*, vol. 3. Elsevier.

6

Structure Formation

A D R I A N N E   S L Y Z

Abstract

In this chapter, I will describe how the galaxies populating the present-day Universe are believed to have originated from extremely small fluctuations in baryons, dark matter, and radiation in the early Universe. I will very briefly outline the currently favoured explanation as to the origin of these fluctuations before discussing their growth and evolution via linear perturbation theory. Finally, to complete the picture, I will sketch how the virialised dark matter halos hosting the galaxies we observe emerged from the cosmological density field, by combining linear theory and a simplified spherically symmetric model to follow the evolution of these perturbations once they have detached from the Universe's expansion and entered the non-linear regime.

6.1 The Emergence of Structure in the Universe

6.1.1 Introduction

Maps of the positions of galaxies in the sky reveal a complex, inhomogeneous distribution. Galaxies are organised in what is called a cosmic web, characterised by sheets, filaments, and voids (see Figure 6.1). The largest among the hierarchy of over-dense structures seen in such cosmic maps are superclusters containing one or more clusters of galaxies undergoing collapse under their own self-gravity. On the opposite end are voids, the most under-dense regions of the Universe. The development of this inhomogeneous matter distribution is called structure formation.

In the framework of the standard cosmological model, we possess a remarkably successful theory for the formation of the observed large-scale structures in the Universe. According to the Big Bang model, the Universe emerged from an initially singular state of infinite density, from which space progressively expanded. This theory is underpinned by three main pieces of observational evidence. First, support for an expanding universe came in 1930 when astronomer Edwin Hubble discovered that galaxies move away from each other at a speed proportional to the distance separating them. Second, the theory of primordial nucleosynthesis stemming from the thermal history of the Big Bang predicts a universal helium abundance consistent with the measured one (about 25% in mass). Finally,

Figure 6.1 Spatial distribution of galaxies with stellar masses in excess of $2 \times 10^{10} M_{\odot}$ in the Horizon-AGN simulated Universe between redshifts 0 (bottom-left corner of the image) and 1 (top-right corner of the image); see www.horizon-simulation.org for detail. The observer, looking to the right and located at the apex of the bottom-most wedge, thus, sees the left hand side of the wedge situated immediately above when the bottommost wedge ends on the right-hand side and so on and so forth, all the way to the end of the uppermost wedge. In other words, the image represents the projection of a unique wedge which subtends a 2 square degree angle on the sky (about 10 times the area covered by the Moon as seen from Earth). Note how the galaxies are not distributed at random but organised in a cosmic web.

the uniform and isotropic black-body radiation prophesied to permeate the entire sky was detected by Penzias and Wilson in 1964, lending further support to the high densities and temperatures reigning in the primordial Universe.

This extreme isotropy and homogeneity of the microwave background radiation immediately raises a puzzle concerning the formation of galaxies. How can galaxies populating the present-day Universe originate from such a ball of incredibly smooth cosmic fire? The answer to this question is believed to rest with the theory of inflation, a very brief phase of exponential expansion of space itself, which occurred immediately after the Big Bang. During this phase, quantum fluctuations in density were stretched to macroscopic scales, while the observable Universe maintained a constant size. However, even though inflation extended these quantum fluctuations to sizes larger than the cosmological horizon, their amplitude, as imprinted in the black-body spectrum of the cosmic microwave background radiation remained very weak, at a level of only 1 part in 10^5. Following inflation, gravitational instability is thus invoked to amplify primordial fluctuations.

Detailed analysis of the cosmic microwave background radiation rules out structure formation in a universe dominated by baryonic matter because the time required for gravitational instability to magnify the density contrasts to present-day values from the weak level of anisotropies at the moment of photon-matter decoupling is too large. In contrast, should matter exist which interacts only weakly with photons, it could decouple at an earlier epoch and could thereby accelerate the later growth of baryonic fluctuations, up to the level of the characteristic density contrast of galaxies.

There are other observational reasons which suggest that a more exotic form of matter dominates the matter density of the Universe. For example, the rotation curves of galaxies remain flat beyond the distances where visible matter can be detected. The mass of clusters of galaxies, deduced from dynamical measurements or gravitational lensing, is higher than their luminous mass comprised of galaxies and hot gas. From the point of view of structure formation, the different possibilities for the nature of this dark matter can roughly be divided into three categories which depend on particle mass. Cold dark matter is characterised by particles with masses ~ 1 GeV or greater, warm dark matter has intermediate mass particles with masses between ~ 100 eV and ~ 100 MeV, and hot dark matter is composed of light, ultra-relativistic particles with masses smaller than 1 eV. The common characteristic of all these particles is that they interact very weakly with the rest of matter.

The nature of dark matter constitutes a very active research subject and is discussed in Chapter 9.

6.1.2 Dynamics of the Universe

The story of structure formation begins from the moment when Einstein's equations become a valid description of the Universe – i.e. a little after the Planck time, a few 10^{-43} seconds after the birth of the Universe. From this moment onwards, it is possible to describe the dynamical evolution of the Universe with the laws of physics as we know

$$\frac{\partial \mathbf{v}}{\partial t} + \frac{1}{a}\left(\mathbf{v} \cdot \frac{\partial}{\partial \mathbf{x}}\right)\mathbf{v} + \frac{1}{a}\frac{da}{dt}\mathbf{v} = -\frac{1}{\rho a}\frac{\partial p}{\partial \mathbf{x}} - \frac{1}{a}\frac{\partial \Phi'}{\partial \mathbf{x}} \qquad (6.11)$$

$$\left(\frac{1}{a}\frac{\partial}{\partial \mathbf{x}}\right)^2 \Phi' = 4\pi G\rho - \Lambda + \frac{3}{a}\frac{d^2 a}{dt^2}, \qquad (6.12)$$

where we have split the gravitational potential in two, writing $\Phi = \Phi' - \frac{a}{2}\frac{d^2 a}{dt^2}\mathbf{x}^2$ to somewhat simplify the form of the Euler equation.[7]

As we wish to describe all the physical quantities appearing in the dynamical evolution equations as perturbations imprinted on the homogeneous and isotropic background Universe previously discussed, we introduce the density contrast of the perturbations, δ, in these equations, by recasting the density as $\rho(\mathbf{x}, t) \equiv \bar{\rho}(t)(1 + \delta(\mathbf{x}, t))$. Note that the velocity $\mathbf{u}$ and the gravitational potential Φ have already undergone a similar procedure, as $\mathbf{v}$ is the deviation from the velocity of the homogeneous and isotropic background component by definition and Φ' is the perturbed part of the term $-\frac{a}{2}\frac{d^2 a}{dt^2}\mathbf{x}^2$, which we can identify with the gravitational potential of this very same background.

If perturbations in all the physical quantities are small compared to their corresponding background values – i.e., $\delta \ll 1$, $\mathbf{v}^2(t_e/d)^2 \ll 1$ and $\Phi'(t_e/d^2)^2/\bar{\rho} \ll 1$, where d is the coherence length of the spatial variation of δ and t_e is the expansion time $\sim (G\bar{\rho})^{-\frac{1}{2}}$ – we can linearise the evolution equations (that is to say, only keep the terms where the small perturbations appear at first order) to obtain

$$\frac{\partial \delta}{\partial t} + \frac{1}{a}\frac{\partial \mathbf{v}}{\partial \mathbf{x}} = 0 \qquad (6.13)$$

$$\frac{\partial \mathbf{v}}{\partial t} + \frac{1}{a}\frac{da}{dt}\mathbf{v} = -\frac{c_s^2}{a}\frac{\partial \delta}{\partial \mathbf{x}} - \frac{1}{a}\frac{\partial \Phi'}{\partial \mathbf{x}} \qquad (6.14)$$

$$\left(\frac{\partial}{\partial \mathbf{x}}\right)^2 \Phi' = 4\pi G\bar{\rho}a^2\delta, \qquad (6.15)$$

where we have used Eqs. 6.4 and 6.5 (dropping the $\bar{p}$ terms as the fluid is assumed to be non-relativistic), which govern the evolution of the homogeneous and isotropic Universe, to get rid of the background zeroth order terms and defined the sound speed in the polytropic[8] fluid as $c_s^2 \equiv dp/d\rho$.

Taking the gradient of Eq. (6.14), and substituting $\mathbf{v}$ using Eq. 6.13 and Φ' using Eq. 6.15 in it, we obtain the following second-order differential equation for the density contrast

$$\frac{\partial^2 \delta}{\partial t^2} + \frac{2}{a}\frac{da}{dt}\frac{\partial \delta}{\partial t} = 4\pi G\bar{\rho}\delta + \frac{c_s^2}{a^2}\left(\frac{\partial}{\partial \mathbf{x}}\right)^2 \delta, \qquad (6.16)$$

---

[7] We are allowed to do that as the second term only introduces a total time derivative in the Poisson equation, which does not change the action.

[8] By imposing a polytropic equation of state for the fluid, we force the perturbations to be adiabatic – i.e., we neglect entropy perturbations.

which fully describes the growth of density perturbations, in the regime where they remain small compared to the background density. Before discussing the solutions to this equation, we describe it qualitatively. Recalling that the Hubble expansion rate is $H \equiv \frac{1}{a}\frac{da}{dt}$, the second term on the left-hand side can be expressed as $2H\frac{\partial \delta}{\partial t}$. Called the 'Hubble drag' term, this term captures how the expansion of the Universe suppresses perturbation growth. The terms on the right-hand side, in contrast, determine how gravity enhances and spatial gradients influence perturbation growth.

The evolution equation for the density contrast is more easily solved by decomposing $\delta(\mathbf{x},t)$ into a set of Fourier modes:

$$\delta(\mathbf{x},t) = \sum_k \delta(\mathbf{k},t)e^{i\mathbf{k}\cdot\mathbf{x}} = \sum_k \delta_{\mathbf{k}}(t)e^{i\mathbf{k}\cdot\mathbf{x}}. \tag{6.17}$$

The amplitude of each Fourier mode then obeys the equation

$$\frac{d^2\delta_{\mathbf{k}}}{dt^2} + 2H\frac{d\delta_{\mathbf{k}}}{dt} = \left(4\pi G\bar{\rho} - \frac{c_s^2 k^2}{a^2}\right)\delta_{\mathbf{k}}, \tag{6.18}$$

as, in the linear regime, we do not have to worry about coupling between the different modes: each Fourier mode $\delta_{\mathbf{k}}$ evolves independently.

Now, because the Universe is composed of radiation (photons, neutrinos), different matter components (baryons, collisionless dark matter), and dark energy (cosmological constant), which dominate at different epochs of the Universe evolution, solutions for the growth of the Fourier modes depend on the cosmological model, the equation of state for the different components, and the epoch considered.

6.2.2 Perturbations in a Matter-Dominated Universe

Let us start by looking at a simple universe mainly composed of ordinary matter – i.e., baryons (by opposition to pressureless dark matter). At first, this may seem like a gross oversimplification, but it constitutes an extremely useful example to understand key features of perturbation growth. Moreover, matter (cold dark matter plus baryons in that case) dominates the energy density throughout most of the evolution of the Universe. Arguably the easiest way to see this last point is to plug Eq. (6.5) into Eq. (6.4), multiply the result by $2da/dt$, and integrate it w.r.t. time to the present day, explicitly breaking up the energy density $\bar{\rho}$ into its total matter and radiation components $\bar{\rho}_m$ and $\bar{\rho}_\gamma$ in the process. This yields the following equation for the evolution of the expansion rate of the homogeneous and isotropic Universe:

$$H^2 = H_0^2\left(\Omega_{\gamma,0}\,a^{-4} + \Omega_{m,0}\,a^{-3} + \Omega_{\Lambda,0}\right), \tag{6.19}$$

where we have omitted the curvature contribution, since the Universe is measured to be flat to a high level of accuracy, and introduced the dimensionless parameters $\Omega_\gamma = \bar{\rho}_\gamma/\rho_c$, $\Omega_m = \bar{\rho}_m/\rho_c$, $\Omega_\Lambda = \bar{\rho}_\Lambda/\rho_c$, with the critical density $\rho_c = 3H^2/(8\pi G)$ and $\bar{\rho}_\Lambda = \Lambda/(8\pi G)$. Given the present-day values of $\Omega_{\gamma,0} \simeq 9.4 \times 10^{-5}$, $\Omega_{m,0} \simeq 0.31$, and $\Omega_{\Lambda,0} \simeq 0.69$ (taken from [10]), and given that Eq. (6.19) constrains their sum to remain equal to 1 at all times, one easily sees that we switch from a cosmological constant dominated expansion rate at the present epoch to one dominated by matter at $a_{\Lambda \mathrm{eq}} = (\Omega_{m,0}/\Omega_{\Lambda,0})^{1/3} \simeq 0.77$ – i.e., a redshift $z_{\Lambda \mathrm{eq}} \equiv a_0/a_{\Lambda \mathrm{eq}} - 1 \simeq 0.3$ – and then to a radiation dominated one at $a_{\mathrm{eq}} = \Omega_{\gamma,0}/\Omega_{m,0} = 3 \times 10^{-4}$ – i.e., $z_{\mathrm{eq}} \simeq 3,300$. Overall, the Universe is thus matter dominated for more than 10 billion years out of a total age of 13.8 billion years.

Going back to our fully baryonic Universe, a solution to Eq. (6.18) depends on the sign of the right-hand side. It is positive and leads to growing perturbations only if wave numbers satisfy the criterion

$$k < k_J = \frac{2\sqrt{\pi G \bar{\rho}}}{c_s} \tag{6.20}$$

or, in terms of wavelength, if the wavelength, λ, of a perturbation is greater than a critical wavelength, λ_J, called the Jeans length

$$\lambda_J = \frac{2\pi}{k_J} = c_s \sqrt{\frac{\pi}{G\bar{\rho}}}. \tag{6.21}$$

Physically, the Jeans length encapsulates the battle between gravity and pressure. Compressing a parcel of baryons adiabatically will cause its density and temperature to rise, creating an overpressured region relative to its surroundings. If the self-gravity of the perturbation is weak, then pressure will iron out gradients, thereby suppressing the growth of perturbations, leaving them to oscillate as acoustic waves. Only if self-gravity overwhelms pressure will the perturbation grow.

Since perturbations smaller than λ_J do not grow, we consider solutions to Eq. (6.18) in the long-wavelength, gravity-dominated limit (i.e., $k \to 0$), in which case this equation becomes[9]

$$\frac{d^2\delta_{\mathbf{k}}}{dt^2} + 2H\frac{d\delta_{\mathbf{k}}}{dt} = 4\pi G\bar{\rho}\delta_{\mathbf{k}}. \tag{6.22}$$

As previously discussed, a good approximation at $0.3 < z < 3,300$ is that $\Omega_m = 1$, in which case $\bar{\rho} = \rho_c$, and Eq. (6.22) becomes

$$\frac{d^2\delta_{\mathbf{k}}}{dt^2} + 2H\frac{d\delta_{\mathbf{k}}}{dt} = \frac{3}{2}H^2\delta_{\mathbf{k}}. \tag{6.23}$$

In a non-expanding universe with $a = 1$, $H = 0$, the solutions to Eq. (6.22) would grow exponentially. The 'Hubble drag' term causes them to grow more slowly in a manner well approximated by a power-law $\delta_{\mathbf{k}} \propto t^n$, where the power law index depends on cosmology, epoch, and equation of state. The solution to Eq. (6.23) in our $\Omega_m = 1$ Einstein–de Sitter cosmology, where $H = \frac{2}{3t}$, is a linear combination of a simple growing mode $D_+ \propto t^{\frac{2}{3}} \propto a$ and a decaying mode $D_- \propto t^{-1} \propto a^{-3/2}$. Since we are interested in forming cosmological structures, we will focus our attention on the growing mode and ignore the decaying part.

In general, for other cosmological models ($\Omega_m \neq 1, \Omega_\Lambda \neq 0$), expressions for the growing mode are more complicated (not analytic) and usually expressed in terms of a

---

[9] Note that since we neglect pressure, this equation also applies to dark matter.

linear growth factor $g(z) \equiv D_+(z) \times (1 + z)$. An approximation for the linear growth factor is given by, e.g., [4] as

$$g(z) \approx \frac{5}{2}\Omega_m \left[\Omega_m^{\frac{4}{7}} - \Omega_\Lambda + \left(1 + \frac{\Omega_m}{2}\right)\left(1 + \frac{\Omega_\Lambda}{70}\right) \right]^{-1}. \qquad (6.24)$$

Perturbations in an open universe or in a universe with a non-zero cosmological constant grow even more slowly than in an Einstein–de Sitter universe because the expansion rate is larger, leading to a stronger Hubble drag.

Taking the Einstein–de Sitter growth rates as an upper limit for the growth rates of perturbations in a matter-dominated universe, one sees a compelling argument against the case of a universe composed mainly of baryons. Given that CMB observations show that the amplitude of the baryonic perturbations at redshift $\sim 1,000$ are a few 10^{-5}, they will have grown by, at most, a factor of 1,000 by redshift 0 according to linear theory, as $D_+ \propto (1 + z)^{-1}$, which brings their amplitude only as high as a few percent, still safely in the regime where our linear solution is valid and far from the $\delta_{\mathbf{k}} \sim 1$ level at which non-linear evolution should take over and form the galaxies we observe today.

A resolution to this puzzle is to invoke dark matter. Because dark matter interacts with radiation only gravitationally, rather than through the scattering processes which affect baryons, larger-amplitude perturbations in the dark matter at high redshifts can be achieved without changing the CMB power spectrum. Larger dark matter perturbations at earlier times allow them to reach larger amplitudes at later times and provide gravitational potential wells for baryons to fall into, thus accelerating the growth of baryonic perturbations. We will now discuss in detail how this proceeds.

6.2.3 *Perturbations in a Radiation-Dominated Universe*

Given that the seed perturbations for structure formation are believed to originate during the inflationary epoch, re-entering the observed Universe around $z \sim 10^6$ for the smallest of them – i.e., galaxy-sized fluctuations – their early growth phase occurs during the radiation-dominated epoch of the Universe at $z > 3,300$. The fluid equations we derived for the evolution of the density perturbations were based on a non-relativistic, Newtonian approach. In principle, to describe perturbation growth during the radiation-dominated epoch, we should solve for the evolution of a photon-baryon fluid using a full relativistic treatment for the fluid in a perturbed space-time metric. In practice, we can get a good sense of the physics using the Newtonian treatment, provided we consider fluctuations in the dark matter fluid only and approximate the photon-baryon fluid as a *smooth* background dominant component which only affects the expansion rate of the Universe,[10] that is to say, we write[11]: $\rho = \bar{\rho}_m(1 + \delta) + \bar{\rho}_\gamma$.

---

[10] This approach only works to describe density fluctuations – i.e., if we are not interested in describing CMB anisotropies.

[11] Technically speaking, to be consistent with our notations, the baryon density, ρ_b, should be removed from the matter density – i.e., ρ_m should become $\rho_m - \rho_b$ – and added to the energy density of the photons ρ_γ. However, since $\rho_b \ll \rho_m$ at all times, and $\rho_m < \rho_\gamma$ before matter radiation equality (but $\delta = \delta_m > \delta_\gamma$, as perturbations in the dark matter do not oscillate), we

From the point of view of structure formation, this halt of perturbation growth related to the present-day value of the cosmological constant arguably constitutes the most clear-cut case of fine-tuning. Indeed, $\bar{\rho}_\Lambda$ will quickly dominate the energy density in the near future as $\bar{\rho}_m$ decays away like a^{-3}. If the value of Λ was only 43% larger than it is measured to be[15] – which is not as large an increase as it seems, given that Λ is 122 orders of magnitude smaller than it should if we interpret it as vacuum energy! – then we would have $a_{\Lambda eq} = (0.01/0.99)^{1/3} \simeq 0.2$, which is small enough to wipe any galaxy cluster out of the night sky and, thus, force galaxies to live in isolation.

6.2.6 The Linear Power Spectrum

After describing the linear growth of primordial fluctuations in the matter density field through different epochs of the Universe, we now turn to the question of the evolution of the spectrum of fluctuations characterised by an initially scale-independent power spectrum widely believed to result from an early inflationary phase. As inflation generates fluctuations on all super-horizon scales, different sized fluctuations will re-enter the horizon at different epochs, which will determine the physical processes that they will be subjected to. The primordial spectrum of fluctuations $P(k, t_{\text{bis}})$ defined by Eq. (6.6), at time t_{bis} when the co-moving scale $\lambda = 2\pi/k$ re-enters the horizon, gets altered by linear perturbation theory to a spectrum $P(k, t)$, at a later time t: $P(k, t) = P(k, t_{\text{bis}}) D_+^2(t)/D_+^2(t_{\text{bis}})$.

Now the horizon scale itself evolves with the expansion of the Universe: if it increases with time, an observer is able to see objects at increasingly further distances. So we first need to calculate how this evolution proceeds. We will denote this co-moving particle horizon – i.e., the co-moving distance to the furthest observable point in the Universe by s_H. How s_H evolves with the scale factor a can be derived from the FLRW metric (Eq. (6.1)). Photons travel along null geodesics $ds^2 = 0$. We can choose our coordinate system to consider radially travelling photons ($d\theta^2 = d\phi^2 = 0$), place the observer at $r = 0$, and define the proper particle horizon for the photon emitted at time t to be at r_{hor}. The FLRW metric, for a flat universe, then gives

$$\int_0^t \frac{dt}{a(t)} = \frac{1}{c} \int_0^{r_{\text{hor}}} dr. \tag{6.36}$$

We can also define the distance between two events measured in a reference frame where the two events happen at the same time ($dt = 0$). This distance, called the proper distance, for an object at $r = 0$ to r_{hor} is

$$s_H^{\text{prop}} = a(t) \int_0^{r_{\text{hor}}} dr, \tag{6.37}$$

---

[15] We assume that other conditions remain the same and, in particular, that the Universe remains flat.

which means that the co-moving particle horizon can be expressed as

$$s_H = \int_0^t \frac{c \, dt}{a(t)}. \tag{6.38}$$

Using $dt = da/\frac{da}{dt}$ and the Friedmann equation (6.19) to get an expression for $da/dt = aH$, the dependence of the co-moving particle horizon on a is

$$s_H = c \int_0^a \frac{da}{a^2 H} = \frac{c}{H_0} \int_0^a \left(\Omega_{\gamma,0} + \Omega_{m,0} a + \Omega_{\Lambda,0} a^4 \right)^{-1/2} da, \tag{6.39}$$

which yields $s_H \propto a$ in the radiation dominated epoch, $s_H \propto a^{1/2}$ in the matter dominated epoch, and $s_H \propto 1/a$ when Λ dominates.

Consider a fluctuation mode with a wavelength λ so large[16] that it re-enters the horizon only during the matter-dominated era. At the time of re-entry, $\lambda = 2\pi/k = s_H \propto a^{1/2}(t_{\text{bis}})$ and so $P(k,t) = P(k, t_{\text{bis}}) \, a^2(t)/a^2(t_{\text{bis}}) \propto k a^2$. In other words, scale-independent perturbations during inflation produce $P(k,t) \propto k$ – i.e., a scale-invariant power spectrum called Harrison-Zel'dovich – on scales that re-enter the Hubble radius during the matter dominated epoch. Applying the same reasoning to a mode that re-enters the horizon well into the radiation-dominated era, we get $1/k \propto a(t_{\text{bis}})$ but $P(k,t) = P(k, t_{\text{bis}}) \propto k^{-3}$ since perturbations do not grow (Mészáros effect). The transition between modes whose last wavelengths re-enter the horizon during the radiation-dominated era and modes whose first wavelengths re-enter the horizon during the matter-dominated era is smooth, so the change in shape of $P(k,t)$ from $\propto k^{-3}$ to $\propto k$ is gradual.

Cosmologists encapsulate this change in the power spectrum shape through a transfer function $T(k)$, defined such as $P(k,t) \propto k T^2(k) D_+(t)$, that depends on the energy densities of the various constituents of the Universe (dark matter, baryons, photons, neutrino, dark energy, etc.) but applies regardless of the power spectrum of initial fluctuations, inflation generated or not.[17] From our previous simple analysis of the shape of $P(k,t)$, we immediately get $T(k) \simeq 1$ on large scales – i.e., for $k \ll k_{\text{eq}}$, where k_{eq} is the wave number corresponding to $s_H(a_{\text{eq}})$ – the co-moving particle horizon size at matter-radiation equality. To get a sense of scales, $k_{\text{eq}} = 2\pi/\lambda_{\text{eq}}$, where $\lambda_{\text{eq}} \sim 100$ Mpc. At the other end of the spectrum, on small scales – i.e., for $k \gg k_{\text{eq}} - T(k) \propto 1/k^2$. The transition where $T(k)$ (and $P(k,t)$) turns over, between the two regimes of scales, happens around k_{eq}.

Importantly, the dependence of $T(k)$ on wave number encodes the non-linear mapping between the primordial power spectrum of fluctuations to the spectrum at later times. This is the main consequence of smaller perturbations re-entering the horizon at earlier times. If they do so during the radiation-dominated phase, then their growth will be suppressed.

---

[16] Note that the reason why we ignore the Λ-dominated epoch is that s_H shrinks during this epoch, which means that perturbations with larger wavelengths than those which re-entered the horizon at the end of the matter-dominated phase will always remain super-horizon.

[17] For completeness sake, even though we have only discussed adiabatic perturbations up to now, we note that $T(k)$ would differ in the case of isocurvature perturbations.

Only when matter dominates will their growth resume. A major ramification of different sized modes growing at different rates at different epochs of the Universe is that structures will form from a power spectrum that is no longer scale invariant. Furthermore, the fact that the scale k_{eq} at which the transfer function bends from unity depends on cosmological parameters indicates that the galaxy distribution can be used to determine them.

We sketched out the derivation of the transfer function for dark matter and radiation focusing on the Mészáros effect, but in practice, $T(k)$ needs to account for all types of matter and radiation and the diverse physical processes that can alter the growth of fluctuations once they penetrate the observable Universe. For example, we saw that baryonic fluctuations oscillate as sound waves during recombination and that they can be damped by photons diffusing out of density peaks after recombination (Silk damping). Fluctuations can also be smoothed out if they enter the horizon while their non-collisional component is still relativistic (free-streaming). These processes will leave imprints on the transfer function.

As a result, an approximation to the transfer function for a cold dark matter–dominated universe will look like

$$T(k) = \frac{\ln(1 + 2.34q)}{2.34q}(1 + 3.89q + (16.1q)^2 + (5.46q)^3 + (6.71q)^4)^{-\frac{1}{4}}, \qquad (6.40)$$

where

$$q = \frac{k}{\Gamma} = \frac{k}{\Omega_{m,0}\, h \exp\left(-\Omega_{b,0}\left[1 + (2h)^{1/2}\,\Omega_{m,0}^{-1}\right]\right)}, \qquad (6.41)$$

where $h = H_0/100$ [12]. This approach can be generalised to other types of matter – e.g., hot dark matter composed of very light neutrinos, warm dark matter composed of intermediate mass particles, and a mixture of cold and hot dark matter. More accurate calculations of the transfer function are now routinely done numerically using Boltzmann codes (e.g., [7]) for any set of cosmological parameters and mix of matter constituents.

6.3 The Final Stretch: Non-linear Growth

6.3.1 The Spherical Top-Hat Model

The present-day density contrast of galaxies and clusters is orders of magnitude above the regime where linear perturbation theory can describe their formation (see Figure 6.2). Knowing the evolution of the Universe background density, back-of-the-envelope calculations of the density contrast give $\delta \sim 10^6$ for galaxies and $\delta \sim 10^3$ for clusters. While in the linear regime – i.e., when $\delta \ll 1$, we evolved different perturbation modes independently with linear perturbation theory. Once the density contrast of a perturbation increases to $\delta \sim 1$, over-densities run away to very large densities via gravitational instability, only prevented from reaching infinity by the random motions of collisionless dark matter and gas pressure, and perturbation modes become correlated. Thus the most accurate description of the growth of structures in the non-linear regime is left to computer simulations (such as that presented in Figure 6.2), but important insights can be gained from simple analytic models.

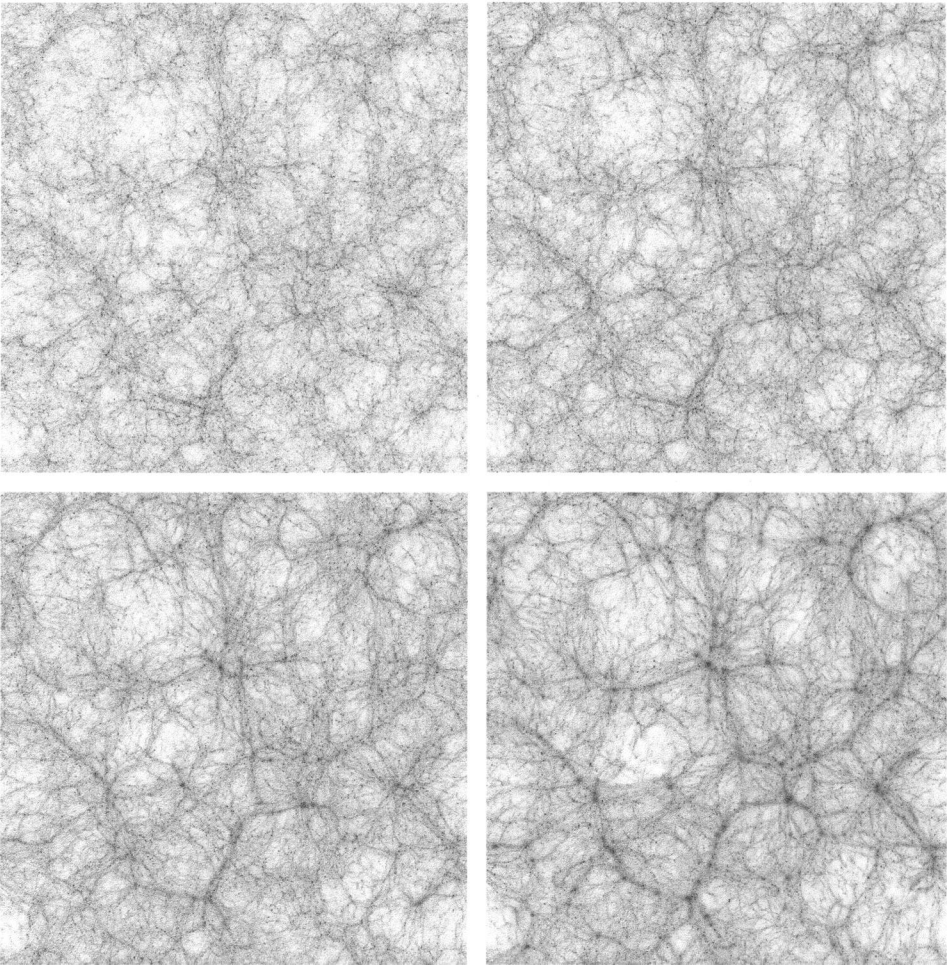

Figure 6.2 Evolution of the gas density in a 20 Mpc/h thick slice of the $(100 \text{ Mpc/h})^3$ volume of the Horizon-AGN simulated Universe between redshifts 0 and 3 (see www.horizon-simulation.org for detail). Top-left panel corresponds to $z = 3$, top-right panel to $z = 2$, bottom-left panel to $z = 1$, and bottom-right panel to $z = 0$. The scale is logarithmic to showcase the enormous range spanned by the density contrast δ, which extends well into the non-linear regime in the densest regions (in dark).

To arrive at an analytic non-linear theory, we must make drastic assumptions about the geometry and density profile of the fluctuations. The simplest model that comes to mind is that of a uniform and isolated spherical density perturbation of collisionless matter evolving in an otherwise completely homogeneous Einstein-de Sitter universe.[18] Commonly called

[18] This latter assumption of an $\Omega_m = 1$ universe is not necessary but yields the simplest analytic solutions while still capturing the essential physics, which is the reason why we adopt it here.

the 'top-hat' perturbation because of the form of its density profile, this very simple model is routinely used because its non-linear evolution can be completely described analytically. Moreover, its evolution provides an understanding of the key stages in the non-linear evolution of a perturbation, namely its growth at a slightly slower rate than the Hubble expansion at early times, its detachment from the Hubble flow owing to its self-gravity once it reaches a certain size, and, finally, its collapse onto itself.

Birkoff's theorem tells us that such a perturbation behaves as if it was an independent universe, with the matter exterior to it having no influence on its evolution. If we split up this perturbation into several concentric spherical shells, as long as different shells do not cross, the mass interior to each shell stays constant. We can then write the equation of motion of a shell located at a radius r inside the top hat perturbation as a function of time t as

$$\frac{d^2 r}{dt^2} = -\frac{GM}{r^2}, \tag{6.42}$$

where the mass M interior to r is

$$M \equiv M(<r) = \frac{4}{3}\pi r^3 \bar{\rho}_m (1 + \delta), \tag{6.43}$$

with δ defined as the top-hat over-density contrast and $\bar{\rho}_m$ as the mean density of the background Universe.

Integrating this equation of motion once yields the following equation of energy conservation:

$$\frac{1}{2}\left(\frac{dr}{dt}\right)^2 - \frac{GM}{r} = E, \tag{6.44}$$

where E is the specific energy the shell. For $E = 0$, r evolves at the same rate as the expansion of the Universe – i.e., $r \propto t^{2/3}$, $\propto a$ – so that, as a result, the top-hat perturbation does not grow. For $E < 0$ the shell is bound and the parametric solution for the value of its radius r at time t is given by

$$\frac{r[\eta]}{r_m} = \frac{1}{2}(1 - \cos \eta)$$

$$\frac{t[\eta]}{t_m} = \frac{1}{\pi}(\eta - \sin \eta). \tag{6.45}$$

Starting from $r[0] = 0$ at $t[0] = 0$, the shell reaches a maximum radius $r_m = r[\pi]$ at time $t_m = t[\pi]$ corresponding to turnaround. By symmetry, the shell collapses back to $r[2\pi] = 0$ and virialises at time $t[2\pi] = 2t_m$. Solutions for r_m and t_m are

$$r_m = \frac{GM}{|E|}$$

$$t_m = \frac{\pi GM}{(2|E|)^{3/2}} \tag{6.46}$$

so that $r_m^3 = (8GM/\pi^2)\, t_m^2$.

From these parametric solutions for r and t, we can deduce the solution for the evolution of δ. Indeed, the density ρ of the top-hat perturbation is given by

$$\rho = \frac{3M}{4\pi r^3} = \frac{6M}{4\pi r_m^3}(1 - \cos\eta)^{-3} \tag{6.47}$$

while that of the background Universe $\bar{\rho}_m$ follows

$$\bar{\rho}_m = \frac{1}{6\pi G t^2} = \frac{\pi}{6 G t_m^2}(\eta - \sin\eta)^{-2}. \tag{6.48}$$

Hence, using $r_m^3 = (8GM/\pi^2)\, t_m^2$, the density contrast of the spherical top-hat perturbation evolves as

$$1 + \delta = \frac{\rho}{\bar{\rho}_m} = \frac{9}{2}\frac{(\eta - \sin\eta)^2}{(1 - \cos\eta)^3}. \tag{6.49}$$

Small values of η correspond to the early evolution of the spherical top-hat perturbation. In that case, the solution for δ reduces to the linear regime solution. More specifically, for $\eta \ll 1$, Taylor expanding $\sin\eta$ and $\cos\eta$ around $\eta = 0$ gives

$$1 + \delta \simeq 1 + 3\frac{\eta^2}{20} \tag{6.50}$$

to second order in η, meaning $\delta \simeq 3\eta^2/20$. We can express δ in terms of t in the small η limit by first Taylor expanding t around $\eta = 0$ to get

$$t = \frac{t_m}{\pi}(\eta - \sin\eta) \simeq \frac{t_m}{\pi}\frac{\eta^3}{6}, \tag{6.51}$$

which gives

$$\eta \simeq \left(\frac{6\pi t}{t_m}\right)^{1/3}, \tag{6.52}$$

and, finally, an expression for the evolution of the density contrast in the linear regime for the Einstein-de Sitter universe:

$$\delta = \frac{3}{20}\left(\frac{6\pi t}{t_m}\right)^{2/3}. \tag{6.53}$$

We can now calculate the difference between the linear regime prediction for δ at various times and its value derived from the spherical top-hat collapse model. Key stages in the evolution of the perturbation occur at the point of maximum expansion (turnaround) and collapse. From Eq. (6.53), linear theory predicts that the density contrast at turnaround when $t = t_m$ is

$$\delta(t_m) = \frac{3}{20}(6\pi)^{2/3} \simeq 1.062 \tag{6.54}$$

and that at collapse, when $t = 2\,t_m$,

$$\delta(2\,t_m) = \frac{3}{20}(12\pi)^{2/3} \simeq 1.686. \tag{6.55}$$

Meanwhile, the spherical top-hat collapse model predicts, from Eq. (6.49), that at turnaround ($\eta = \pi$),

$$\delta(t_m) = \frac{9\pi^2}{16} - 1 \simeq 4.6, \tag{6.56}$$

while at collapse ($\eta = 2\pi$),

$$\delta(2\,t_m) = \infty. \tag{6.57}$$

Of course, the prediction from the spherical collapse model breaks down some point after turnaround when the shells inevitably cross because in reality the perturbation embedded in a universe with structures will not be exactly homogeneous and will experience torques from its neighbours, giving it angular momentum and causing it to deviate from perfect spherical symmetry. Hence, non-spherical motions will develop. Virialisation takes over as oscillating shells interact gravitationally and exchange energy to ultimately halt the collapse.

Neither linear perturbation theory nor the simple spherical top-hat model can give us the final end state of the perturbation, but we can determine it from energy conservation arguments. To estimate the final density of the collapsed, virialised perturbation, we use the virial theorem to estimate its radius after collapse has ended. Neglecting the surface pressure term, the perturbation will reach virial equilibrium when its final kinetic K_f and potential W_f energies obey

$$2K_f + W_f = 0. \tag{6.58}$$

Energy conservation relates the total final energy of the perturbation, E_f, to its total energy at turnaround, E_{ta}:

$$E_f = K_f + W_f = E_{\text{ta}}. \tag{6.59}$$

Since the velocity of the perturbation comes to a halt at turnaround, all the energy at this point is in potential form so that $E_{\text{ta}} = -GM/r_{\text{ta}}$. Combining the virial theorem and energy conservation, we can show that $E_f = W_f/2 = -GM/(2\,r_{\text{vir}})$ as $W_f = -GM/r_{\text{vir}}$, where r_{vir} is the virial radius of the perturbation. Hence, $r_{\text{vir}} = r_{\text{ta}}/2$, indicating that after it has virialised, the perturbation has collapsed to half the radius it had at turnaround.

Now we are in a position to estimate the over-density, Δ_{vir}, of the collapsed perturbation:

$$1 + \Delta_{\text{vir}} = \frac{\rho(t_{\text{coll}})}{\bar{\rho}_m(t_{\text{coll}})}, \tag{6.60}$$

where $\rho(t_{\text{coll}})$ is the density of the top-hat perturbation at collapse and $\bar{\rho}_m(t_{\text{coll}})$ is the density of the background Universe at the same time. It is evident that the collapsed perturbation acquires its density because it increases during the collapse, while at the same time, the density of the background Universe in which it is embedded decreases due to the expansion of the Universe, thereby enhancing the density contrast of the perturbation relative to a collapse during which the background remains constant. Because the background density in a matter-dominated, Einstein–de Sitter universe scales as $a^{-3} \sim t^{-2}$ and $t_{\text{coll}} = 2\,t_{\text{ta}}$, this means that $\bar{\rho}_m(t_{\text{coll}}) = \frac{1}{4}\bar{\rho}_m(t_{\text{ta}})$. Furthermore, as the radius of the collapsed perturbation is half the radius of the perturbation at turnaround, its density at collapse is a factor 8 greater than its density at turnaround, $\rho(t_{\text{coll}}) = 8\,\rho(t_{\text{ta}})$. As a result,

$$1 + \Delta_{\text{vir}} = 32\frac{\rho(t_{\text{ta}})}{\bar{\rho}_m(t_{\text{ta}})} = 32 \times (1 + \delta(t_m)) \tag{6.61}$$

and given that we showed earlier that $1 + \delta(t_m) = 9\pi^2/16$, we obtain

$$1 + \Delta_{\text{vir}} = 18\pi^2 \simeq 178. \tag{6.62}$$

So although the spherical top-hat collapse model predicted that the density at collapse would shoot to infinity giving a singularity, virialisation intervenes when shells cross, and the spherical top-hat model breaks down, leading to an over-density of $\simeq 178$. This is two orders of magnitude higher than the prediction for the value of the over-density ($\delta_{\text{coll,lin}} \sim 1.686$) at $t_{\text{coll}} = 2\,t_{\text{ta}}$ if it had continued to evolve according to linear theory. For a low-Ω_m universe such as ours, the mean background density is lower than in an $\Omega_m = 1$, and so the virialised over-density is higher yet.

Despite its simplicity, the spherical top-hat model not only highlights the key stages in the evolution of a density perturbation (turnaround and collapse), but it also gives insight to the hierarchy of collapsing perturbations. Arguments based on energy conservation reveal that in an Einstein–de Sitter universe, all over-densities will eventually collapse. The argument goes as follows. Approximating the velocity v of a mass shell with radius r_{init} by the Hubble flow velocity $v = H\,r_{\text{init}}$, the energy of the shell can be expressed as

$$E = \frac{1}{2}H^2 r_{\text{init}}^2 - \frac{GM}{r_{\text{init}}}. \tag{6.63}$$

Using $M = 4\pi/3\,r_{\text{init}}^3\bar{\rho}_m\,(1 + \delta(t_{\text{init}}))$ and $\bar{\rho}_m = \rho_c = 3H^2/(8\pi G)$, we have

$$E = \frac{1}{2}H^2 r_{\text{init}}^2 - \frac{1}{2}H^2 r_{\text{init}}^2(1 + \delta), \tag{6.64}$$

which gives $E < 0$ for all over-dense perturbations ($\delta > 0$) in an Einstein–de Sitter universe.

Energy conservation arguments also show that the radius at turnaround only depends on the initial over-density of the perturbation. By equating the energy at turnaround,

$$E(t_{\text{ta}}) = -\frac{GM}{r_{\text{ta}}} = -\frac{H^2 r_{\text{init}}^3}{2\,r_{\text{ta}}}\,(1 + \delta(t_{\text{init}})), \tag{6.65}$$

to the initial energy,

$$E(t_{\text{init}}) = -\frac{H^2 r_{\text{init}}^2}{2}\delta(t_{\text{init}}), \tag{6.66}$$

one gets

$$\frac{r_{\text{ta}}}{r_{\text{init}}} = \frac{1 + \delta(t_{\text{init}})}{\delta(t_{\text{init}})}. \tag{6.67}$$

[10] Planck Collaboration *et al.* 2016. 'Planck 2015 Results. XIII. Cosmological Parameters'. *Astronomy and Astrophysics*, **594**(Sept.), A13.

[11] Sheth, R. K., and Tormen, G. 2002. 'An Excursion Set Model of Hierarchical Clustering: Ellipsoidal Collapse and the Moving Barrier'. *Monthly Notices of the Royal Astronomical Society*, **329**(Jan.), 61–75.

[12] Sugiyama, N. 1995. 'Cosmic Background Anisotropies in Cold Dark Matter Cosmology'. *Astrophysical Journal Supplements*, **100**(Oct.), 281.

[13] Vilenkin, A., and Shellard, E. P. S. 2000. *Cosmic Strings and Other Topological Defects*. Cambridge University Press.

[14] Weinberg, S. 1972. *Gravitation and Cosmology: Principles and Applications of the General Theory of Relativity*. Wiley VCH.

[15] Liddle, A., and Lyth, D. 2000. *Cosmological Inflation and Large-Scale Structure*. Cambridge University Press.

[16] Press, W. H., and Schechter, P. 'Formation of Galaxies and Clusters of Galaxies by Self-Similar Gravitational Condensation'. *Astrophysical Journal*, 187, 425–438.

Part III
Fine-Tuning in Particle and Nuclear Physics

7

Nuclear Physics and Its Impact on Primordial and Stellar Nucleosynthesis

JEAN-PHILIPPE UZAN

Abstract

Nuclear physics is at work in cosmology both during Big Bang nucleosynthesis, during which light elements are formed, and during stellar evolution, where heavier elements (actually, all nuclei but hydrogen, helium, and lithium) are synthetised. It means that a successful stellar nucleosynthesis, in particular to form carbon and oxygen, is a key step in the emergence of complexity in our Universe. Besides, stars provide a long-lived, low-entropy source of energy also necessary for planetary life to emerge. Nuclear processes depend on many fundamental constants, the values of which affect cross sections, binding energies, reaction rates, and masses of particles. They also affect the stability of those nuclei. In our Universe, the Mendeleev table summarises the variety of nuclei and their isotopes that are then involved to form molecules. The goal of this chapter is to detail the connections between the Standard Model of particle physics and nuclear physics in order to understand how the fundamental constants affect the stability and production of the nuclei in the Universe and how it could limit the apparition of chemical complexity and eventually of life.

7.1 Introduction

Nature confronts us with structures at different scales, complexity, and properties, from fundamental particles to molecules and cells to planets, stars, and galaxies. Each of them can be described by a scientific theory, with its own ontology and structures. They form a hierarchy of theories organised in modules in interaction (see References [45, 97, 111] for a discussion). Indeed, those theories are not independent. Higher levels are built on lower levels and more fundamental theories that define a space of possibility for the former. And higher theories set the context in which the dynamics of the lower-level theories develop. This is related to both bottom-up action and top-down causation [45].

Hierarchy of physical theories. The fact that we can understand the Universe and its laws has a deep implication on this structure of theories. At each step in our construction of physical theories, we have been dealing with phenomena below a typical energy scale, mostly for technological constraints, and it turned out (empirically) that we have always been able to design a consistent theory, valid in such a restricted regime. This is not expected a priori, and this empirical fact is deeply rooted in the mathematical structure of the theories that describe nature. In particular, they have to enjoy a *scale-decoupling principle* in the sense that there exist energy scales below which effective theories are sufficient to understand a set of physical phenomena that can be observed. *Effective theories* are then the most fundamental concepts in the scientific approach to the understanding of nature, and they always come with a domain of validity inside which they are efficient to describe all related phenomena [111]. They offer a successful explanation at a given level of complexity based on concepts of that particular level.

This implies that the structure of the theories is such that there is a kind of stability and independence of higher levels with respect to more fundamental ones. It follows that various disciplines have developed independently in almost quasi-autonomous domains, each of them having its own ontology and dynamics that are independent of our ability to formulate a theory explaining these concepts. On the one hand, we can hope to relate the concepts and constants of a given level to those of an underlying level. For instance, we understand that the proton is a composite structure of three quarks, and we may try to determine its physical characteristics (charge, mass, gyromagnetic factor, quantum numbers) in terms of these more fundamental entities [79]. We know that this reductionist approach is limited and can only be achieved for some structures since there exist emergent phenomena (information, life, consciousness) that cannot be reduced to the concepts of a lower level.

Goal of this chapter. This chapter investigates the first layer of this hierarchy, namely the one involving the properties of nuclear matter and, in particular, its stability and its production. In our Universe, the diversity of the atomic nuclei opens a large possibility space of combinations for atoms and then molecules. These properties are first roughly summarised in the Mendeleev table, and one success of the standard cosmological model and nuclear astrophysics is to understand how the Mendeleev table is populated – i.e., in which sites each nuclei is produced and with which abundance.

As can be seen from Figure 7.1, only helium is significantly synthesised during Big Bang nucleosynthesis; some lithium, beryllium, and boron and slight traces of carbon, nitrogen, and oxygen are synthesised. It is understood that most stable beryllium and boron were created in the interstellar medium when cosmic rays induced fission in heavier elements found in interstellar gas and dust. Elements lighter than iron are formed in small and large stars while heavier elements are formed mostly in explosive nucleosynthesis in supernovae. In terms of nuclear processes, one usually distinguishes different processes:

- The s-process (slow neutron-capture process). It occurs at low neutron density and intermediate temperature conditions in stars. Heavier nuclei are created by neutron capture, increasing the atomic mass of the nucleus by one, followed by a β^- decay, leading to a nucleus of higher atomic number.
- The r-process (rapid neutron-capture process). It occurs during core-collapse supernovae and is responsible for the creation of approximately half of the neutron-rich atomic nuclei

Figure 7.1 Simplified summary of the main sites of formation of the different atomic nuclei. Besides helium mostly produced during Big Bang nucleosynthesis, all nuclei are formed during the stellar evolution. It is important to note that some elements are only known to be human made and have not been found in any natural environment. A more detailed summary, included the different isotopes is presented in Figure 7.17 in Appendix A. We emphasise that the existence of isotopes, is also important, wether they are stable or not. Note that some nuclei have never been observed to be produced in natural environments. They do not seem to have any importance for the emergence of complexity but they may be a way to characterise an intelligent civilisation.

heavier than iron. The process entails a succession of rapid neutron captures (hence the name r-process) by heavy seed nuclei. It creates very neutron-rich heavy isotopes, which can then decay to the first stable isotope.

- The p-process (proton process). It refers to a proton capture process which is the source of certain naturally occurring, proton-rich isotopes of the elements

Let us also note that it is important to distinguish stable isotopes from unstable isotopes. For instance, radioactive beryllium-10 is produced in the atmosphere of the Earth by the cosmic-ray spallation of oxygen. We shall not describe all these routes here. A summary of all the sites of production, as understood today, is given in Appendix A.

Some heavy elements, for which we only have traces, are also important for the emergence of life. Our discussion mostly focuses on carbon, nitrogen, and oxygen. They are, indeed, the building blocks of life as we know them and will enable a complex chemistry. This is indeed not a sufficient condition for life to emerge. Many other elements are necessary for life as we know it to exist. But it is almost impossible to quantify their effect and whether they can be replaced by other mechanisms if they did not exist.

This raises different questions. First, one would like to determine if (and which) nuclei are stable and then if they can be produced with a large enough abundance. To answer these questions, we would need to connect quantum chromodynamics (QCD) to nuclear physics, which is a difficult task and for which many links are still not fully understood. This means that one will have to rely on different approaches: theoretical, numerical, and modelisation. It also means that we will have to embed nuclear physics in a cosmological context for Big Bang nucleosynthesis (BBN) and in an astrophysical model for stellar evolution.

Fundamental physics. To begin, let us define our starting point concerning fundamental physics. Today, gravitation is well described by general relativity, and the most fundamental (experimentally tested) theory of matter is the Standard Model of particle physics. It is a theoretical construction based on an action and many choices such as the mathematical description of the matter fields (This is not completely arbitrary. It is based on the representation of the Poincaré group, which allows one to define scalar, spinor, vector, etc., structures, but we still have to decide to identify one kind of particle with a mathematical structure [113]), symmetries, and constants. None of them can be explained by the theory at hand. In particular, it is important for the constants to be measurable. For this model, they actually are and have been measurable, as we shall describe later. There is, indeed, no way to express them in terms of more fundamental quantities, and there is no equation for them. By testing their constancy, one actually shows that at the level of accuracy of the experiments and their timescales, it is a good hypothesis. In case of a disagreement, one could promote them to a dynamical field but would have to explain why they are almost frozen today [107–110] – i.e., of a stabilisation mechanism.

It is also important to remember that any measurement is just a comparison between two physical systems, one usually defining a system of units. It follows that only dimensionless constant can be measured [46], and only the change of these constants would change the physics. We are considering only these parameters. Given a list of N constants, one can always pick up three of them to define units so that one is left with $N - 3$ fundamental parameters that affect the magnitude of any physical process.

Coming back to the Standard Model of particle physics, we assume it is our fundamental theory, even though it does not incorporate massive neutrinos and dark matter so that we know it calls for an extension. This theory offers the space of possibility for higher levels complexity to emerge. Changing the values of the fundamental constants may result in the technical impossibility for nuclei to be stable. How fine-tuned is the Universe? This is what we shall now illustrate, keeping in mind that we would like to estimate how far in the chain of physical theories of higher complexity level they propagate.

Content. This chapter focuses on the relation between QCD and nuclear physics. There, one needs to determine how cross sections, binding energies, lifetimes of unstable nuclei (or of the neutron), or simply characteristics such as the mass of the proton depend on the fundamental constants listed in Table 7.1. This has been intensively investigated, and we refer to References [107–110] for reviews. It is of huge importance for the description of Big Bang nucleosynthesis and for stellar nucleosynthesis.

This chapter will first provide some generalities on the study of the influence of fundamental constants (Section 7.2). Section 7.3 then details different approaches to relate masses, gyromagnetic factors, binding energies, and resonance energies to the parameters

Table 7.1 *List of the fundamental constants of our Standard Model. See Reference [83]*
for further details on the measurements.

Constant	Symbol	Value
Speed of light	c	$299{,}792{,}458$ m s^{-1}
Planck constant (reduced)	$\hbar$	$1.054\,571\,628(53) \times 10^{-34}$ J s
Newton constant	G	$6.674\,28(67) \times 10^{-11}$ m^2 kg^{-1} s^{-2}
Weak coupling constant (at m_Z)	$g_2(m_Z)$	0.6520 ± 0.0001
Strong coupling constant (at m_Z)	$g_3(m_Z)$	1.221 ± 0.022
Weinberg angle	$\sin^2\theta_W(91.2\text{ GeV})_{\overline{MS}}$	0.23120 ± 0.00015
Electron Yukawa coupling	h_e	2.94×10^{-6}
Muon Yukawa coupling	h_μ	0.000607
Tauon Yukawa coupling	h_τ	0.0102156
Up Yukawa coupling	h_u	0.000016 ± 0.000007
Down Yukawa coupling	h_d	0.00003 ± 0.00002
Charm Yukawa coupling	h_c	0.0072 ± 0.0006
Strange Yukawa coupling	h_s	0.0006 ± 0.0002
Top Yukawa coupling	h_t	1.002 ± 0.029
Bottom Yukawa coupling	h_b	0.026 ± 0.003
Quark CKM matrix angle	$\sin\theta_{12}$	0.2243 ± 0.0016
	$\sin\theta_{23}$	0.0413 ± 0.0015
	$\sin\theta_{13}$	0.0037 ± 0.0005
Quark CKM matrix phase	δ_{CKM}	1.05 ± 0.24
Higgs potential quadratic coefficient	$\hat{\mu}^2$	$-(250.6 \pm 1.2)$ GeV^2
Higgs potential quartic coefficient	λ	1.015 ± 0.05
QCD vacuum phase	θ_{QCD}	10^{-9}

of the Standard Model of particle physics, which means that it starts with a short summary of the formulation of the Standard Model of particle physics in order for the read to understand the nature of its constants (this is complementary to Chapter 8). Masses, cross sections, and binding energies are the key quantities that appear in nuclear physics. Once we have seen how they can be related to the fundamental constants, we can give the two main astrophysical applications of these computations: the synthesis of light nuclei during Big Bang nucleosythesis in Section 7.4 (see also Chapter 5) and the production of heavier elements during stellar evolution in Section 7.5. We will pay special attention to the production of carbon-12 and oxygen-16 since they are key nuclei for life as we know it on Earth. Section 7.6 will summarise the implications for fine-tuning.

7.2 Strategy

In order to study the effects of a variation of the fundamental constants on the prediction of nuclear physics, we proceed as follows. First, we only consider a local variation; that is, we assume some changes of the fundamental constants in a domain close to the actual values observed in our Universe and, more important, we assume that the structure of the QCD theory remains unchanged. General studies on the constraints on the time variation of fundamental constants [107–110] provide upper bounds on possible variations. This means that these constraints give a measure of a universe that will appear similar to ours to observers similar to us. Our observational ability to constrain such variations provide a coarse-graining scale on the space of fundamental constants, under which two universes cannot be distinguished.

Under this hypothesis, we can define sensitivity coefficients that characterise the effect of the change of a fundamental constant. Given an observable O, the value of which depends on a set of primary parameters G_k, the sensitivity of the measured value of O to these parameters is

$$\frac{\mathrm{d}\ln O}{\mathrm{d}\ln G_k} = c_k. \tag{7.1}$$

The primary parameters can indeed be fundamental constants, such as the gravitational constant, the fine-structure constant, or combination of them, such as masses of composite particles, cross sections, binding energies, etc. The computation of the coefficient c_k requires a physical description of the system. Indeed, the observation O depends on both the parameters G_k and on external parameters that we shall call X, such as temperature, magnetic fields, etc.

Any observation of the system provides a value of O with some errors bars – i.e., $O_{\mathrm{obs}} \pm \Delta O_{\mathrm{obs}}$. Usually, the values of the primary parameters are assumed to be constant and take the value derived locally in the solar system, as tabulated in Table 7.1. Two different approaches shall be distinguished:

- One can use the observations to set constraints on the space-time variations of the primary parameters. The constraints will depend on the accuracy of the observations (ΔO_{obs}) as well as our knowledge of the effects of the external parameters X.
- One can let the constants vary in order to quantify the change that may result in the fact that the phenomena O does not happen or happens in a very different way.

The values of the parameters c_k offer a way to either choose the system O that will best constrain a set of parameters or determine the constants that play a drastic role in the phenomena O and that can thus be subject to some fine-tuning. At this stage, the approach is completely standard from a physics point of view, since it just assumes that the values of the G_k are not known a priori and that we try to measure them on the system O.

As we have already emphasised, the parameters G_k may not be fundamental constants. They need to be related to a set of fundamental constant α_i, and we define

$$\frac{\mathrm{d}\ln G_k}{\mathrm{d}\ln \alpha_i} = d_{ki}. \tag{7.2}$$

The computation of the coefficients d_{ki} requires one to specify the theoretical framework and depends heavily on our knowledge of nuclear physics and the assumptions of unification. This shall be discussed in the next section. In the following, a particular set of parameters d_{ki} has been singled out for the sensitivity of the mass of a body A to a variation of the fundamental constants

$$\frac{d \ln m_A}{d\alpha_i} = f_{Ai}. \tag{7.3}$$

As we have emphasised, we restrict our analysis to local variations of the fundamental constants, simply for technical reasons. In a general setting – such as in most multiverse scenarios, fundamental constants, and even the structures of the theory – can radically change. Among this range of parameters lies a subset that we shall call the *anthropic range*, which allows for the Universe to support the existence of observers (but not necessarily similar to us). The local analysis is indeed very restrictive since the mathematical form of the laws of physics may as well change so that what we are restricting to a local analysis in the neighbourhood of our observed Universe (in the space of theories). The determination of the anthropic region is not a prediction but just a characterisation of the sensitivity of 'our' Universe to a change of the fundamental constants ceteris paribus. Once this range is determined, one can ask the general question of quantifying the probability that we observe a universe as ours, hence providing a probabilistic prediction. This involves the use of the Anthropic Principle, which expresses the fact that we observe not just observations but observations made by us, and, requires us to state what an observer actually is (see References [8, 110] for a detailed discussion on fine-tuning and anthropic arguments). We shall not discuss this in detail in this chapter.

7.3 From the Standard Model of Particle Physics to Nuclear Physics

In order to detail the connection between these two theories, we start by recalling the basics of the Standard Model of particle physics. The main goal is to define the structures and the constants that appear at that level. We shall then discuss the running of the coupling constants with energy and the possibility of unification. It has deep implications on fine-tuning since it can tie the values of different constants together, hence suppressing some parts of the parameter space. We follow with a general discussion on the masses and gyromagnetic factors of composite particles. We then describe the cluster approach to model nuclei. It is an important tool that will allow us to compute binding energies and resonance energies. We finish with a summary of the tuning issues related to the stability of the nuclei.

7.3.1 *The Standard Model of Particle Physics in a Nutshell*

The Standard Model of non-gravitational interactions accounts for all the interactions except gravity, classified depending on the representations of the groups $SU(3)_c$ for the strong interaction, and $SU(2)_L \times U(1)_Y$ for the electroweak ones. This last symmery is broken by the Higgs mechanism along the symmetry-broken process $SU(2)_L \times U(1)_Y \to U(1)_{\text{elec}}$.

The differences between the shell model predicted g-factors and the experimental values can be attributed to the effects of the polarisation of the non-valence nucleons and spin-spin interaction [13, 51, 54]. Taking these effects into account, the refined formula relevant is

$$g = 2 \left[g_n \, b \, \langle s_z \rangle^o + (g_p - 1)(1 - b) \langle s_z \rangle^o + j \right], \tag{7.42}$$

where $g_n = -3.826$, $\langle s_z \rangle^o$ is the spin expectation value of the single-valence proton in the shell model and is one-half of the coefficient of g_s in Eq. (7.41), and b is determined by the spin-spin interaction and appears in the expressions for the spin expectation value of the valence proton $\langle s_{z_p} \rangle = (1 - b) \langle s_z \rangle^o$ and non-valence neutrons $\langle s_{z_n} \rangle = b \langle s_z \rangle^o$. Following the preferred method in References [13, 51], it is found

$$\langle s_{z_n} \rangle = \frac{\frac{g}{2} - j - (g_p - 1) \langle s_z \rangle^o}{g_n + 1 - g_p}, \quad \text{and} \quad \langle s_{z_p} \rangle = \langle s_z \rangle^o - \langle s_{z_n} \rangle. \tag{7.43}$$

Therefore, the variation of the g-factor can be written as

$$\frac{\delta g}{g} = \frac{\delta g_p}{g_p} \frac{2 g_p \langle s_{z_p} \rangle}{g} + \frac{\delta g_n}{g_n} \frac{2 g_n \langle s_{z_n} \rangle}{g} + \frac{\delta b}{b} \frac{2(g_n - g_p + 1) \langle s_{z_n} \rangle}{g}. \tag{7.44}$$

The main step is, thus, the computation of the gyromagnetic factors of the proton and neutron. In Reference [79], the dependence of the g-factors was expressed as

$$\frac{\delta g_p}{g_p} = \kappa_{u_p} \frac{\delta m_u}{m_u} + \kappa_{d_p} \frac{\delta m_d}{m_d} + \kappa_{s_p} \frac{\delta m_s}{m_s} + \kappa_{QCD_p} \frac{\delta \Lambda_{QCD}}{\Lambda_{QCD}}, \tag{7.45}$$

$$\frac{\delta g_n}{g_n} = \kappa_{u_n} \frac{\delta m_u}{m_u} + \kappa_{d_n} \frac{\delta m_d}{m_d} + \kappa_{s_n} \frac{\delta m_s}{m_s} + \kappa_{QCD_n} \frac{\delta \Lambda_{QCD}}{\Lambda_{QCD}}, \tag{7.46}$$

where the coefficients κ_i have been calculated by three methods: constituent quark model, chiral perturbation theory, and based on lattice results. The results can vary according to the method by a factor 10 for some of those coefficients. This illustrates the difficulty to relate nuclear physics to QCD.

7.3.5 Effect of the Strength Nuclear Force

As just illustrated, it is sometimes difficult to relate nuclear physics parameters. In the following, we will use a microscopic model [74, 115] that proved to be useful and efficient to analyse the effect of the variation of the strength of the nuclear interaction (NN) and electromagnetic interaction on cross sections, binding energies, and energy levels. In such an approach, the wave function of a nucleus with atomic number A, spin J, and total parity π is a solution of a Schrödinger equation with a Hamiltonian given by

$$H = \sum_i^A T(\mathbf{r}_i) + \sum_{i > j = 1}^A V(\mathbf{r}_{ij}). \tag{7.47}$$

$T(\mathbf{r}_i)$ is the kinetic energy of nucleon i. The nucleon-nucleon interaction $V(\mathbf{r}_{ij})$ depends only on the set of relative distances $\mathbf{r}_{ij} = \mathbf{r}_i - \mathbf{r}_j$. It can be decomposed as

$$V(\mathbf{r}_{ij}) = V_C(\mathbf{r}_{ij}) + V_N(\mathbf{r}_{ij}), \tag{7.48}$$

where the potential $V_C(\mathbf{r})$ arises from the electromagnetic interaction and $V_N(\mathbf{r})$ from the nuclear interaction. The expression for V_N is detailed in Appendix B. The eigenstates $\Psi^{JM\pi}$ with energy $E^{J\pi}$ of the system are solutions, as usual, of the Schrödinger equation associated with the Hamiltonian given in Eq. (7.47),

$$H\Psi^{JM\pi} = E^{J\pi}\Psi^{JM\pi}. \tag{7.49}$$

The total wave function $\Psi^{JM\pi}$ is a function of the $A - 1$ coordinates $\mathbf{r}_{ij}$.

When $A > 4$, no exact solutions of Eq. (7.49) can be found, and approximate solutions have to be constructed. For those cases, we use a cluster approximation in which $\Psi^{JM\pi}$ is written in terms of α-nucleus wave functions. Because the binding energy of the α particle is large, this approach has been shown to be well adapted to cluster states; and, in particular, to ^{8}Be and ^{12}C [71, 102]. In the particular case of these two nuclei, the wave functions are respectively expressed as

$$\Psi^{JM\pi}_{^8Be} = \mathcal{A}\phi_\alpha\phi_\alpha g_2^{JM\pi}(\boldsymbol{\rho})$$

$$\Psi^{JM\pi}_{^{12}C} = \mathcal{A}\phi_\alpha\phi_\alpha\phi_\alpha g_3^{JM\pi}(\boldsymbol{\rho},\mathbf{R}), \tag{7.50}$$

where ϕ_α is the α wave function, defined in the $0s$ shell model with an oscillator parameter b; $\mathcal{A}$ is the antisymmetrisation operator between the A nucleons of the system. For two-cluster systems, the wave function $g_2^{JM\pi}(\boldsymbol{\rho})$ depends on the relative coordinate $\boldsymbol{\rho}$ between the two α particles. For three-cluster systems, $\mathbf{R}$ is the relative distance between two α particles, and $\boldsymbol{\rho}$ is the relative coordinate between the third α particle and the ^{8}Be centre of mass. The relative wave functions, g_2 and g_3, are obtained by solving the Schrödinger equation (7.49).

One then needs to specify the nucleon-nucleon potential $V_N(\mathbf{r}_{ij})$. We shall use the microscopic interaction model [105], which contains one linear parameter (admixture parameter u) whose standard value is $u = 1$. It can be slightly modified to reproduce important inputs such as the resonance energy of the Hoyle state. The binding energies of the deuteron (-2.22 MeV) and the α particle (-24.28 MeV) do not depend on u. For the deuteron, the Schrödinger equation is solved exactly.

To take into account the variation of the fundamental constants, we introduce the parameters δ_α and δ_{NN} to characterise the change of the strength of the electromagnetic and nucleon-nucleon interaction respectively. This is implemented by modifying the interaction potential (7.48) so that

$$V(\mathbf{r}_{ij}) = (1 + \delta_\alpha)V_C(\mathbf{r}_{ij}) + (1 + \delta_{NN})V_N(\mathbf{r}_{ij}). \tag{7.51}$$

Such a modification will affect binding energies and resonant energies simultaneously.

where η_S and η_V are the strength of respectively the scalar (attractive) and vector (repulsive) nuclear contact interactions normalised to their actual value. These two parameters need to be related to the QCD parameters [41]. We also refer to [55] for the study of the dependence of the binding of light ($A \leq 8$) nuclei on possible variations of hadronic masses – including meson, nucleon, and nucleon-resonance masses.

7.3.7 Resonance Energies

The formalism developed in Section 7.3.5 allows one to compute the resonance energies of different nuclear reactions. It is, indeed, out of the scope of this text to investigate all possible reactions. We shall focus of beryllium-8 and carbon-12, which are of primary importance for nucleosynthesis.

For each set of values ($\delta_\alpha, \delta_{NN}$), one can solve Eq. (7.49) with the interaction potential (7.51). We emphasise that the parameter u is determined from the experimental ^{8}Be and ^{12}C(0_2^+) energies ($u = 0.954$). We assume that δ_{NN} varies in the range $[-0.015, 0.015]$. Following Reference [44], one obtains the results depicted in Figure 7.2. The sensitivities of $E_R(^8$Be$)$ and $E_R(^{12}$C$)$ to δ_{NN} are scaling as

$$E_R(^8\text{Be}) \equiv -B_8 \, (0.09184 - 12.208 \times \delta_{NN}) \text{ MeV}, \qquad (7.64)$$

B_8 being the ^{8}Be binding energy with respect to two-alpha break-up (with this convention, $B_8 < 0$ for unbound ^{8}Be), and

$$E_R(^{12}\text{C}) = (0.2876 - 20.412 \times \delta_{NN}) \text{ MeV}. \qquad (7.65)$$

The numerical results for the sensitivities of $E_R(^8$Be$)$ and $E_R(^{12}$C$)$ to δ_{NN} as well as the preceding linear fits are shown in Figure 7.2. The effect of δ_α on these quantities is negligible. Note that for $\delta_{NN} \gtrsim 0.007$, $E_R(^8$Be$)$ is negative and ^{8}Be becomes stable. Using the bijective relation (7.57) between B_D and δ_{NN}, we can also express our results as

$$E_R(^8\text{Be}) = (0.09184 - 12.208 \, \Delta B_D / B_D) \text{ MeV}, \qquad (7.66)$$

$$E_R(^{12}\text{C}) = (0.2876 - 3.570 \, \Delta B_D / B_D) \text{ MeV}. \qquad (7.67)$$

To estimate the effect of δ_α in Eq. (7.51), we can approximate the Coulomb energy by $(3/5)Z(Z-1)\alpha\hbar c / R_c$, where $R_c = 1.3 A^{1/3}$ fm, which gives 9 MeV for ^{12}C and 0.9 MeV for ^{4}He. The variation of $Q_{\alpha\alpha\alpha}$ is thus of the order of $+6$ MeV $\times \delta_\alpha$. The direct effect of δ_α is thus of opposite sign but considerably less important. This is in qualitative agreement with References [88, 89].

It is appropriate at this point to note that within the limits of variation of δ_{NN} that we are considering here, the effect on promoting the stability of dineutron or diproton states is negligible. Working within the context of the same nuclear model, we estimate that a value of $\delta_{NN} \geq 0.15$ (for the dineutron) or ≥ 0.35 (for the diproton) would be required in order to induce stability for the dineutron or diproton, respectively.

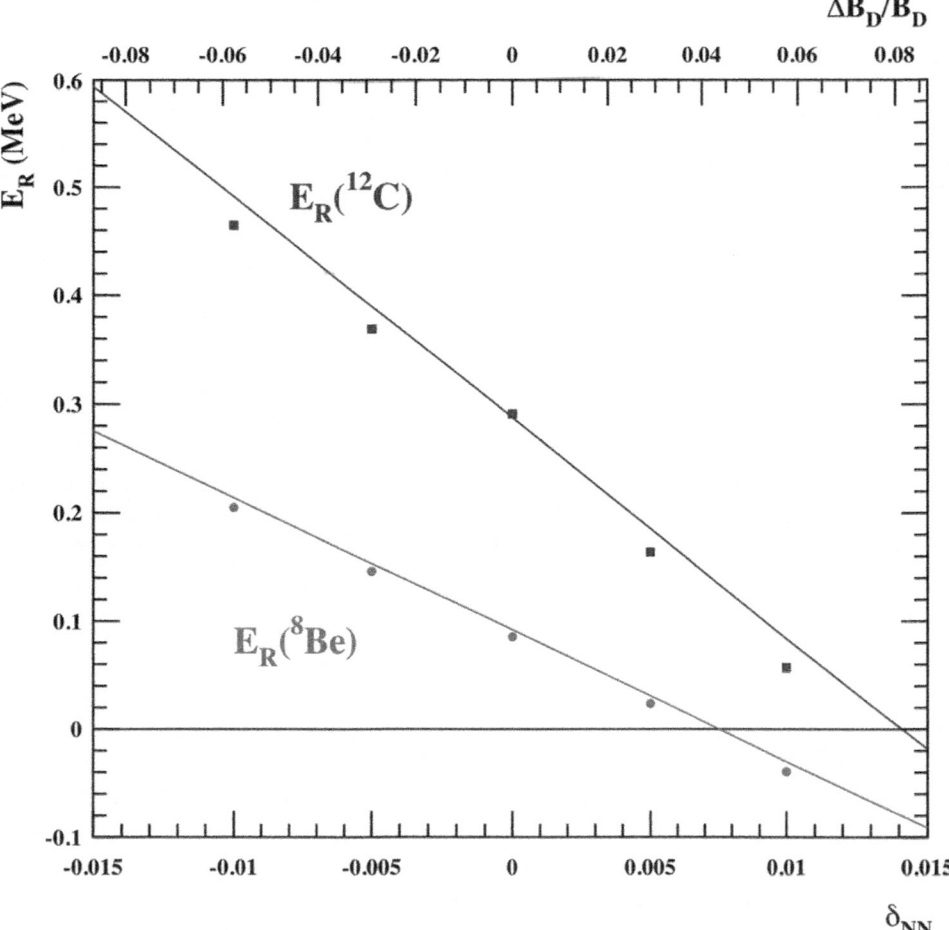

Figure 7.2 Variation of the resonance energies as a function of δ_{NN}. The symbols represent the results of the microscopic calculation while the lines correspond to the adopted linear relationship between E_R and δ_{NN}. From Reference [44].

7.3.8 Implications for Fine-Tuning

From the previous sections, it can be deduced that most low-energy nuclear properties will depend on the values of the gauge couplings and on the masses of the quarks. In particular, changing these constants may affect the stability of some particles (we mentioned the fact that beryliium-8 may become bound).

Several attempts to study the parameter space of the Standard Model of particle physics have been made.

They are often model dependent and restricted to a subset of parameters. For instance, Reference [9] considered a model in which up and down fermions get their masses from

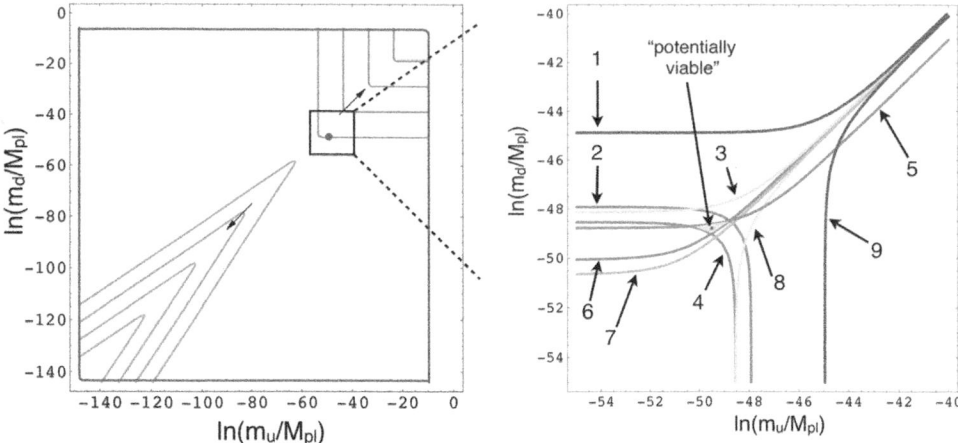

Figure 7.3 Parameter space of the masses of the up and down quarks. The lines show the limits of different life-permitting criteria, as calculated by Reference [9] and summarised in the text. The small green region marked 'potentially viable' shows where all these constraints are satisfied. From Reference [9].

different Higgs doublets. The result of their analysis in the plane of the up and down quarks masses is depicted in Figure 7.3, where they have indicated the thresholds defined by different criteria: (1) above the blue line, there exists a single stable element with chemistry similar to helium-4; (2) above the red line, the deuteron is unstable and decays via the strong force so that nucleosynthesis in hydrogen-burning stars would fail; (3) above the green line, neutrons in nuclei decay so that hydrogen is the only stable element; (4) below this red curve, the diproton is stable; two protons can fuse to helium-2 via a very fast electromagnetic reaction rather than the much slower, weak nuclear pp-chain; (5) above this red line, the production of deuterium in stars absorbs energy rather than releasing it. Also, the deuterium is unstable to weak decay; (6) below this red line, atoms are unstable since a proton in a nucleus can capture an orbiting electron to form a neutron; (7) below the orange curve, isolated protons are unstable, leaving no hydrogen left over from the early Universe to power long-lived stars and play a crucial role in organic chemistry; (8) below this green line, protons in nuclei decay so that any atoms that formed would disintegrate into a cloud of neutrons; (9) below this blue line, there is only a single stable particle, which can combine with a positron to produce an element with the chemistry of hydrogen. A handful of chemical reactions are possible, with their most complex product being (an analogue of) H_2.

Another example in the case of unification is provided by the analysis of Reference [65] in the case of the grand unified theory, in the parameter space of the mass of the electron and the difference between the masses of the down and up quarks, taking into account the stability of hydrogen and deuterium. This slice of the parameter space is depicted in Figure 7.4.

Figure 7.4 Constraints in the plane of the mass of the electron and the difference between the masses of the down and up quarks. Figure taken from Reference [65].

The effect of the QCD parameters on the strength of the interactions and masses have been extensively discussed; see, e.g., References [8, 11, 20, 21, 104]. A short summary of those constraints is the following (see Reference [8] for an extensive review on these questions).

- The existence of stable atoms requires the radius of the electron orbit to be significantly larger than the nuclear radius – i.e., $\alpha\mu/\alpha_s \ll 1$ [11].
- The existence of hydrogen – in particular, to power stars and form water – implies $m_e < m_n - m_p$. Otherwise, the electron will be captured by the proton to form a neutron [37].
- To ensure that the atomic constituents of chemical species maintain their identity in chemical reactions, one may require that the typical energy of chemical reactions is much smaller than the typical energy of nuclear reactions – i.e., $\alpha^2\mu/\alpha_s^2 \ll 1$ [11].

- Stable ordered molecular structures are not stable only if $\mu^{1/4} \ll 1$ [11].
- The stability of the proton requires $\alpha \lesssim (m_d - m_u)/141$ MeV so that the extra electro-magnetic mass-energy of a proton relative to a neutron is more than counterbalanced by the bare quark masses [61, 64].
- Unless $\alpha \ll 1$, the electrons in atoms and molecules are unstable to pair creation [11].
- Unless $\alpha_s \lesssim 0.3\alpha^{1/2}$, carbon and all larger elements are unstable [11].
- Unless $\alpha_s/\alpha_{s,0} \gtrsim 0.91$ [40], the deuteron is unstable, and the main nuclear reaction in stars (pp) does not proceed.
- The grey stripe on the left of each plot shows where $\alpha < \alpha_G$, rendering electric forces weaker than gravitational ones.

Many other constraints exist [8, 11, 20, 21, 104], but they also involve cosmological or astrophysical parameters – that is, the environment in which the Standard Model of particle physics evolves. This bounds were summarised mostly for the reader to be aware that many conditions exist for the basic building blocks of nuclear physics to exist.

In the following sections of this chapter, we only consider the effect on BBN and stellar nucleosynthesis. The range of variation of the parameters will be assumed to satisfy the constraints cited earlier.

7.4 Primordial Nucleosynthesis

Big Bang nucleosynthesis describes the synthesis of light nuclei in the primordial Universe. It is considered as the *second* pillar of the Big Bang model. It is worth noting that BBN has been essential in the past, first to estimate the baryonic density of the Universe and give an upper limit on the number of neutrino families, as was later confirmed from the measurement of the Z^0 width by LEP experiments at CERN.

This section starts by describing the evolution of the Universe during the radiation era in Section 7.4.1 and then the basics of the BBN mechanism in Section 7.4.2. The observational status of the abundances of light elements is summarised in Section 7.4.3. Section 7.4.4 describes the dependence of BBN of the different constants, and Section 7.4.5 investigates the effect of a change of these parameters.

7.4.1 Description of the Universe in Its Early Phases

Cosmological Dynamics

In the standard cosmological model [92], which we shall assume here, the Universe is described by a Friedmann-Lemaître (FL) space-time with metric

$$ds^2 = -(u_\mu dx^\mu)^2 + (g_{\mu\nu} + u_\mu u_\nu)dx^\mu dx^\nu, \tag{7.68}$$

which clearly shows that the cosmic time t is the proper time measured by these fundamental observers. As a second consequence, this symmetry implies that the most general form of the stress-energy tensor is the one of a perfect fluid

$$T_{\mu\nu} = \rho u_\mu u_\nu + P(g_{\mu\nu} + u_\mu u_\nu), \tag{7.69}$$

with ρ and P being the energy density and isotropic pressure measured by the fundamental observers. The Einstein equations with the stress-energy tensor (7.69) reduce to the Friedmann equations

$$H^2 = \frac{8\pi G}{3}\rho - \frac{K}{a^2} + \frac{\Lambda}{3}, \tag{7.70}$$

$$\frac{\ddot{a}}{a} = -\frac{4\pi G}{3}(\rho + 3P) + \frac{\Lambda}{3}, \tag{7.71}$$

together with the conservation equation $(\nabla_\mu T^{\mu\nu} = 0)$

$$\dot{\rho} + 3H(\rho + P) = 0. \tag{7.72}$$

This gives two independent equations for three variables (a, ρ, P) that requires the choice of an equation of state

$$P = w\rho \tag{7.73}$$

to be integrated. It is convenient to use the conformal time defined by $dt = a(\eta)d\eta$ and the normalised density parameters

$$\Omega_i = 8\pi G\rho_i/3H_0^2, \qquad \Omega_\Lambda = \Lambda/3H_0^2, \qquad \Omega_K = -K/a_0^2 H_0^2, \tag{7.74}$$

that satisfy, from Eq. (7.70), $\sum_i \Omega_i + \Omega_\Lambda + \Omega_K = 1$, so that the Friedmann equation takes the form

$$\frac{H^2}{H_0^2} = \sum_i \Omega_i(1+z)^{3(1+w_i)} + \Omega_K(1+z)^2 + \Omega_\Lambda, \tag{7.75}$$

where the redshift z has been defined as $1 + z = a_0/a$.

Thermal History

Since for radiation $\rho \propto a^4$ and for pressureless matter $\rho \propto a^3$, it can be concluded that the Universe was dominated by radiation in its early phase. The density of radiation today is mostly determined by the temperature of the cosmic microwave background so that equality takes place at a redshift

$$z_{eq} \simeq 3612\,\Theta_{2.7}^{-4}\left(\frac{\Omega_{m0}h^2}{0.15}\right), \tag{7.76}$$

obtained by equating the matter and radiation energy densities and where $\Theta_{2.7} \equiv T_{CMB}/2.725\,\mathrm{mK}$. Since the temperature scales as $(1 + z)$, the temperature at which the matter and radiation densities were equal is $T_{eq} = T_{CMB}(1 + z_{eq})$, which is of order

$$T_{eq} \simeq 5.65\,\Theta_{2.7}^{-3}\,\Omega_{m0}h^2\,\mathrm{eV}. \tag{7.77}$$

Above this energy, the matter content of the Universe is in a very different form from that of today. As it expands, the photon bath cools down, which implies a *thermal history*. In particular:

- When the temperature T becomes larger than twice the rest mass m of a charged particle, the energy of a photon is large enough to produce particle-antiparticle pairs. Thus, when $T \gg m_e$, both electrons and positrons were present in the Universe, so that the particle content of the Universe changes during its evolution while it cools down.
- Symmetries can be spontaneously broken.
- Some interactions may be efficient only above a temperature, typically as long as the interaction rate Γ is larger than the Hubble expansion rate.
- The freeze-out of some interaction can lead to the existence of relic particles.

Radiation Era at Thermodynamical Equilibrium

Particles interaction are mainly characterised by a reaction rate Γ. If this reaction rate is much larger than the Hubble expansion rate, then it can maintain these particles in *thermodynamic equilibrium* at a temperature T. Particles can thus be treated as perfect Fermi-Dirac and Bose-Einstein gases with distribution[1]

$$F_i(E, T) = \frac{g_i}{(2\pi)^3} \frac{1}{\exp\left[(E - \mu_i)/T_i(t)\right] \pm 1} \equiv \frac{g_i}{(2\pi)^3} f_i(E, T), \qquad (7.78)$$

where g_i is the degeneracy factor, μ_i is the chemical potential, and $E^2 = p^2 + m^2$. The normalisation of f_i is such that $f_i = 1$ for the maximum phase space density allowed by the Pauli principle for a fermion. T_i is the temperature associated with the given species, and, by symmetry, it is a function of t alone, $T_i(t)$. Interacting species have the same temperature. Among these particles, the Universe contains an electrodynamic radiation with black-body spectrum. Any species interacting with photons will, hence, have the same temperature as these photons as long as $\Gamma_i \gg H$. The photon temperature $T_\gamma = T$ will thus be called the *temperature of the Universe*.

As long as thermal equilibrium holds, one can define thermodynamical quantities such as the number density n, energy density ρ, and pressure P as

$$n_i = \int F_i(p, T)\mathrm{d}^3 p \quad \rho_i = \int F_i(p, T)E(p)\mathrm{d}^3 p \quad P_i = \int F_i(p, T)\frac{p^2}{3E}\mathrm{d}^3 p. \qquad (7.79)$$

For ultra-relativistic particles ($m, \mu \ll T$), the density at a given temperature T is then given by

$$\rho_{\mathrm{r}}(T) = g_*(T)\left(\frac{\pi^2}{30}\right) T^4. \qquad (7.80)$$

---

[1] The distribution function depends a priori on $(\boldsymbol{x}, t)$ and $(\boldsymbol{p}, E)$, but the homogeneity hypothesis implies that it does not depend on $\boldsymbol{x}$, and isotropy implies that it is a function of $p^2 = \boldsymbol{p}^2$. Thus, it follows from the cosmological principle that $f(\boldsymbol{x}, t, \boldsymbol{p}, E) = f(E, t) = f[E, T(t)]$.

g_* represents the effective number of relativistic degrees of freedom at this temperature:

$$g_*(T) = \sum_{i=\text{bosons}} g_i \left(\frac{T_i}{T}\right)^4 + \frac{7}{8} \sum_{i=\text{fermions}} g_i \left(\frac{T_i}{T}\right)^4. \tag{7.81}$$

The factor 7/8 arises from the difference between the Fermi and Bose distributions. In the radiation era, the Friedmann equation then takes the simple form

$$H^2 = \frac{8\pi G}{3} \left(\frac{\pi^2}{30}\right) g_* T^4. \tag{7.82}$$

Numerically, this amounts to

$$H(T) \cong 1.66 g_*^{1/2} \frac{T^2}{M_p}, \qquad t(T) \cong 0.3 g_*^{-1/2} \frac{M_p}{T^2} \sim 2.42 g_*^{-1/2} \left(\frac{T}{1\,\text{MeV}}\right)^{-2} \text{s}. \tag{7.83}$$

Description of the Neutrinos

In order to follow the evolution of the matter content of the Universe, it is convenient to have conserved quantities such as the entropy. It can be shown [92] to be defined as $S = sa^3$ in terms of the entropy density s as

$$s \equiv \frac{\rho + P - n\mu}{T}. \tag{7.84}$$

It satisfies $d(sa^3) = -(\mu/T)d(na^3)$ and is, hence, constant (1) as long as matter is neither destroyed nor created, since then na^3 is constant, or (2) for non-degenerate relativistic matter, $\mu/T \ll 1$. In the cases relevant for cosmology, $d(sa^3) = 0$. It can be expressed in terms of the temperature of the photon bath as

$$s = \frac{2\pi^2}{45} q_* T^3, \qquad \text{with} \qquad q_*(T) = \sum_{i=\text{bosons}} g_i \left(\frac{T_i}{T}\right)^3 + \frac{7}{8} \sum_{i=\text{fermions}} g_i \left(\frac{T_i}{T}\right)^3. \tag{7.85}$$

If all relativistic particles are at the same temperature, $T_i = T$, then $q_* = g_*$. Note also that $s = q_*\pi^4/45\zeta(3)n_\gamma \sim 1.8 q_* n_\gamma$, so that the photon number density gives a measure of the entropy.

The standard example of the use of entropy is the determination of the temperature of the cosmic neutrino background. Neutrinos are in equilibrium with the cosmic plasma as long as the reactions $\nu + \bar{\nu} \longleftrightarrow e + \bar{e}$ and $\nu + e \longleftrightarrow \nu + e$ can keep them coupled. Since neutrinos are not charged, they do not interact directly with photons. The cross section of weak interactions is given by $\sigma \sim G_F^2 E^2 \propto G_F^2 T^2$ as long as the energy of the neutrinos is in the range $m_e \ll E \ll m_W$. The interaction rate is thus of the order of $\Gamma = n\langle \sigma v \rangle \simeq G_F^2 T^5$. We obtain that $\Gamma \simeq \left(\frac{T}{1\,\text{MeV}}\right)^3 H$. Thus, close to $T_D \sim 1\,\text{MeV}$, neutrinos decouple from the cosmic plasma. For $T < T_D$, the neutrino temperature decreases as $T_\nu \propto a^{-1}$ and remains equal to the photon temperature.

Slightly after decoupling, the temperature becomes smaller than m_e. Between T_D and $T = m_e$ there are four fermionic states (e^-, e^+, each having $g_e = 2$) and two bosonic states (photons with $g_\gamma = 2$) in thermal equilibrium with the photons. Thus, we have that $q_\gamma(T > m_e) = \frac{11}{2}$ while for $T < m_e$ only the photons contribute to q_γ and, hence, $q_\gamma(T < m_e) = 2$. The conservation of entropy implies that after $\bar{e} - e$ annihilation, the temperatures of the neutrinos and the photons are related by

$$T_\gamma = \left(\frac{11}{4}\right)^{1/3} T_\nu. \tag{7.86}$$

Thus, the temperature of the Universe is increased by about 40% compared to the neutrino temperature during the annihilation. Since $n_\nu = (3/11)n_\gamma$, there must exist a cosmic background of neutrinos with a density of 112 neutrinos per cubic centimetre and per family, with a temperature of around 1.95 K today.

7.4.2 Mechanism

The standard BBN scenario [15, 16, 91, 92] proceeds in three main steps:

1. For $T > 1$ MeV, ($t < 1$ s) a first stage during which the neutrons, protons, electrons, positrons, and neutrinos are kept in statistical equilibrium by the (rapid) weak interaction

$$n \longleftrightarrow p + e^- + \bar{\nu}_e, \qquad n + \nu_e \longleftrightarrow p + e^-, \qquad n + e^+ \longleftrightarrow p + \bar{\nu}_e. \tag{7.87}$$

As long as statistical equilibrium holds, the neutron-to-proton ratio is

$$\left(\frac{n}{p}\right) = \exp\left(-\frac{Q_{np}}{k_B T}\right), \tag{7.88}$$

where $Q_{np} \equiv (m_n - m_p)c^2 = 1.29$ MeV. The abundance of the other light elements is given by [92]

$$Y_A = g_A \left(\frac{\zeta(3)}{\sqrt{\pi}}\right)^{A-1} 2^{(3A-5)/2} A^{5/2} \left[\frac{k_B T}{m_N c^2}\right]^{3(A-1)/2} \eta^{A-1} Y_p^Z Y_n^{A-Z} e^{B_A/k_B T}, \tag{7.89}$$

where g_A is the number of degrees of freedom of the nucleus $^A_Z X$, m_N is the nucleon mass, η the baryon-photon ratio, and $B_A \equiv (Z m_p + (A - Z)m_n - m_A)c^2$ is the binding energy.

2. Around $T \sim 0.8$ MeV ($t \sim 2$ s), the weak interactions freeze out at a temperature T_f determined by the competition between the weak interaction rates and the expansion rate of the Universe and, thus, roughly determined by $\Gamma_w(T_f) \sim H(T_f)$; that is,

$$G_F^2(k_B T_f)^5 \sim \sqrt{G N_*}(k_B T_f)^2, \tag{7.90}$$

where G_F is the Fermi constant and N_* the number of relativistic degrees of freedom at T_f. Below T_f, the number of neutrons and protons changes only from the neutron β-decay between T_f to $T_N \sim 0.1$ MeV when $p + n$ reactions proceed faster than their inverse dissociation.

3. For 0.05 MeV$< T <$ 0.6 MeV (3 s $< t <$ 6 min), the synthesis of light elements occurs only by two-body reactions. This requires the deuteron to be synthesised ($p + n \rightarrow D$), and the photon density must be low enough for the photodissociation to be negligible. This happens roughly when

$$\frac{n_d}{n_\gamma} \sim \eta^2 \exp(-B_D/T_N) \sim 1 \tag{7.91}$$

with $\eta \sim 3 \times 10^{-10}$. The abundance of ^{4}He by mass, Y_p, is then well estimated by

$$Y_p \simeq 2\frac{(n/p)_N}{1 + (n/p)_N} \tag{7.92}$$

with

$$(n/p)_N = (n/p)_f \exp(-t_N/\tau_n) \tag{7.93}$$

with $t_N \propto G^{-1/2}T_N^{-2}$ and $\tau_n^{-1} = 1.636G_F^2(1 + 3g_A^2)m_e^5/(2\pi^3)$, with $g_A \simeq 1.26$ being the axial-vector coupling of the nucleon.

4. The abundances of the light element abundances, Y_i, are then obtained by solving a series of nuclear reactions

$$\dot{Y}_i = J - \Gamma Y_i,$$

where J and Γ are time-dependent source and sink terms (see Figure 7.5).

5. Today, BBN codes include up to 424 nuclear reaction network [28] with up-to-date nuclear physics. In standard BBN, only D, ^{3}He, ^{4}He, and ^{7}Li are significantly produced as well as traces of ^{6}Li, ^{9}Be, ^{10}B, ^{11}B, and CNO. The most recent up-to-date predictions are discussed in References [30, 33].

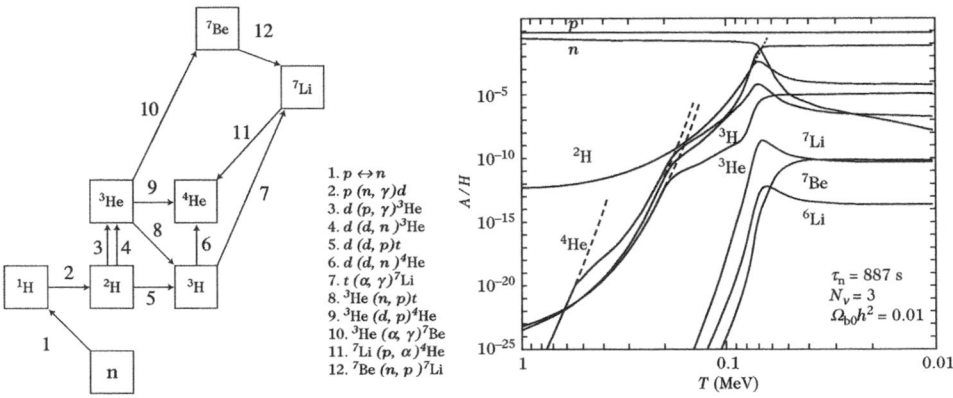

Figure 7.5 Left: The minimal 12 reactions network needed to compute the abundances up to lithium. Right: The evolution of the abundances of neutron, proton, and the lightest elements as a function of temperature (i.e., time). Below 0.01 MeV, the abundances are frozen and can be considered as the primordial abundances. Figure taken from Reference [92].

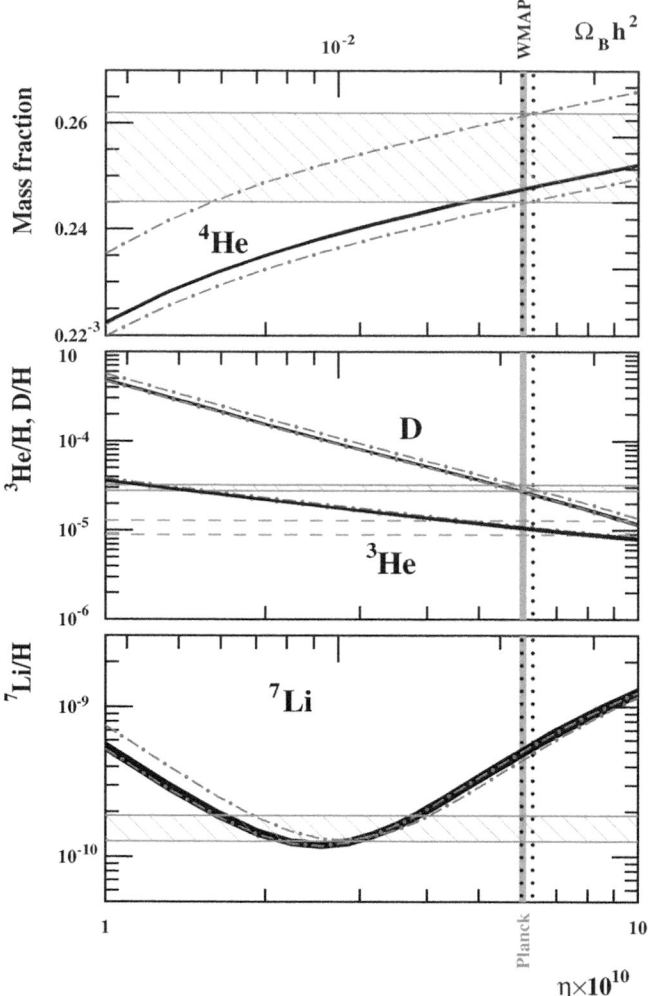

Figure 7.6 Abundances of ^{4}He D, ^{3}He, and ^{7}Li (thick solid curve) as a function of the baryon-to-photon ratio (bottom) or baryonic density (top). The vertical areas correspond to the WMAP (dot, black) and Planck (solid, grey) baryonic densities while the horizontal hatched areas represent the adopted observational abundances. The dot-dashed lines correspond to the extreme values of the *effective* neutrino families coming from CMB Planck study, $N_{\text{eff}} = (3.02, 3.70)$. Taken from Reference [33].

7.4.3 Observations

These predictions need to be compared to the observation of the abundances of the different nuclei (Figure 7.6).

Deuterium is a very fragile isotope, easily destroyed after BBN. Its most primitive abundance is determined from the observation of clouds at high redshift, on the line of sight of distant quasars. Recently, more precise observations of damped Lyman-α systems at high redshift have led to provide [35, 93] the mean value

$$D/H = (2.53 \pm 0.04) \times 10^{-5}. \qquad (7.94)$$

After BBN, ^{4}He is still produced by stars, essentially during the main-sequence phase. Its primitive abundance is deduced from observations in H$_{II}$ (ionised hydrogen) regions of compact blue galaxies. The primordial ^{4}He mass fraction, Y_p, is obtained from the extrapolation to zero metallicity but is affected by systematic uncertainties. Recently, references [5, 6] have determined that

$$Y_p = 0.2465 \pm 0.0097. \qquad (7.95)$$

Contrary to ^{4}He, ^{3}He is both produced and destroyed in stars all along its galactic evolution so that the evolution of its abundance as a function of time is subject to large uncertainties. Moreover, ^{3}He has been observed in our galaxy [7], and one only gets a local constraint

$$^3\text{He/H} = (1.1 \pm 0.2) \times 10^{-5}. \qquad (7.96)$$

Consequently, the baryometric status of ^{3}He is not firmly established [112].

Primitive lithium abundance is deduced from observations of low-metallicity stars in the halo of our galaxy where the lithium abundance is almost independent of metallicity, displaying the so-called Spite plateau [101]. This interpretation assumes that lithium has not been depleted at the surface of these stars, so the presently observed abundance can be assumed to be equal to the primitive one. The small scatter of values around the Spite plateau is indeed an indication that depletion may not have been very efficient. However, there is a discrepancy between the value (1) deduced from these observed spectroscopic abundances and (2) the BBN theoretical predictions assuming Ω_b is determined by the CMB observations. Many studies have been devoted to the resolution of this so-called Lithium problem and many possible 'solutions', none fully satisfactory, have been proposed. For a detailed analysis, see the proceedings of the meeting 'Lithium in the Cosmos' [70]. Note that the idea according to which introducing neutrons during BBN may solve the problem has today been shown [31, 32] to be generically inconsistent with lithium and deuterium observations. Astronomical observations of these metal poor halo stars [98] have thus led to a relative primordial abundance of

$$\text{Li/H} = (1.58 \pm 0.31) \times 10^{-10}. \qquad (7.97)$$

The origin of the light elements lithium, beryllium, and boron is a crossing point between optical and gamma spectroscopy, non-thermal nucleosynthesis (via spallation with galactic cosmic ray), stellar evolution, and Big Bang nucleosynthesis. We shall not discuss them in detail but just mention that, typically, ^{6}Li/H $\sim 10^{-11}$. Beryllium is a fragile nucleus formed in the vicinity of Type II supernovae by non-thermal process (spallation). The observations in metal-poor stars provide a primitive abundance at very low metallicity, of the order of Be/H $= 3. \times 10^{-14}$ at [Fe/H] $= -4$. This observation has to be compared to the typical primordial Be abundance, $Be/H = 10^{-18}$. Boron has two isotopes: ^{10}B and ^{11}B, and is also synthesised by non-thermal processes. The most recent observations give $B/H = 1.7 \times 10^{-12}$, to be compared to the typical primordial B abundance $B/H = 3 \times 10^{-16}$. For a general review of these light elements, see Reference [103].

Finally, CNO elements are observed in the lowest-metal poor stars (around [Fe/H] = −5). The observed abundance of CNO is typically [CNO/H] = −4, relatively to the solar abundance – i.e., primordial CNO/H $< 10^{-7}$. For a review, see Reference [56] and references therein.

7.4.4 Parameters

As can be seen in the description, the general BBN mechanism depends on

- External parameters (mostly the number of families of neutrinos or, more generally, the number of relativistic degrees of freedom, and the baryonic density)
- Several primary parameters, such as the neutron lifetime (That dictates the free neutron decay and appears in the normalisation of the proton-neutron reaction rates. It is the only weak interaction parameter, and it is related to the Higgs vev.) and the deuterium binding energies.
- Nuclear constants, such as the neutron-to-proton mass difference (it enters in the neutron-proton ratio)
- Fundamental constants, such as the Newton constant (which will affect the Hubble expansion rate at the time of nucleosynthesis in the same way as extra-relativistic degrees of freedom do so that it modifies the freeze-out time T_f), the fine-structure constant (which enters in the Coulomb barriers of the reaction rates through the Gamow factor in all the binding energies), or the Higgs vev (via the Fermi constant).

In full generality, the effect of these constants on the BBN predictions is difficult to model because of the intricate structure of QCD and its role in low-energy nuclear reactions. Thus, a solution is to proceed in *two steps*: first by determining the dependencies of the light element abundances on the BBN parameters and then by relating those parameters to the fundamental constants.

Sensitivities. In order to evaluate to these parameters, we vary them independently and compute numerically the abundances of the light elements, assuming that the cosmological parameter η is fixed to the value determined by the analysis of the cosmic microwave background anisotropies by the Planck satellite. The result is depicted on Figure 7.7 and shows that the most sensitive parameter is B_D, mostly because it controls the deuterium bottleneck.

Expression of the nuclear parameters. Q_{np} can be expressed in terms of the mass on the quarks u and d and the fine-structure constant as

$$Q_{np} = a\alpha \Lambda_{QCD} + (m_d - m_u), \tag{7.98}$$

m_e, B_D, Q_{np} and τ_n variations

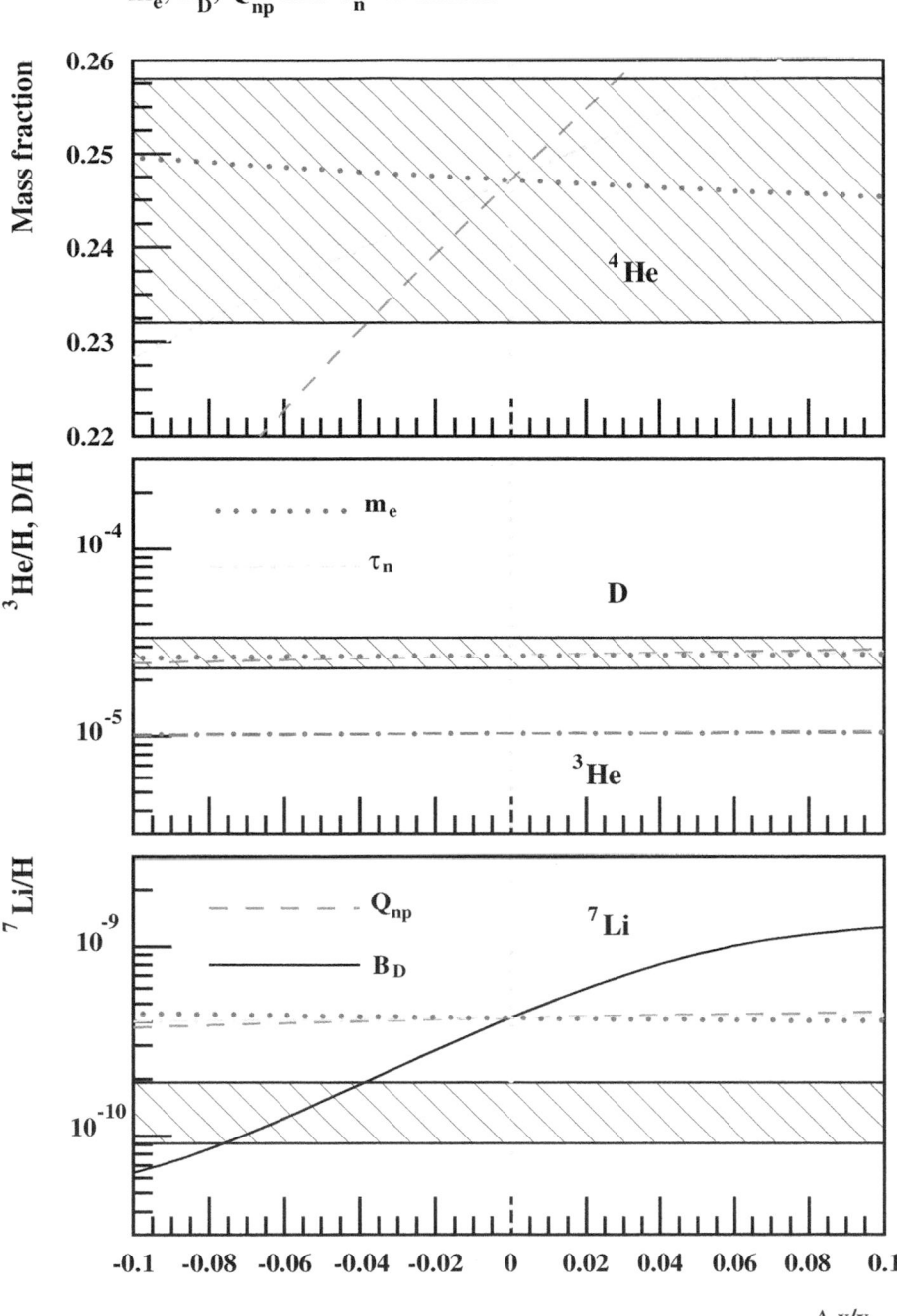

Figure 7.7 Dependence of the light element abundances on the independent variation of the main BBN parameters, assuming that η is fixed by the analysis of the Planck satellite. Figure taken from Reference [26].

Figure 7.8 The relative variation of the $^4\mathrm{He}(\alpha, \gamma)^8\mathrm{Be}$ reaction rate assuming that $^8\mathrm{Be}$ is bound by 10, 50, and 100 keV, relative to the 50 keV rate. Figure taken from Reference [29].

neglect the difference in the rates once $B_8 > 0$. The reaction rate is essentially given by the radiative capture cross section at the Gamow energy $E_0(T)$, which is proportional to E_γ^5, where the photon energy is $E_\gamma = E_{\mathrm{cm}} + B_8$ and E_{cm} is the α-α centre-of-mass energy.

Hoyle state. The computation of the effect of a change of the nuclear interaction on the Hoyle state is described in Section 7.5, and the effects on the cross sections and reaction rates are detailed in Appendix B.

Taking that modifications in our BBN code, one can compute the CNO yields as a function of δ_{NN}, as displayed in Figure 7.9(a). The carbon abundance shows a maximum at $\delta_{NN} \approx 0.006$, C/H $\approx 10^{-21}$, which is *six orders of magnitude* below the carbon abundance in standard BBN. This can be understood from Figure 7.9(b), that displays the $^4\mathrm{He}(\alpha\alpha, \gamma)^{12}\mathrm{C}$ rate as a function of δ_{NN} for temperatures relevant to BBN – i.e., from 0.1 to 1 GK. Clearly, the variation of the rate with δ_{NN} is limited at the highest temperatures where BBN production occurs so that the amplification of $^{12}\mathrm{C}$ production does not exceed a few orders of magnitudes. Indeed, while stars can process CNO at 0.1 GK over billions of years, in BBN, the optimal temperature range for producing CNO is passed through in a matter of minutes. This is not sufficient for $^{12}\mathrm{C}$ (CNO) nucleosynthesis in BBN. Furthermore, the baryon density during BBN remains in the range 10^{-5}–0.1 g/cm^3 between 1.0 and 0.1 GK. This makes three-body reactions like $^4\mathrm{He}(\alpha\alpha, \gamma)^{12}\mathrm{C}$ much less efficient compared to two-body reactions.

The maximum of the $^{12}\mathrm{C}$ production as a function of δ_{NN} in Figure 7.9 thus reflects the maxima in the $^8\mathrm{Be}(\alpha, \gamma)^{12}\mathrm{C}$ and $^4\mathrm{He}(\alpha\alpha, \gamma)^{12}\mathrm{C}$ rates. They are due to the effect of

Figure 7.9 Left: ^{12}C production (in number of atoms relative to H) through the $^4\text{He}(\alpha\alpha, \gamma)^{12}\text{C}$ reaction (left) or via a stable ^{8}Be (right) as a function of the N-N interaction. For clarity, the rates of all other CNO producing reactions are set to zero to study these specific channels. The dotted line just connects the results of the two types of calculations: via $^4\text{He}(\alpha\alpha, \gamma)^{12}\text{C}$, as in Reference [44], or via $^4\text{He}(\alpha, \gamma)^8\text{Be}^{\text{stable}}(\alpha, \gamma)^{12}\text{C}$. Right: The $^4\text{He}(\alpha\alpha, \gamma)^{12}\text{C}$ rate as a function of δ_{NN} at constant temperature relevant for BBN from 0.1 GK (lower curve) to 1 GK (upper curve), in steps of 0.1 GK. Taken from Reference [29].

the sharp resonances (in both the ^{8}Be ground state and the ^{12}C 'Hoyle state' that dominate the cross section). The sensitivity of the reaction rates of a sharp resonance is detailed in Appendix D.

To summarise, for $\delta_{NN} \lesssim 0.006$, the rate decreases (increases) as a function of E_R (δ_{NN}) because of the dominating exponential factor, $\exp(-E_R/kT)$, while for $\delta_{NN} \gtrsim 0.006$, it increases (decreases) because of the penetrability. This evolution is followed by the ^{12}C production displayed in Figure 7.9. For $\delta_{NN} \geq 0.00752$, when ^{8}Be is bound, ^{12}C production drops to C/H $\approx 5 \times 10^{-23}$ for $B_8 = 10$. For larger B_8, the abundance drops sharply, as seen in Figure 7.9. For $B_8 = 50$ keV, C/H $\approx 5 \times 10^{-29}$ and is no longer in the range shown in the figure. For $B_8 = 100$ keV, corresponding to $\delta_{NN} = 0.0156$, the Hoyle state is even below threshold, and the production is vanishingly small. If ^{8}Be is bound, reactions that normally produce two α-particles could form ^{8}Be instead. We considered the following reactions

$$^7\text{Be}(n, \gamma)2\alpha \qquad ^7\text{Li}(p, \gamma)2\alpha$$
$$^7\text{Li}(n, \gamma)^8\text{Li}(\beta^+)2\alpha \qquad ^7\text{Be}(d, p)2\alpha$$
$$^7\text{Li}(d, n)2\alpha \qquad ^7\text{Be}(t, np)2\alpha$$
$$^7\text{Be}(^3\text{He}, 2p)2\alpha \qquad ^7\text{Be}(n, \gamma)2\alpha$$
$$^7\text{Li}(^3\text{He}, d)2\alpha \qquad ^7\text{Be}(t, d)2\alpha$$
$$^7\text{Li}(t, 2n)2\alpha \qquad ^7\text{Li}(^3\text{He}, np)2\alpha$$

using the same rates as in Reference [28] but replacing 2α with ^{8}Be. The only significant enhancement comes from the ^{7}Li(d,n)2α reaction, but even in the most favourable case ($B_8 = 10$ keV), C/H reaches $\approx 10^{-21}$. This is still *six orders of magnitude* below the SBBN yield [28] that proceeds via the reactions listed in Eqs. (7.105) and (7.106).

^{3}H(d,n)^{4}He and ^{3}He(d,p)^{4}He reactions. Both reactions proceed through the $A = 5$ nuclei ^{5}He and ^{5}Li and are dominated by a low-energy $\frac{3}{2}^+$ resonance (at $E_R^{\text{exp}} = 0.048$ MeV for ^{5}He and $E_R^{\text{exp}} = 0.21$ MeV for ^{5}Li) and whose properties can be calculated within the same microscopic model that we used for ^{4}He($\alpha\alpha, \gamma$)^{12}C, but with ^{3}H+d and ^{3}He+d cluster structures. Unlike the case for ^{8}Be, the lifetime of the ^{5}He and ^{5}Li states is extremely short (the width of the ^{8}Be ground state is 6 eV, whereas the widths of the $\frac{3}{2}^+$ resonances in ^{5}He and ^{5}Li are of the order of 1 MeV). Therefore, the issue of producing $A = 5$ bound states, or even shifting their ground state energy down to the Gamow window, is not relevant. Even a two-step process, like the 3α reaction, where ^{5}He or ^{5}Li in thermal equilibrium would capture a subsequent nucleon to form ^{6}Li is completely negligible because they are unbound by ~ 1 MeV compared with the 92 keV of ^{8}Be. Hence, no significant equilibrium abundance of $A = 5$ nuclei can be reached.

The computation is similar as the one for the Hoyle state; i.e., (1) determine from the microscopic model the variation of the resonance energy with δ_{NN} and then (2) deduce the effect on the reaction rate.

Using the parameterisation (7.51) for the nucleon-nucleon interaction, we modify the resonance energy. Both the excitation energies of the $\frac{3}{2}^+$ resonance and of the thresholds vary. It is found that

$$\Delta E_R = -0.327 \times \delta_{NN} \tag{7.107}$$

for ^{3}H(d,n)^{4}He and

$$\Delta E_R = -0.453 \times \delta_{NN} \tag{7.108}$$

for ^{3}He(d,p)^{4}He (units are MeV). These energy dependences are much weaker ($\sim$ 20–30 keV for $|\delta_{NN}| \leq 0.03$) than for ^{8}Be and ^{12}C (see Eqs. (7.64) and (7.65)). This is expected for broad resonances which are weakly sensitive to the nuclear interaction [115].

The reaction rates are shown in Figure 7.10. As expected from Eqs. (7.107) and (7.108), they are only slightly affected by variations of δ_{NN} (less than 5%). From the sensitivity study of Reference [25], we deduce that ^{3}H(d,n)^{4}He rate variations have no effect on BBN while ^{3}He(d,p)^{4}He rate variations induce only very small ($\leq$4%) changes in the ^{7}Li and ^{3}He abundances. Because the change is so small, we can make a linear approximation to the sensitivity, as shown in Table 7.2, displaying $(\delta Y/Y)/\delta_{NN}$ values for both reactions.

Constraints on δ_{NN}. These results have been implemented in a BBN code in order to compute the primordial abundances of the light elements as a function of δ_{NN}. Reference [26] concluded that, in terms of δ_{NN}, $-0.7\% < \delta_{NN} < +0.5\%$. To test the importance

Table 7.2 *Abundance sensitivity, $\partial \log Y / \delta_{NN}$, to a variation of the N–N interaction at WMAP baryon density. Blank entries correspond to negligible values.*

Reaction	Y_p	D/H	^{3}He/H	^{7}Li/H
^{3}H(d,n)^{4}He				-0.015
^{3}He(d,p)^{4}He		-0.027	-1.14	-1.10

Figure 7.10 Left: Relative variation of the ^{3}H(d,n)^{4}He (solid line) and ^{3}He(d,p)^{4}He (dashed line) rates for δ_{NN} = -0.030, -0.015, 0.015, and 0.030. Right: Effect of the variation of the N-N interaction induced solely by the modification of the nuclear rates of ^{3}He(d,p)^{4}He and ^{3}H(d,n)^{4}He on the primordial abundances of the light element compared to the constraints obtained in Reference [26]. Figure taken from Reference [29].

of the variations in the $A = 5$ rates, ^{3}He(d,p)^{4}He and ^{3}H(d,n)^{4}He were first considered. We emphasise that a 30% variation in δ_{NN} is unrealistic since it corresponds to a 175% variation on B_D. Given the D and ^{4}He primordial abundances, one gets the constraint

$$-0.0025 < \delta_{NN} < 0.0006. \tag{7.109}$$

Those allowed variations in δ_{NN} are too small to reconcile ^{7}Li abundances with observations, where $\delta_{NN} \approx -0.01$ is required. We can easily extend our analysis by allowing both η_{10} (the most important external parameter) and δ_{NN} to vary. This allows one to set a joint constraint on the two parameters δ_{NN} and baryonic density, as depicted on

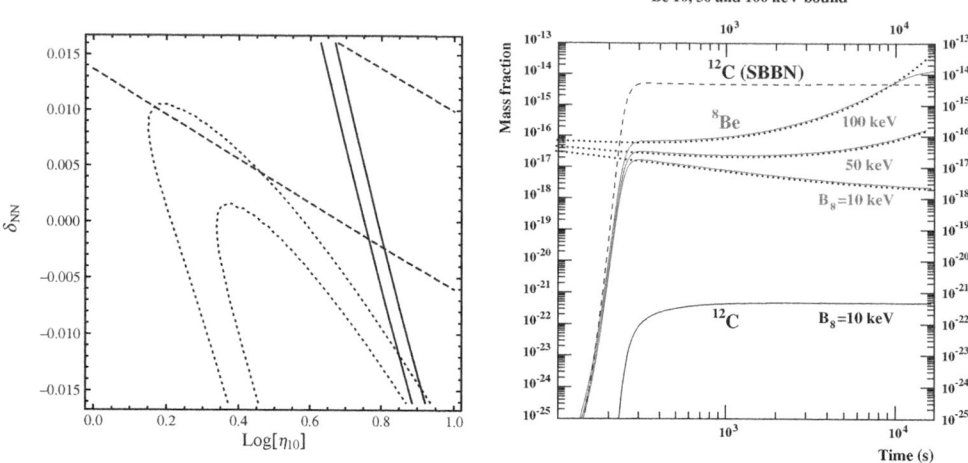

Figure 7.11 Left: Limits on of η_{10} (the number of baryons per 10^{10} photons) and δ_{NN} provided by observational constraints on D (solid) ^{4}He (dash) and ^{7}Li (dot). Right: ^{12}C and ^{8}Be mass fractions as a function of time, assuming ^{8}Be is bound by 100, 50, and 10 keV, as shown by the upper to lower solid curves, respectively [29]. (Only the ^{12}C mass fraction curve, for B_8 = 10 keV, is shown; others are far below the scale shown). The dotted lines correspond to the computation at thermal equilibrium and the dashed line to the SBBN [28] production. Figure taken from Reference [27].

Figure 7.11. No combination of values allow for the simultaneous fulfilment of the ^{4}He, D, and ^{7}Li observational constraints.

Note that the most influential reaction on ^{7}Li is surprisingly [26, 54] n(p,γ)d, as it affects the neutron abundance and the ^{7}Be destruction by neutron capture. The production of ^{12}C by the ^{4}He($\alpha\alpha$, γ)^{12}C, or the ^{4}He(α, γ)^{8}Be and ^{8}Be(α, γ)^{12}C reactions as a function of δ_{NN}, has been investigated. This is to be compared to the CNO (mostly ^{12}C) standard BBN production, CNO/H = (0.5 − 3) × 10^{-15}, in number of atoms relative to hydrogen. A network of ≈ 400 reactions was used, but the main nuclear path to CNO was found to proceed from ^{7}Li(n,γ)^{8}Li(α,n)^{11}B, followed by ^{11}B(p,γ)^{12}C, ^{11}B(d,n)^{12}C, ^{11}B(d,p)^{12}B, and ^{11}B(n,γ)^{12}B reactions. To disentangle the ^{12}C production through the ^{4}He→^{8}Be→^{12}C link, from the standard ^{7}Li→^{8}Li→^{11}B→^{12}C paths, we reduced the network to the reactions involved in A < 8 plus the ^{4}He($\alpha\alpha$, γ)^{12}C, or the ^{4}He(α, γ)^{8}Be and ^{8}Be(α, γ)^{12}C reactions, depending whether or not ^{8}Be would be stable for a peculiar value of δ_{NN}. The carbon abundance shows a maximum at δ_{NN} ≈ 0.006, C/H ≈ 10^{-21} [29], which is *six orders of magnitude* below the carbon abundance in SBBN [28]. This can be understood as the baryon density during BBN remains in the range 10^{-5}–0.1 g/cm^3 between 1.0 and 0.1 GK, substantially lower than in stars (e.g., 30–3,000 g/cm^3 in stars considered by Ekström *et al.* [44]). This makes three-body reactions like ^{4}He($\alpha\alpha$, γ)^{12}C much less efficient compared to two-body reactions. In addition, while stars can produce CNO at 0.1 GK over billions of years, in BBN, the optimal temperature range for producing CNO is passed through in a matter of minutes. Finally, in stars, ^{4}He($\alpha\alpha$, γ)^{12}C operates during the helium-burning phase without significant sources of ^{7}Li, d, p, and n to allow the ^{7}Li$\to$^{8}Li$\to$^{11}B$\to$^{12}C, $A = 8$, bypass process.

Note that the maximum is achieved for $\delta_{NN} \approx 0.006$ when ^{8}Be is still unbound, so contrary to a common belief, a stable ^{8}Be would not have allowed the build-up of heavy elements during BBN. This is illustrated in Figure 7.11 which displays the evolution of the ^{12}C and ^{8}Be mass fractions as a function of time when ^{8}Be is supposed to be bound by 10, 50, and 100 keV (solid lines). They both increase with time until equilibrium between two α-particle fusion and ^{8}Be photodissociation prevails, as shown by the dotted lines. For the highest values of B_8, the ^{8}Be mass fraction increases until, due to the expansion, equilibrium drops out, as shown by the late time behaviour of the upper curve ($B_8 = 100$ keV) in Figure 7.11 (right). For $B_8 \gtrsim 10$ keV, the ^{12}C production falls well below, out of the frame, because the ^{8}Be$(\alpha, \gamma)^{12}$C reaction rate decreases dramatically due to the downward shift of the Hoyle state. For comparison, the SBBN ≈ 400 reactions network (essentially the ^{7}Li$\to$^{8}Li$\to$^{11}B$\to$^{12}C chain) result [28] is plotted (dashed line) in Figure 7.11. It shows that for the ^{4}He$\to$^{8}Be$\to$^{12}C path to give a significant contribution, not only ^{8}Be should have been bound by much more than 100 keV, but also the ^{8}Be$(\alpha, \gamma)^{12}$C rate should have been much higher in order to transform most ^{8}Be in ^{12}C.

7.4.6 Summary

This section has described the dependence of the production of light nuclei during BBN on the fundamental parameters of nuclear physics. The most sensitive parameters are $(Q_{np}, B_D, \tau_n, m_e, \alpha)$.

It can be first concluded (see Figure 7.7) that to be compatible with existing observations, an independent variation of these parameters has to satisfy

$$-8.2 \times 10^{-2} \lesssim \frac{\Delta \tau_n}{\tau_n} \lesssim 6 \times 10^{-2}, \quad -4 \times 10^{-2} \lesssim \frac{\Delta Q_{np}}{Q_{np}} \lesssim 2.7 \times 10^{-2}, \quad (7.110)$$

and

$$-7.5 \times 10^{-2} \lesssim \frac{\Delta B_D}{B_D} \lesssim 6.5 \times 10^{-2}, \quad (7.111)$$

at a 2σ level. The deuterium data set the tighter constraint $-4 \times 10^{-2} \lesssim \Delta \ln B_D \lesssim 3 \times 10^{-2}$.

Similarly, the constraint on the parameter δ_{NN} has been obtained in Figure 7.11. It has also been shown that a stable ^{8}Be would not have allowed the build-up of heavy elements during BBN, and, thus, there is a large abundance of carbon-12.

There exist several public codes to compute BBN abundances: PArthENope,[2] FASTBBN [50, 77],[3] Kawano-Wagoner [72, 114], and AlterBBN.[4] No public code includes a modification of the standard physics.

---

[2] http://parthenope.na.infn.it.
[3] www-thphys.physics.ox.ac.uk/users/SubirSarkar/bbn/fastbbn.f.
[4] http://superiso.in2p3.fr/relic/alterbbn/.

differences arise from the fact that the analytical expression for the reaction rate is no longer accurate (see Appendix D).

The sensitivity to a variation of the intensity of the N-N interaction arises from the fact that $dQ_{\alpha\alpha\alpha}/d\delta_{NN} \sim 10^2 Q_{\alpha\alpha\alpha}$. The fact that the typical correction to the resonant energies is of order 10 MeV ($\times \delta_{NN}$) – compared to the resonant energies themselves, which are of order 0.1 MeV – allows one to put relatively strong constraints on any variation.

Implications on the Stellar Evolution

The Geneva stellar code was adapted to take into account the reaction rates computed earlier. The version of the code we use is the one described in Reference [43]. Here, we only consider models of $15M_\odot$ and $60M_\odot$ without rotation and assume an initial chemical composition given by $X = 0.7514$, $Y = 0.2486$ and $Z = 0$. This corresponds to the BBN abundance of He at the baryon density determined by WMAP [73] and at zero metallicity as is expected to be appropriate for Population III stars. For 16 values of the free parameter δ_{NN} in the range $-0.009 \leq \delta_{NN} \leq +0.006$, we computed a stellar model which was followed up to the end of core He burning (CHeB). As we will see, beyond this range in δ_{NN}, stellar nucleosynthesis is unacceptably altered. Note that for some of the most extreme cases, the set of nuclear reactions now implemented in the code should probably be adapted for a computation of the advanced evolutionary phases.

Focusing on the limited range in δ_{NN} will allow us to study the impact of a change of the fundamental constants on the production of carbon and oxygen in Population III massive stars. In this context, we recall that the observations of the most iron-poor stars in the halo offer a wonderful tool to probe the nucleosynthetic impact of the first massive stars in the Universe. Indeed, these halo stars are believed to form from material enriched by the ejecta of the first stellar generations in the Universe. Their surface chemical composition (at least on the main sequence) still bears the mark of the chemical composition of the cloud from which they formed and, thus, allows us to probe the nucleosynthetic signature of the first stellar generations. Any variation of the fundamental constants – which, for instance, would prevent the synthesis of carbon and/or oxygen – would be very hard to conciliate with present-day observations of the most iron-poor stars. For instance, the two most iron-poor stars [23, 57] both show strong overabundances of carbon and oxygen with respect to iron.

Figure 7.15 (left) shows the HR diagram for the models with δ_{NN} between -0.009 and $+0.006$ in increments of 0.001 (from left to right). Once the CNO cycle has been triggered (as we are about to discuss), the main sequence (MS) tracks are shifted towards cooler T_{eff} for increasing δ_{NN}. There is a difference of about 0.20 dex between the two extreme models. Figure 7.15 (right) shows the central temperature at the moment of the CNO-cycle ignition (lower curve). On the zero-age main sequence (ZAMS), the standard ($\delta_{NN} = 0$) model has not yet produced enough ^{12}C to be able to rely on the CNO cycle, so it starts by continuing its initial contraction until the CNO cycle ignites. In this model, CNO ignition occurs when the central H mass fraction reaches 0.724 – i.e., when less than 3% of the initial H has been burned. Models with $\delta_{NN} < 0$ (i.e., a lower 3α rate) yield a phase of

Figure 7.15 Left: HR diagrams for $15M_\odot$ models with $\delta_{NN} = 0$ (thick black) and ranges from left to right from -0.009 to $+0.006$ in steps of 0.001 (using same colour code as Figure 7.14). Right: The central temperature at CNO ignition (circles) and at the beginning of CHeB (squares) as a function of δ_{NN}. The labels on the CNO-ignition curve show the central H mass fraction at that moment. Note that above $\delta_{NN} \sim +0.001$, CNO ignition occurs on or before the ZAMS. Figure taken from Reference [44].

contraction which is longer for lower δ_{NN} (i.e., larger $|\delta_{NN}|$): in these models, the less-efficient 3α rates need a higher T_c to produce enough ^{12}C for triggering the CNO cycle. Models with $\delta_{NN} > 0$ (i.e., a higher 3α rate) are directly sustained by the CNO cycle on the ZAMS: the star can more easily counteract its own gravity and the initial contraction is stopped earlier, so H burning occurs at lower T_c and ρ_c (Figure 7.15, right), i.e., at a slower pace. The MS lifetime, τ_{MS}, is sensitive to the pace at which H is burned, so it increases with δ_{NN}. The relative difference between the Standard Model MS lifetime τ_{MS} at $\delta_{NN} = 0$ and τ_{MS} at $\delta_{NN} = -0.009$ $(+0.006)$ amounts to -17% $(+19\%)$.

Implications on the Nuclear Constants

While the differences in the 3α rates do not lead to strong effects in the evolution characteristics on the MS, the CHeB phase amplifies the differences between the models. The upper curve of Figure 7.15 (right) shows the central temperature at the beginning of CHeB. There is a factor of 2.8 in temperature between the models with $\delta_{NN} = -0.009$ and $+0.006$. To get an idea of what this difference represents, we can relate these temperatures to the grid of Population III models computed by [80]. The $15M_\odot$ model with $\delta_{NN} = -0.009$ starts its CHeB at a higher temperature than a standard $100\ M_\odot$ of the same stage. In contrast, the model with $\delta_{NN} = +0.006$ starts its CHeB phase with a lower temperature than a standard $12M_\odot$ star at CNO ignition. Table 7.6 presents the characteristics of the models for each value of δ_{NN} at the end of CHeB. From these characteristics, we distinguish four different cases:

[61] Hall, L. J., and Nomura, Y. 2008. 'Evidence for the Multiverse in the Standard Model and Beyond'. *Physical Review D*, **78**(3), 035001.

[62] Harnik, R., Kribs, G. D., and Perez, G. 2006. 'A Universe without Weak Interactions'. *Physical Review D*, **74**(3), 035006.

[63] Heger, A., Langer, N., and Woosley, S. E. 2000. 'Presupernova Evolution of Rotating Massive Stars: I. Numerical Method and Evolution of the Internal Stellar Structure'. *Astrophysical Journal*, **528**(Jan.), 368–396.

[64] Hogan, C. J. 2000. 'Why the Universe Is Just So'. *Reviews of Modern Physics*, **72**(Oct.), 1149–1161.

[65] Hogan, C. J. 2007. *Quarks, Electrons and Atoms in Closely Related Universes*. Cambridge University Press. Page 221–230.

[66] Hoyle, F. 1954. 'On Nuclear Reactions Occuring in Very Hot Stars: I. The Synthesis of Elements from Carbon to Nickel'. *Astrophysical Journal Supplements*, **1**(Sept.), 121.

[67] Ichikawa, K., and Kawasaki, M. 2004. 'Big Bang Nucleosynthesis with a Varying Fine Structure Constant and Nonstandard Expansion Rate'. *Physical Review D*, **69**(12), 123506.

[68] Iliadis, C. 2007. *Nuclear Physics of Stars*. Wiley-VCH Verlag.

[69] Iocco, F., *et al.* 2007. 'Path to Metallicity: Synthesis of CNO Elements in Standard BBN'. *Physical Review D*, **75**(8), 087304.

[70] Iocco, F., Bonifacio, P., and Vangioni, E. 2012. 'Lithium in the Cosmos'. *Memorie della Societa Astronomica Italiana Supplementi*, **22**, 3.

[71] Kamimura, M. 1981. 'Transition Densities between the $0_1^+, 2_1^+, 4_1^+, 0_2^+, 2_2^+, 1_1^-$ and 3_1^- States in ^{12}C Derived from the Three-Alpha Resonating-Group Wave Functions'. *Nuclear Physics A*, **351**(Jan.), 456–480.

[72] Kawano, L. 1992 (Jan.). *Let's Go: Early Universe 2: Primordial Nucleosynthesis the Computer Way*. Fermilab. https://ntrs.nasa.gov/search.jsp?R=19920015920.

[73] Komatsu, E., *et al.* 2009. 'Five-Year Wilkinson Microwave Anisotropy Probe Observations: Cosmological Interpretation'. *Astrophysical Journal Supplements*, **180**(Feb.), 330–376.

[74] Korennov, S., and Descouvemont, P. 2004. 'A Microscopic Three-Cluster Model in the Hyperspherical Formalism'. *Nuclear Physics A*, **740**(Aug.), 249–267.

[75] Langacker, P., Segrè, G., and Strassler, M. J. 2002. 'Implications of Gauge Unification for Time Variation of the Fine Structure Constant'. *Physics Letters B*, **528**(Feb.), 121–128.

[76] Langanke, K., Wiescher, M., and Thielemann, F. K. 1986. 'The Triple-Alpha-Reaction at Low Temperatures'. *Zeitschrift fur Physik A Hadrons and Nuclei*, **324**, 147–152.

[77] Lisi, E., Sarkar, S., and Villante, F. L. 1999. 'Big Bang Nucleosynthesis Limit on N_ν'. *Physical Review D*, **59**(12), 123520.

[78] Livio, M., *et al.* 1989. 'The Anthropic Significance of the Existence of an Excited State of C-12'. *Nature*, **340**(July), 281–284.

[79] Luo, F., Olive, K. A., and Uzan, J. P. 2011. 'Gyromagnetic Factors and Atomic Clock Constraints on the Variation of Fundamental Constants'. *Physical Review D*, **84**(9), 096004.

[80] Marigo, P., *et al.* 2001. 'Zero-Metallicity Stars: I. Evolution at Constant Mass'. *Astronomy and Astrophysics*, **371**(May), 152–173.

[81] Morel, P., and Lebreton, Y. 2008. 'CESAM: A Free Code for Stellar Evolution Calculations'. *Astrophysics and Space Science*, **316**(Aug.), 61–73.

[82] Müller, C. M., Schäfer, G., and Wetterich, C. 2004. 'Nucleosynthesis and the Variation of Fundamental Couplings'. *Physical Review*, **70**(Oct), 083504.

[83] Nakamura, K., and Particle Data Group. 2010. 'Review of Particle Physics'. *Journal of Physics G: Nuclear Physics*, **37**(7), 075021.

[84] Navrátil, P., *et al.* 2009. 'Topical Review: Recent Developments in No-Core Shell-Model Calculations'. *Journal of Physics G: Nuclear Physics*, **36**(8), 083101.

[85] Newton, J. R., *et al.* 2007. 'Gamow peak in Thermonuclear Reactions at High Temperatures'. *Physical Review C*, **75**(4), 045801.

[86] Nollett, K. M., and Lopez, R. E. 2002. 'Primordial Nucleosynthesis with a Varying Fine Structure Constant: An Improved Estimate'. *Physical Review D*, **66**(6), 063507.

[87] Nomoto, K., Thielemann, F. K., and Miyaji, S. 1985. 'The Triple Alpha Reaction at Low Temperatures in Accreting White Dwarfs and Neutron Stars'. *Astronomy and Astrophysics*, **149**(Aug.), 239–245.

[88] Oberhummer, H., Csótó, A., and Schlattl, H. 2000. 'Stellar Production Rates of Carbon and Its Abundance in the Universe'. *Science*, **289**(July), 88–90.

[89] Oberhummer, H., Csótó, A., and Schlattl, H. 2001. 'Bridging the Mass Gaps at /A=5 and /A=8 in Nucleosynthesis'. *Nuclear Physics A*, **689**(June), 269–279.

[90] Oberhummer, H., *et al.* 2003. 'Temporal Variation of Coupling Constants and Nucleosynthesis'. *Nuclear Physics A*, **719**(May), C283–C286.

[91] Peebles, P. J. E. 1966. 'Primordial Helium Abundance and the Primordial Fireball. II'. *Astrophysical Journal*, **146**(Nov.), 542.

[92] Peter, P., and Uzan, J. 2009. *Primordial Cosmology*. Oxford Graduate Texts. Oxford University Press.

[93] Pettini, M. and Cooke, R. 2012. 'A New, Precise Measurement of the Primordial Abundance of Deuterium'. *Monthly Notices of the Royal Astronomical Society*, **425**(Oct.), 2477–2486.

[94] Reeves, H. 1968. *Stellar Evolution and Nucleosynthesis*. Cambridge University Press.

[95] Rohlf, J. 1994. *Modern Physics from [α] to Z*. Wiley.

[96] Rozental, I. L. 1988. *Big Bang, Big Bounce: How Particles and Fields Drive Cosmic Evolution*. Springer.

[97] Sánchez-Cañizares, J. 2017. 'How Can Physics Underlie the Mind? Top-Down Causation in the Human Context, by G. Ellis'. *Contemporary Physics*, **58**(Oct.), 356–356.

[98] Sbordone, L., *et al.* 2010. 'The Metal-Poor End of the Spite Plateau: I. Stellar Parameters, Metallicities, and Lithium Abundances'. *Astronomy and Astrophysics*, **522**(Nov.), A26.

[99] Schlattl, H., *et al.* 2004. 'Sensitivity of the C and O Production on the 3α Rate'. *Astrophysics and Space Science*, **291**(Apr.), 27–56.

[100] Serot, B. D., and Walecka, J. D. 1997. 'Recent Progress in Quantum Hadrodynamics'. *International Journal of Modern Physics E*, **6**, 515–631.

[101] Spite, F. and Spite, M. 1982. 'Abundance of Lithium in Unevolved Halo Stars and Old Disk Stars: Interpretation and Consequences'. *Astronomy and Astrophysics*, **115**(Nov.), 357–366.

[102] Suzuki, Y., *et al.* 2008. 'Local versus Nonlocal $\alpha\alpha$ Interactions in a 3 α Description of ^{12}C'. *Physics Letters B*, **659**(Jan.), 160–164.

[103] Symposium, I. A. U., *et al.* 2010. *Light Elements in the Universe (IAU S268)*. IAU Symposium and Colloquium Proceedings Series. Cambridge University Press.

[104] Tegmark, M. 1998. 'Is "the Theory of Everything" Merely the Ultimate Ensemble Theory?' *Annals of Physics*, **270**(Nov.), 1–51.

[105] Thompson, D. R., Lemere, M., and Tang, Y. C. 1977. 'Systematic Investigation of Scattering Problems with the Resonating-Group Method'. *Nuclear Physics A*, **286**(Aug.), 53–66.

[106] Tilley, D., *et al.* 2004. 'Energy Levels of Light Nuclei A=8,9,10'. *Nuclear Physics A*, **745**(3), 155–362.

[107] Uzan, J. P. 2003a. 'Tests of Gravity on Astrophysical Scales and Variation of the Constants'. *Annales Henri Poincaré*, **4**(Dec.), 347–369.

[108] Uzan, J. P. 2003b. 'The Fundamental Constants and Their Variation: Observational and Theoretical Status'. *Reviews of Modern Physics*, **75**(Apr.), 403–455.

[109] Uzan, J. P. 2009. 'Fundamental Constants and Tests of General Relativity: Theoretical and Cosmological Considerations'. *Space Science Reviews*, **148**(Dec.), 249–265.

[110] Uzan, J. P. 2011. 'Varying Constants, Gravitation and Cosmology'. *Living Reviews in Relativity*, **14**(Mar.), 2.

[111] Uzan, J. P. 2013. 'Models of the Cosmos and Emergence of Complexity'. In *The Causal Universe*. Copernicus Center Press. Pages 93–120.

[112] Vangioni-Flam, E., *et al.* 2003. 'On the Baryometric Status of ^{3}He'. *Astrophysical Journal*, **585**(Mar.), 611–616.

[113] Villani, C., Uzan, J., and Moncorgé, V. 2014. *La maison des mathématiques*. Beaux livres. Le Cherche Midi.

[114] Wagoner, R. V. 1973. 'Big-Bang Nucleosynthesis Revisited'. *Astrophysical Journal*, **179**(Jan.), 343–360.

[115] Wildermuth, K., and T'ang, Y. C. 1977. *A Unified Theory of the Nucleus*. Vieweg Braunschweig.

[116] Woosley, S. E., and Heger, A. 2015. 'The Remarkable Deaths of 9-11 Solar Mass Stars'. *Astrophysical Journal*, **810**(Sept.), 34.

[117] Woosley, S. E., Heger, A., and Weaver, T. A. 2002. 'The Evolution and Explosion of Massive Stars'. *Reviews of Modern Physics*, **74**(Nov.), 1015–1071.

[118] Yoo, J. J., and Scherrer, R. J. 2003. 'Big Bang Nucleosynthesis and Cosmic Microwave Background Constraints on the Time Variation of the Higgs Vacuum Expectation Value'. *Physical Review D*, **67**(4), 043517.

Appendix A: **Summary of the Formation Sites of Atomic Isotopes**

200

E. Anders and N. Grevesse

Table 3. Abundance of the Nuclides (Atoms/10^6 Si)

Element, A		Atom Percent	Process*	Abund.†
1 H	1	99.9966		2.79×10^{10}
	2	0.0034	U	9.49×10^{5}
2 He	3	0.0142	U,h?	3.86×10^{5}
	4	99.9858	U,h	2.72×10^{9}
3 Li	6	7.5	X	4.28
	7	92.5	U,x,h	52.82
4 Be	9	100	X	0.73
5 B	10	19.9	X	4.22
	11	80.1	X	16.98
6 C	12	98.90	He	9.99×10^{6}
	13	1.10	H,N	1.11×10^{5}
7 N	14	99.634	H,N	3.12×10^{6}
	15	0.366	H,N	1.15×10^{4}
8 O	16	99.762	He	2.37×10^{7}
	17	0.038	N,H	9.04×10^{3}
	18	0.200	He,N	4.76×10^{4}
9 F	19	100	N	843
10 Ne	20	92.99	C	3.20×10^{6}
	21	0.226	C,Ex	7.77×10^{3}
	22	6.79	He,N	2.34×10^{5}
11 Na	23	100	CNeEx	5.74×10^{4}
12 Mg	24	78.99	N,O	8.48×10^{5}
	25	10.00	NeExC	1.07×10^{5}
	26	11.01	NeExC	1.18×10^{5}
13 Al	27	100	Ne,Ex	8.49×10^{4}
14 Si	28	92.23	O,Ex	9.22×10^{5}
	29	4.67	Ne,Ex	4.67×10^{4}
	30	3.10	Ne,Ex	3.10×10^{4}
15 P	31	100	Ne,Ex	1.04×10^{4}
16 S	32	95.02	O,Ex	4.89×10^{5}
	33	0.75	Ex	3.86×10^{3}
	34	4.21	O,Ex	2.17×10^{4}
	36	0.02	Ex,Ne,S	1.03×10^{2}
17 Cl	35	75.77	Ex	2860
	37	24.23	Ex,C,S	913
18 Ar	36	84.2	Ex	8.50×10^{4}
	38	15.8	O,Ex	1.60×10^{4}
	40		S,Ne	26
	40			*25±14*
19 K	39	93.2581	Ex	3516
	40	0.01167	S,Ex,Ne	0.440
	40			*5.48*
	41	6.7302	Ex	253.7
20 Ca	40	96.941	Ex	5.92×10^{4}
	42	0.647	Ex,O	395
	43	0.135	Ex,C,S	82.5
	44	2.086	Ex,S	1275
	46	0.004	ExCNe	2.4
	48	0.187	E,Ex	114
21 Sc	45	100	ExNeE	34.2
22 Ti	46	8.0	Ex	192
	47	7.3	Ex	175
	48	73.8	Ex	1771
	49	5.5	Ex	132
	50	5.4	E	130
23 V	50	0.250	Ex,E	0.732
	51	99.750	Ex	292
24 Cr	50	4.345	Ex	587
	52	83.789	Ex	1.131×10^{4}
	53	9.501	Ex	1283
	54	2.365	E	319
25 Mn	55	100	Ex,E	9550
26 Fe	54	5.8	Ex	5.22×10^{4}
	56	91.72	Ex,E	8.25×10^{5}
	57	2.2	E,Ex	1.98×10^{4}
	58	0.28	He,E,C	2.52×10^{3}
27 Co	59	100	E,C	2250
28 Ni	58	68.27	E,Ex	3.37×10^{4}
	60	26.10	E	1.29×10^{4}
	61	1.13	E,Ex,C	557
	62	3.59	E,Ex,C	1770
	64	0.91	Ex	449
29 Cu	63	69.17	Ex,C	361
	65	30.83	Ex	161
30 Zn	64	48.63	Ex,E	613
	66	27.90	E,S	352
	67	4.10	E,S	51.7
	68	18.75	E,S	236
	70	0.62	E,S	7.8
31 Ga	69	60.108	S,e,r	22.7
	71	39.892	S,e,r	15.1
32 Ge	70	20.5	S,e	24.4
	72	27.4	S,e,r	32.6
	73	7.8	e,s,r	9.28
	74	36.5	e,s,r	43.4
	76	7.8	E	9.28
33 As	75	100	R,s	6.56
34 Se	74	0.88	P	0.55
	76	9.0	S,p	5.6
	77	7.6	R,s	4.7
	78	23.6	R,s	14.7
	80	49.7	R,s	30.9
	82	9.2	R	5.7
35 Br	79	50.69	R,s	5.98
	81	49.31	R,s	5.82
36 Kr	78	0.339	P	0.153
	80	2.22	S,p	0.999
	82	11.45	S	5.15
	83	11.47	R,s	5.16
	84	57.11	R,s	25.70
	86	17.42	S,r	7.84
37 Rb	85	72.165	R,s	5.12
	87	27.835	S	1.97
	87			*2.11*
38 Sr	84	0.56	P	0.132
	86	9.86	S	2.32
	87	7.00	S	1.64
	87			*1.51*
	88	82.58	S,r	19.41
39 Y	89	100	S	4.64
40 Zr	90	51.45	S	5.87
	91	11.22	S	1.28
	92	17.15	S	1.96
	94	17.38	S	1.98
	96	2.80	R	0.320
41 Nb	93	100	S	0.698
42 Mo	92	14.84	P	0.378
	94	9.25	P	0.236
	95	15.92	R,s	0.406
	96	16.68	S	0.425
	97	9.55	R,s	0.244
	98	24.13	R,s	0.615
	100	9.63	R	0.246
44 Ru	96	5.52	P	0.103
	98	1.88	P	0.0350
	99	12.7	R,s	0.236
	100	12.6	S	0.234
	101	17.0	R,s	0.316
	102	31.6	R,S	0.588
	104	18.7	R	0.348
45 Rh	103	100	R,s	0.344
46 Pd	102	1.020	P	0.0142
	104	11.14	S	0.155
	105	22.33	R,s	0.310
	106	27.33	R,S	0.380
	108	26.46	R,S	0.368
	110	11.72	R	0.163
47 Ag	107	51.839	R,s	0.252
	109	48.161	R,s	0.234
48 Cd	106	1.25	P	0.0201
	108	0.89	P	0.0143
	110	12.49	S	0.201
	111	12.80	R,S	0.206
	112	24.13	S,R	0.388
	113	12.22	R,S	0.197
	114	28.73	R,S	0.463
	116	7.49	R	0.121
49 In	113	4.3	p,s,r	0.0079
	115	95.7	R,S	0.176
50 Sn	112	0.973	P	0.0372
	114	0.659	P,s	0.0252
	115	0.339	p,s,r	0.0129
	116	14.538	S,r	0.555
	117	7.672	R,S	0.293
	118	24.217	S,r	0.925
	119	8.587	S,R	0.328
	120	32.596	S,R	1.245
	122	4.632	R	0.177
	124	5.787	R	0.221
51 Sb	121	57.362	R,s	0.177
	123	42.638	R	0.132
52 Te	120	0.09	P	0.0043
	122	2.57	S	0.124
	123	0.89	S	0.0428
	124	4.76	S	0.229
	125	7.10	R,s	0.342
	126	18.89	R,S	0.909
	128	31.73	R	1.526
	130	33.97	R	1.634
53 I	127	100	R	0.90
54 Xe	124	0.121	P	0.00571
	126	0.108	P	0.00509
	128	2.19	S	0.103
	129	27.34	R	1.28
	130	4.35	S	0.205
	131	21.69	R	1.02
	132	26.50	R,s	1.24
	134	9.76	R	0.459
	136	7.94	R	0.373
55 Cs	133	100	R,s	0.372
56 Ba	130	0.106	P	0.00476
	132	0.101	P	0.00453
	134	2.417	S	0.109
	135	6.592	R,s	0.296
	136	7.854	S	0.353
	137	11.23	S,r	0.504
	138	71.70	S	3.22
57 La	138	0.089	P	0.000397
	138			*0.000409*
	139	99.911	S,r	0.446
58 Ce	136	0.19	P	0.00216
	138	0.25	P	0.00284
	138			*0.00283*
	140	88.48	S,r	1.005
	142	11.08	R	0.126
59 Pr	141	100	R,S	0.167
60 Nd	142	27.13	P	0.225
	143	12.18	R,S	0.101
	143			*0.100*
	144	23.80	S,R	0.197
	145	8.30	R,s	0.0687
	146	17.19	R,S	0.142
	148	5.76	R	0.0477
	150	5.64	R	0.0467
62 Sm	144	3.1	P	0.00800
	147	15.0	R,s	0.0387
	147			*0.0399*
	148	11.3	S	0.0292
	149	13.8	R,S	0.0356
	150	7.4	S	0.0191
	152	26.7	R,S	0.0689
	154	22.7	R	0.0586
63 Eu	151	47.8	R,s	0.0465
	153	52.2	R,s	0.0508
64 Gd	152	0.20	P,s	0.00066
	154	2.18	S	0.00719
	155	14.80	R,s	0.0488
	156	20.47	R,s	0.0676
	157	15.65	R,s	0.0516
	158	24.84	R,s	0.0820
	160	21.86	R	0.0721
65 Tb	159	100	R	0.0603
66 Dy	156	0.056	P	0.000221
	158	0.096	P	0.000378
	160	2.34	S	0.00922
	161	18.91	R	0.0745
	162	25.51	R,s	0.101
	163	24.90	R	0.0982
	164	28.19	R,S	0.111
67 Ho	165	100	R	0.0889
68 Er	162	0.14	P	0.000351
	164	1.61	P,S	0.00404
	166	33.6	R,s	0.0843
	167	22.95	R	0.0576
	168	26.8	R,S	0.0672
	170	14.9	R	0.0374
69 Tm	169	100	R,s	0.0378
70 Yb	168	0.13	P	0.000322
	170	3.05	S	0.00756
	171	14.3	S	0.0354
	172	21.9	R,S	0.0543
	173	16.12	R,s	0.0400
	174	31.8	S,R	0.0788
	176	12.7	R	0.0315
71 Lu	175	97.41	R,s	0.0357
	176	2.59	S	0.000951
	176			*0.001035*
72 Hf	174	0.162	P	0.000249
	176	5.206	S	0.00802
	176			*0.00793*
	177	18.606	R,s	0.0287
	178	27.297	R,s	0.0420
	179	13.629	R,s	0.0210
	180	35.100	S,R	0.0541
73 Ta	180	0.012	p,s,r	2.48×10^{-6}
	181	99.988	R,S	0.0207
74 W	180	0.13	P	0.000173
	182	26.3	R,s	0.0350
	183	14.3	R,s	0.0190
	184	30.67	R,s	0.0408
	186	28.6	R	0.0380
75 Re	185	37.40	R,s	0.0193
	187	62.60	R	0.0324
	187			*0.0351*
76 Os	184	0.018	P	0.000122
	186	1.58	S	0.0107
	187	1.6	S	0.0108
	187			*0.00807*
	188	13.3	R,s	0.0898
	189	16.1	R	0.109
	190	26.4	R	0.178
	192	41.0	R	0.277
77 Ir	191	37.3	R	0.247
	193	62.7	R	0.414
78 Pt	190	0.0127	P	0.000170
	192	0.78	S	0.0105
	194	32.9	R	0.441
	195	33.8	R	0.453
	196	25.2	R	0.338
	198	7.19	R	0.0963
79 Au	197	100	R	0.187
80 Hg	196	0.1534	P	0.00052
	198	9.968	S	0.0339
	199	16.873	R,S	0.0574
	200	23.096	S,r	0.0785
	201	13.181	S,r	0.0448
	202	29.863	S,r	0.1015
	204	6.865	R	0.0233
81 Tl	203	29.524	R,S	0.0543
	205	70.476	S,R	0.1297
82 Pb	204	1.94	S	0.0611
	206	19.12	R,S	0.602
	206			*0.593*
	207	20.62	R,S	0.650
	207			*0.644*
	208	58.31	R,s	1.837
	208			*1.828*
83 Bi	209	100	R,s	0.144
90 Th	232	100	RA	0.0335
	232			*0.0420*
92 U	235	0.7200	RA	6.48×10^{-5}
	235			*0.00573*
	238	99.2745	RA	0.00893
	238			*0.0181*

*Assignments to nucleosynthetic processes are from Cameron (1982), Schramm (private communication, 1982), Walter et al. (1989), Woosley and Hoffman (1986, 1989), and Beer and Penzhorn (1987). Processes are listed in the order of importance, with minor processes (10-30% for r- and s-processes) shown in lower case. See above references for details.

U = cosmological nucleosynthesis
H = hydrogen burning
N = hot or explosive hydrogen burning
He = helium burning
C = carbon burning
O = oxygen burning
Ne = neon burning
Ex = explosive nucleosynthesis
E = nuclear statistical equilibrium
S = s-process
R = r-process
RA = r-process producing actinides
P = p-process
X = cosmic-ray spallation

† Italicized values refer to abundances 4.55×10^{9} yr ago.

Figure 7.17 Abundance of the nucleides (atoms/10^6 Si) with the process of production in our Universe. Figure taken from Reference [2].

Appendix B: **Details of the Microscopic Model**

Here, we provide some technical details about the microscopic calculation used to determine the ^{8}Be and ^{12}C binding energies. This calculation is based on the description of the nucleon-nucleon interaction by the Minnesota (MN) force [105], adapted to low-mass systems.

The nuclear part of the interaction potential V_N between nucleons i and j is given by

$$V_{Nij}(r) = \left[V_R(r) + \frac{1}{2}(1 + P_{ij}^\sigma)V_t(r) + \frac{1}{2}(1 - P_{ij}^\sigma)V_s(r) \right] \left[\frac{1}{2}u + \frac{1}{2}(2 - u)P_{ij}^r \right],$$

(7.131)

where $r = |\mathbf{r}_i - \mathbf{r}_j|$ and P_{ij}^σ and P_{ij}^r are the spin and space exchange operators, respectively. The radial potentials $V_R(r), V_s(r), V_t(r)$ are expressed as Gaussians and have been optimized to reproduce various properties of the nucleon-nucleon system – such as the deuteron binding energy at $\delta_{NN} = 0$, or the low-energy phase shifts. They have been fit as [105]

$$V_R(r) = 200 \ \exp(-1.487r^2)$$
$$V_s(r) = -91.85 \ \exp(-0.465r^2)$$
$$V_t(r) = -178 \ \exp(-0.639r^2),$$

(7.132)

where energies are expressed in MeV and lengths in fm.

In Eq. (7.131), the exchange-admixture parameter u takes standard value $u = 1$ but can be slightly modified to reproduce important properties of the A-nucleon system (for example, the energy of a resonance). This does not affect the physical properties of the interaction. The MN force is an effective interaction, adapted to cluster models. It is not aimed at perfectly reproducing all nucleon-nucleon properties, as realistic forces used in ab initio models [84], where the cluster approximation is not employed. The potentials are expressed as Gaussian factors, well adapted to cluster models, where the nucleon orbitals are also Gaussians [115].

The wave functions (Eq. (7.50)) are written in the resonating group method (RGM), which clearly shows the factorisation of the system wave function in terms of individual cluster wave functions. In practice the radial wave functions are expanded over Gaussians, which provides the generator coordinate method (GCM), fully equivalent to the RGM [115] but better adapted to numerical calculations. Some details are given here for the simpler two cluster case. The radial function $g_2^{JM\pi}(\rho)$ is written as a sum over Gaussian functions centered at different values of the generator coordinate R_n. This allows us to write the ^{8}Be wave function (7.50) as

$$\Psi_{^8Be}^{JM\pi} = \sum_n f^{J\pi}(R_n)\Phi^{JM\pi}(R_n),$$

(7.133)

where $\Phi^{JM\pi}(R_n)$ is a projected Slater determinant.

This development corresponds to a standard expansion on a variational basis. The binding energies $E^{J\pi}$ of the system are obtained by diagonalisation of

$$\sum_n \left[H^{J\pi}(R_n, R_{n'}) - E^{J\pi} N^{J\pi}(R_n, R_{n'}) \right] f^{J\pi}(R_n) = 0, \tag{7.134}$$

where the overlap and Hamiltonian kernels are defined as

$$N^{J\pi}(R_n, R_{n'}) = \langle \Phi^{J\pi}(R_n) | \Phi^{J\pi}(R_{n'}) \rangle,$$
$$H^{J\pi}(R_n, R_{n'}) = \langle \Phi^{J\pi}(R_n) | H | \Phi^{J\pi}(R_{n'}) \rangle. \tag{7.135}$$

The Hamiltonian H is given by Eq. (7.47). Standard techniques exist for the evaluation of these many-body matrix element. The choice of the nucleon-nucleon interaction directly affects the calculation of the hamiltonian kernel and, therefore, of the eigen-energy $E^{J\pi}$.

For three-body wave functions, the theoretical developments are identical, but the presentation is more complicated due to the presence of two relative coordinates (ρ, R). The problem is addressed by using the hyperspherical formalism [74].

Appendix C: **Numerical Integration of the Cross Sections**

To take into account the (energy-dependent) finite widths of the two resonances involved in this two-step process, one has to perform numerical integrations as was done in NACRE following References [76, 87]. Here, the condition of thermal equilibrium is relaxed, but it is assumed that the timescale for alpha capture on ^{8}Be is negligible compared to its lifetime against alpha decay. The rate is calculated as in NACRE for the resonance of interest:

$$N_A^2 \langle \sigma v \rangle^{\alpha\alpha\alpha} = 3N_A \left(\frac{8\pi\hbar}{\mu_{\alpha\alpha}^2} \right) \left(\frac{\mu_{\alpha\alpha}}{2\pi k_B T} \right)^{3/2} \int_0^\infty \frac{\sigma_{\alpha\alpha}(E)}{\Gamma_\alpha(E)} \exp(-E/k_B T) N_A \langle \sigma v \rangle^{\alpha^8 Be} \, E \, dE,$$

where $\mu_{\alpha\alpha}$ is the reduced mass of the $\alpha + \alpha$ system, and E is the energy with respect to the $\alpha + \alpha$ threshold. The elastic cross section of $\alpha + \alpha$ scattering is given by a Breit-Wigner expression:

$$\sigma_{\alpha\alpha}(E) = \frac{\pi}{k^2} \omega \frac{\Gamma_\alpha^2(E)}{(E - E_R)^2 + \Gamma_\alpha^2(E)/4}, \tag{7.136}$$

where k is the wave number, $E_R \equiv E_R(^8Be)$, $\Gamma_\alpha \equiv \Gamma_\alpha(^8Be)$, and ω is a statistical factor (here, equal to 2 to account for identical particles with spin zero).

The $N_A \langle \sigma v \rangle^{\alpha^8 Be}$ rate assumes that ^{8}Be has been formed at an energy E different from $E_{^8Be}$ [76]. This rate is given by

$$N_A \langle \sigma v \rangle^{\alpha^8 Be} = N_A \frac{8\pi}{\mu_{\alpha^8 Be}^2} \left(\frac{\mu_{\alpha^8 Be}}{2\pi k_B T} \right)^{3/2} \int_0^\infty \sigma_{\alpha^8 Be}(E'; E) \exp(-E'/k_B T) \, E' \, dE',$$

$$\tag{7.137}$$

where $\mu_{\alpha^8\text{Be}}$ is the reduced mass of the $\alpha + {}^8\text{Be}$ system, and E' is the energy with respect to its threshold (which varies with the formation energy E). As in References [76, 87], we parametrise $\sigma_{\alpha^8\text{Be}}(E'; E)$ as

$$\sigma_{\alpha^8\text{Be}}(E'; E) = \frac{\pi \hbar^2}{2\mu_{\alpha^8\text{Be}} E'} \frac{\Gamma_\alpha(E')\Gamma_\gamma(E' + E)}{[E' - E_R({}^{12}\text{C}) + E - E_R({}^8\text{Be})]^2 + \frac{1}{4}\Gamma(E'; E)^2}, \quad (7.138)$$

where the partial widths are those of the Hoyle state and, in particular, $\Gamma = \Gamma_\alpha({}^{12}\text{C}) + \Gamma_\gamma({}^{12}\text{C})$. The various integrals are calculated numerically. The experimental widths at resonance energy can be found in Table 7.5.

However, one must include (1) the energy dependence of those widths, away from the resonance energy and (2) the variation of the widths at the resonant energy when this energy changes due to a change in the nuclear interaction.

The energy dependence of the particle widths $\Gamma_\alpha(E)$ is given by

$$\Gamma_\alpha(E) = \Gamma_\alpha(E_R) \frac{P_\ell(E, R_c)}{P_\ell(E_R, R_c)}, \quad (7.139)$$

where P_ℓ is the penetration factor associated with the relative angular momentum, ℓ (0 here), and the channel radius, R_c.[9] The penetration factor is related to the Coulomb functions by

$$P_\ell(E, R) = \frac{\rho}{F_\ell^2(\eta, \rho) + G_\ell^2(\eta, \rho)}, \quad (7.140)$$

where $\rho = kR$ and

$$\eta = \frac{Z_1 Z_2 \alpha}{v/c} \quad (7.141)$$

is the Sommerfeld parameter.

For radiative capture reactions, the energy dependence of the gamma width $\Gamma_\gamma(E)$ is given by

$$\Gamma_\gamma(E) \propto \alpha E^{2\lambda+1}, \quad (7.142)$$

where λ is the multipolarity (here, 2 for $E2$) of the electromagnetic transition.

The relevant widths as a function of δ_{NN} are given in Figure 7.18. They are directly linked to the resulting change of $E_R({}^8\text{Be})$ and $E_R({}^{12}\text{C})$.

The radiative width, $\Gamma_\gamma({}^{12}\text{C})$, with its E^5 energy dependence, shows little evolution. (The energy of the final state at 4.44 MeV is assumed to be constant). In contrast, the ${}^8\text{Be}$ alpha width undergoes large variations due to the effect of Coulomb barrier penetrability. Note that compared to these variations, those induced by a change of α in the Coulomb barrier penetrability (Eqs. (7.141) and (7.140)) and Γ_γ are considerably smaller.

---

[9] We choose $R_c = 1.3\,(A_1^{1/3} + A_2^{1/3})$ fm, for nuclei A_1 and A_2.

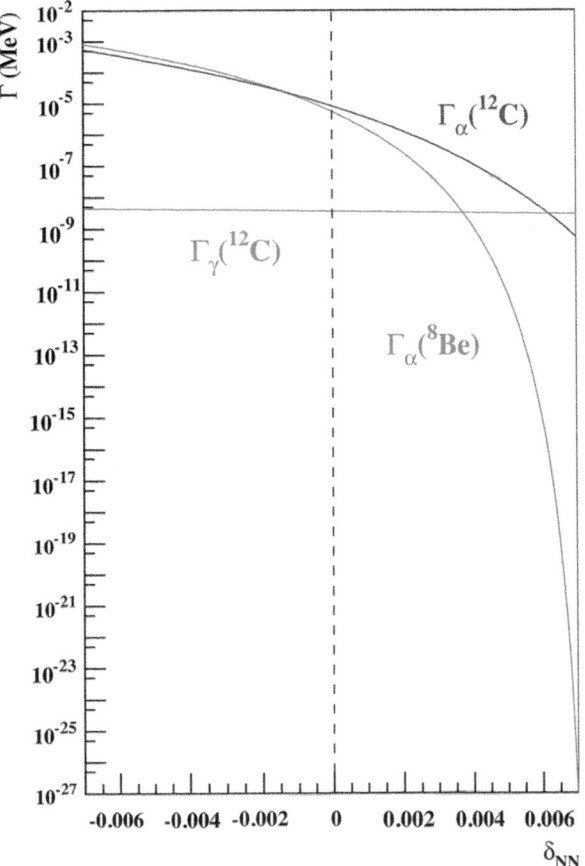

Figure 7.18 The partial widths of ^{8}Be and ^{12}C as a function of δ_{NN}. Figure taken from Reference [44].

Numerical integration is necessary at low temperatures as the reaction takes place through the low-energy wing of resonances. It takes even more relative importance at a given temperature when the resonance energy is shifted upwards. On the other hand, when δ_{NN} increases, the resonance energies decrease, and the $\Gamma_\alpha(^8\text{Be})$ becomes so small that the numerical integration becomes useless and soon gives erroneous results because of the finite numerical resolution. For this reason, when $\Gamma_\alpha(^8\text{Be}) < 10^{-8}$ MeV, we use instead the Saha equation for the first step and the sharp resonance approximation for the second step – i.e., Eq. (7.126) when $\Gamma_\alpha(^{12}\text{C}) < 10^{-8}$ MeV. (Note that for high values of δ_{NN}, the condition $\Gamma_\gamma \ll \Gamma_\alpha$ does not hold anymore and $\gamma \neq \Gamma_\gamma$.)

At temperatures in excess of $T_9 \simeq 2$, one must include the contribution of the higher ^{12}C levels like the one observed by [59]. As this is not of importance for this study; we just

Figure 7.19 The $^4\text{He}(\alpha\alpha, \gamma)^{12}\text{C}$ reaction rate as a function of temperature for different values of δ_{NN}. Solid (dashed) lines represent the result of the numerical calculation (analytical approximation) with $\delta_{NN} = 0$ (centre), $\delta_{NN} > 0$ (above) and $\delta_{NN} < 0$ (below). (For the negative value of δ_{NN}, the larger difference is caused by the failure of the numerical integration and the analytical solution is preferred (see text). Figure taken from Reference [44].

added the contribution given by the last terms in the NACRE analytical approximation and neglected any induced variation.

Figure 7.19 shows the numerically integrated $^4\text{He}(\alpha\alpha, \gamma)^{12}\text{C}$ reaction rates for different values of δ_{NN} compared with the analytical approximation Eq. (7.126). The difference is important at low temperatures and small δ_{NN} values but becomes negligible for $\delta_{NN} \gtrsim 0$. At the highest values of δ_{NN}, we consider the numerical calculation uses the Saha equation for the first step, but the total widths of the ^{12}C level also becomes too small to be accurately numerically calculated; we use Eq. (7.126) instead.

The ^8Be lifetime with respect to alpha decay, $(h/\Gamma_\alpha(^8\text{Be}))$, exhibits the opposite behaviour, indicating that, for large values of δ_{NN}, it becomes stable. Before that, its lifetime is so long that the $^4\text{He}(\alpha\alpha, \gamma)^{12}\text{C}$ reaction should be considered as a real two-step process with ^8Be included in the network, as the assumption that alpha decay is much faster than alpha capture may not hold anymore. Fortunately, our network calculations show that this situation is encountered only for $\delta_{NN} \gtrsim 0.006$ for the temperatures and densities considered in our stellar evolution studies.

Appendix D: Sensitivity of the Nuclear Reaction Rates on the Resonance Energy

The contribution of a sharp resonance in the (α, γ) channel is given by

$$\langle \sigma v \rangle \propto \frac{\Gamma_\alpha(E_R)\Gamma_\gamma(E_R)}{\Gamma(E_R)} \exp\left(-\frac{E_R}{k_B T}\right), \tag{7.143}$$

where Γ_α is the entrance (alpha-)width, Γ_γ is the exit (gamma-)width, and Γ is the total width ($\Gamma = \Gamma_\alpha + \Gamma_\gamma$ if these are the only open channels). Figure 1 in Reference [44] displays the ^{12}C level scheme: the radiative width, Γ_γ, is associated with the decay to the first ^{12}C excited state at 4.44 MeV as the decay to the ground state proceeds only through the much less efficient electron-positron pair emission. The corresponding decay energy is then $E_\gamma(E_R) = 3.21$ MeV $+ \Delta E_R(^{12}$C$)$. Equation (7.143) shows that for a fixed E_R – i.e., δ_{NN} – the contribution increases with temperature, as seen in Figures 7.9. While the radiative width $\Gamma_\gamma(E_R) \propto E_\gamma^{2\ell+1}$ is almost insensitive to $E_R(\delta_{NN})$, $\Gamma_\alpha(E_R)$ is very sensitive to $E_R(\delta_{NN})$ variations because of the Coulomb and centrifugal barriers penetrability, $P_\ell(E)$. The reduced widths γ_x^2, defined by

$$\Gamma_x(E) = 2\gamma_x^2 P_\ell(E) \quad (x \neq \gamma), \tag{7.144}$$

are corrected for these effects so that they reflect the nuclear properties only and are, as a good approximation, independent of δ_{NN}.

Depending on whether $\Gamma_\alpha \ll \Gamma_\gamma$ or $\Gamma_\gamma \ll \Gamma_\alpha$, the sensitivity of $\langle \sigma v \rangle$ (Eq. (7.143)) to E_R or δ_{NN} variations is very different. This is due to the very different energy dependence of Γ_α and Γ_γ, as discussed in detail in Reference [85]. In the latter case, using Eq. (7.143), the sensitivity of the rate to E_R (δ_{NN}) variations is simply given by

$$\frac{\partial \ln\langle \sigma v \rangle}{\partial \ln E_R} = -\frac{E_R}{k_B T} \tag{7.145}$$

as the prefactor in Eq. (7.143) is reduced to Γ_γ, which is almost constant. Since δE_R and δ_{NN} have opposite signs (Eqs. (7.66) and (7.65)), the rate *increases* with δ_{NN}. In the former case, the same factor is reduced to the very energy-dependent Γ_α, and we have

$$\langle \sigma v \rangle \propto \gamma_\alpha^2 \exp\left(-\sqrt{\frac{E_G}{E_R}} - \frac{E_R}{k_B T}\right), \tag{7.146}$$

where the penetrability, $P_\ell(E)$, has been approximated by $\exp(-\sqrt{E_G/E})$ with Gamow energy, E_G. It is well known that the exponential in Eq. (7.146) can be well approximated (see, e.g., Reference [3]) by

$$\exp\left[-\left(\frac{E_R - E_0}{\Delta E_0/2}\right)^2\right], \tag{7.147}$$

with

$$E_0 = \left(\frac{\mu}{2}\right)^{1/3} \left(\frac{\pi e^2 Z_1 Z_2 kT}{\hbar}\right)^{2/3} = 0.1220 \, (Z_1^2 Z_2^2 A)^{1/3} \, T_9^{2/3} \text{ MeV} \tag{7.148}$$

and

$$\Delta E_0 = 4 \, (E_0 kT/3)^{1/2} = 0.2368 \, (Z_1^2 Z_2^2 A)^{1/6} \, T_9^{5/6} \, \text{MeV}, \qquad (7.149)$$

which define the Gamow window. Recalling that the reduced width γ^2 only reflects the nuclear structure and is assumed to be constant, it is straightforward to calculate the sensitivity of the rate to $E_R \, (\delta_{NN})$ variations:

$$\frac{\partial \ln \langle \sigma v \rangle}{\partial \ln E_R} = 4 \left(\frac{E_0(T) - E_R(\delta_{NN})}{\Delta E_0(T)/2} \right). \qquad (7.150)$$

Since, for large δ_{NN} and T, we have $E_0 > E_R$, the rate *decreases* with δ_{NN}.

The condition $\Gamma_\alpha = \Gamma_\gamma$ that marks the boundary between these two opposite evolutions in Eq. (7.143) can be found in figure A.1 of Ekström et al. [44], at $\delta_{NN} \approx 0.006$.

8

Fine-Tunings at Particle Scales

GIULIA ZANDERIGHI

Abstract

Particle physics, like most fundamental sciences, has a number of open questions that are generally perceived as being of utmost importance. Accordingly, trying to address those questions motivates most of the experimental and theoretical activity in the field. In particle physics, these questions can be classified into two main categories. First, there are questions that arise from experimental observations – for instance, what are the constituents of dark matter, or what is the origin of the matter-antimatter asymmetry seen in our Universe? These questions have a well-defined, unique answer which must be understood before a complete understanding of the fundamental theory that governs all interactions can be claimed. Additionally, there are questions that are more related to aesthetic arguments and to our belief that the fundamental theory should obey certain criteria of symmetry and simplicity that we often refer to as the beauty or elegance of a theory. The most common queries in the second category include, for example, why are there three generations of quarks and leptons in the Standard Model (SM) of particle physics, why is the mass spectrum of fermions hierarchical, what is the origin of neutrino masses, what resolves the strong charge-parity (CP) problem, and why are the electroweak (EW) scale and the Planck scale so widely separated? These questions, some which will be discussed in detail in the following, are sometimes considered even more fundamental and interesting than those of the first type. For instance, discovering that dark matter is an axion particle would be a great achievement but would not fundamentally advance the understanding of our ultimate theory. On the other hand, it is not clear whether all questions of the second type have a well-defined answer.

8.1 Introduction

Particle physics, like most fundamental sciences, has a number of open questions that are generally perceived as being of utmost importance. Accordingly, trying to address those questions motivates most of the experimental and theoretical activity in the field. In particle physics, these questions can be classified into two main categories. First, there are questions that arise from experimental observations – for instance, what are the constituents of dark matter, or what is the origin of the matter-antimatter asymmetry seen in our Universe? These questions have a well-defined, unique answer which must be understood before a complete understanding of the fundamental theory that governs all interactions can be claimed.

Additionally, there are questions that are more related to aesthetic arguments and to our belief that the fundamental theory should obey certain criteria of symmetry and simplicity that we often refer to as the beauty or elegance of a theory. The most common queries in the second category include, for example, why are there three generations of quarks and leptons in the Standard Model (SM) of particle physics, why is the mass spectrum of fermions hierarchical, what is the origin of neutrino masses, what resolves the strong charge-parity (CP) problem, or why are the electroweak (EW) scale and the Planck scale so widely separated? These questions, some which will be discussed in detail in the following, are sometimes considered even more fundamental and interesting than those of the first type. For instance, discovering that dark matter is an axion particle would be a great achievement but would not fundamentally advance the understanding of our ultimate theory. On the other hand, it is not clear whether all questions of the second type have a well-defined answer. Hence, while a resolution is not required, if it were to happen, it would most likely lead to the discovery of new particles not present in the SM, to new symmetries of nature, and, ultimately, to the unveiling of new fundamental principles. Prominent questions of the second type address, in particular, the fine-tuning or naturalness problem in the SM that will be discussed in this chapter. In particle physics, we generally say that a parameter is fine-tuned if its observed value is much smaller than the size of corrections that the parameter receives when higher-order quantum effects are accounted for. In particular, fine-tuned parameters typically involve a cancellation between two large, apparently unrelated terms.

In the past decades, a number of situations have been encountered in particle physics when a parameter of the theory seemed fine-tuned. The apparent fine-tuning problem could be resolved by realising that the description used was only suitable at low energies and that extending the theory by including new particles and interactions could remove the alleged fine-tuning problem. A number of notable examples of apparent fine-tuning – namely the energy field of the electron, the charged and neutral pion mass difference, and the kaon transition rates – are presented in Section 8.2. Since, in the past, requiring that a theory was not fine-tuned led to formulating the correct extension of the theory, understanding how to remove the fine-tuning problems in the SM has been used as a guiding principle to construct theories of new physics, beyond the SM (BSM) over the past 40 years.

While we have some intuition about whether a parameter is fine-tuned, in order to use fine-tuning as a solid guiding principle, it is important to quantify how fine-tuned a parameter or a theory is. This is addressed in Section 8.3. The technical naturalness criterion, as formulated by 't Hooft in the 1980s [54] and discussed in more detail in Section 8.3.1, states that a parameter is allowed to be very small if setting it to zero increases the symmetry of the theory. If, however, no new symmetry is recovered and a parameter is unnaturally small, then the theory is considered to be unnatural. If a theory is found to be unnatural, it is useful to quantify the degree of fine-tuning. A number of measures that have been suggested in the past and are used today in particle physics are described in Section 8.3.2.

In contemporary particle physics, three areas face a severe fine-tuning problem: the Higgs mass related to the electroweak symmetry breaking, the strong CP-violating angle θ, and the cosmological constant Λ. The last problem is discussed extensively in Chapter 2 (by Peacock); hence, in this chapter, we will concentrate on the other two problems.

The first problem, discussed in more detail in Section 8.4, has to do with the fact that the Higgs mass takes a value that is several orders of magnitude smaller than the size of quantum corrections that it would receive in extensions of the SM. Hence, if the SM is just a low-energy effective theory, then the Higgs mass suffers from a fine-tuning problem. Section 8.4 also presents a number of possible solutions to the fine-tuning problem in the EW sector in terms of BSM theories.

Section 8.5 describes the fine-tuning of the strong CP angle – i.e., the fact that an angular parameter, usually denoted as θ, that enters the QCD Lagrangian, happens to be very small. The fine-tuning of the CP angle is quite different from the fine-tuning in the EW sector: while in the latter case the Higgs mass receives very large quantum corrections of the order of the ultraviolet (UV) cut-off of the theory, this is not the case for the strong CP angle. The problem is rather that θ is an angle that can take any value between 0 and 2π in the SM. Experimental measurements constrain θ to be very close to zero, but there is no explanation of why this angle should vanish or be so small. Section 8.5 also describes the most popular solution to the problem in terms of a new symmetry of the Lagrangian, the Peccei-Quinn symmetry. The latter leads to the existence of a new particle, the so-called QCD axion.

Finally, in Section 8.6, we will discuss anthropic arguments as possible approaches to fine-tuning problems. Anthropic arguments in their original and weaker formulation, as applied first to the cosmological constant, rely on the observation that, if the cosmological constant had a value that is much different from the one we observe, galaxies would not have formed. These arguments can be made much stronger by requiring that our life as human beings on the Earth should be possible and, hence, that light and heavy elements should have formed in the right abundance for our existence to be possible. When extended to requiring such conditions, anthropic arguments can be used to explain the fine-tuning in the EW sector, but these approaches remain controversial.

8.2 Historical Examples

It is useful to first discuss a few examples where fine-tuning and naturalness arguments provided an indication about the presence of new states and interactions. We will see that these considerations can be, and in some cases even have been, used to predict the energy scale at which new degrees of freedom are expected to appear.

one finds

$$M_u = \frac{G_F^2}{2\pi^2} g_u \Lambda^2 \left(\bar{s}\gamma_\mu P_L d\right)\left(\bar{\mu}\gamma^\mu P_L \mu\right), \qquad (8.22)$$

which is – like the result in Eq. (8.13) – quadratically divergent.

However, in the diagram shown in Figure 8.4, both an up quark and a charm quark can circulate in the loop. In order to take the charm quark into account, one should replace g_u with $\sum_{q=u,c} g_q$ in Eq. (8.22). The form of the coupling g_c then follows from the observation that, in analogy to Eq. (8.17), the charm quark also couples to a linear combination of the two down-type quarks d and s, namely to

$$s' = \cos\theta_c\, s - \sin\theta_c\, d. \qquad (8.23)$$

From this, it follows that

$$g_c = -\sin\theta_c \cos\theta_c, \qquad (8.24)$$

and as a result, $\sum_{q=u,c} g_q = 0$. In physical terms, the latter result means that the quadratic divergence of the up-quark loop is exactly cancelled by the charm-quark contribution. This cancellation is called GIM mechanism. In the case of the up-quark and charm-quark contribution to the $K_L \rightarrow \mu^+\mu^-$ decay, the perfect GIM cancellation is broken by terms $(m_c^2 - m_u^2)/m_W^2 \simeq m_c^2/m_W^2 \ll 1$, which implies that the physical charm-quark mass acts as an effective UV cut-off.

Similarly, the presence of the charm quark can explain the observed suppression of K_0–$\bar{K}_0$ mixing. In fact, as illustrated in Figure 8.5, in this case, four diagrams contribute if we restrict ourselves to the exchange of up quarks and charm quarks. In the limit of $m_u = m_c = 0$, the individual contributions to the K_0–$\bar{K}_0$ amplitude with an exchange of a quark of flavour i and a quark of flavour j take the form

$$M_{ij} \propto G_F^2 g_i g_j, \qquad (8.25)$$

where $i, j = u, c$. In the total amplitude $M = \sum_{i,j=u,c} M_{ij}$, hence, all Λ^2 terms cancel as a result of the GIM mechanism. Like in the case of $K_L \rightarrow \mu^+\mu^-$, finite-mass effects again spoil the complete GIM cancellation but the residual mixing effects are very small being both Cabibbo-Kobayashi-Maskawa (CKM) suppressed ($\propto \sin^2\theta_c$) and suppressed by the small charm-quark mass ($\propto m_c^2/m_W^2$). In the SM, the top quark also enters the loops that describe K_0–$\bar{K}_0$ mixing (see Figure 8.5); the discussed GIM cancellation, however, persists as a result of the unitarity of the CKM matrix [39].

Like for the mass difference between charged and neutral pions, one can also extract in the case of K_0–$\bar{K}_0$ mixing a bound on the effective cut-off scale $\Lambda \simeq m_c$. One just has to calculate the K_L–K_S mass difference $\Delta M_K = M_{K_L} - M_{K_S}$ and compare it to experiment. A calculation similar to the one leading to Eq. (8.22) gives

$$\frac{\Delta M_K}{M_K} \simeq \frac{G_F^2}{6\pi^2} f_K^2 \sin^2\theta_c\, m_c^2, \qquad (8.26)$$

Figure 8.5 The one-loop box diagrams describing K_0–$\bar{K}_0$ mixing. To obtain the full amplitude the contributions of all internal up-type quarks have to be included.

where $M_K = (M_{K_L} + M_{K_S})/2$ and f_K denotes the kaon decay constant. Using the values $\Delta M_K / M_K \simeq 7 \cdot 10^{-15}$ and $f_K \simeq 0.1$ GeV, one finds

$$m_c \simeq 1.4 \, \text{GeV}, \tag{8.27}$$

in very good agreement with the actual mass of the charm quark as determined today by lattice QCD. In fact, the very same argument presented here has been used already more than 40 years ago by Gaillard and Lee in Reference [26] to predict that the charm quark should be lighter than 10 GeV. This is one of the first examples in the history of particle physics where the presence of a quadratic divergence in a theoretical prediction, together with an accurate measurement, has been used to predict the correct mass-scale of a new particle.

8.3 Quantifying Fine-Tuning

8.3.1 Technical Naturalness

We have seen in Section 8.2.1 that despite the fact that the electron mass is very small – in particular, much smaller than the Planck scale – this is considered natural. In fact, no fine-tuning is involved since setting the electron mass to zero one recovers an additional symmetry, the chiral symmetry. This symmetry guarantees that corrections to the electron mass remain proportional to the electron mass itself, and hence, it cannot receive large radiative corrections. This idea was formalised by 't Hooft [54] and elevated to a principle of naturalness which states that if a quantity in nature is small but a new symmetry is recovered were the quantity to vanish, then the smallness of the quantity is natural. This principle is usually referred to as technical naturalness. 't Hooft, however, made an even stronger statement, saying that a quantity in nature should be small only if the underlying theory recovers an additional symmetry when the quantity vanishes. Quantities that do not respect this principle are considered fine-tuned. Let us now consider a few examples of 'small' quantities and see if they are technically natural in the 't Hooft sense.

A first example is given by fermion masses in the SM, such as quark or neutrino masses, which exhibit a large spread in values, from less than 1 eV to more than 100 GeV. If we consider, for instance, the light quarks u, d, and s, the corresponding QCD Lagrangian contains a term of the form

$$\mathcal{L} \supset \bar{\psi} \left(i \gamma^\mu D_\mu - m \right) \psi, \tag{8.28}$$

If m is a physical mass of the order of the EW scale – i.e., $\mathcal{O}(100\,\text{GeV})$ – and Λ is taken to be the Planck scale $\mathcal{O}(10^{19}\,\text{GeV})$, then

$$r_0 = 10^{-34} - g_0^2. \tag{8.38}$$

This equation implies that r_0 must be adjusted to the 34th decimal digit, since, otherwise, the mass m will become of order $\mathcal{O}(10^{19}\,\text{GeV})$ rather than $\mathcal{O}(100\,\text{GeV})$. This high sensitivity of physical parameters to tiny variations of fundamental parameters is precisely what Wilson deemed to be unnatural.

A quantitative formalisation of naturalness, following Wilson's idea, was formulated by Barbieri and Giudice in Reference [13] in the context of quantifying the naturalness of various low-energy supersymmetry models that arise from supergravity theories. Let us consider an observable $\mathcal{O}$, which could be, for example, a mass or a mass difference, and is a function of a number of parameters a_i – for instance, the masses and couplings of new states. The fine-tuning measure Δ_{BG} of Reference [13] is defined as follows. One requires that for every a_i, the following relation holds:

$$\Delta_{BG}(\mathcal{O}; a_i) \equiv \left| \frac{a_i}{\mathcal{O}} \frac{\partial \mathcal{O}(\{a_i\})}{\partial a_i} \right| < \Delta, \tag{8.39}$$

which means that a relative variation of each of the parameters a_i does not produce a relative variation of $\mathcal{O}$ that is more than Δ times larger. So a value of Δ of 10 means that there is at most an order of magnitude cancellation between the parameters.

If we go back to the example in Eq. (8.36) and use now the measure in Eq. (8.39), we obtain

$$\Delta_{BG}(m^2, g_0) = \frac{g_0}{m^2} \frac{\partial m^2(\{g_0\})}{\partial g_0} = 2g_0^2 \frac{\Lambda^2}{m^2}. \tag{8.40}$$

This implies that unless the UV cut-off is close to the physical, observed mass m, this mass is unnatural, since $\Delta_{BG}(m^2, g_0)$ will become very large. For instance, taking $g_0 = 1$, Λ to be the Planck mass and m to be the Higgs mass, one obtains $\Delta_{BG}(m^2, g_0) = \mathcal{O}(10^{34})$.

Since in model building one typically wants to avoid models that are fine-tuned, Eq. (8.39) can be used at fixed Δ to constrain the parameter space of a theory. For instance, one can look at the allowed parameter space so that a theory is not fine-tuned by more than one or two orders of magnitude. Alternatively, given a new-physics theory, one can first exclude regions in parameter space where predictions are in conflict with experimental observations. For those regions of parameter space that remain not excluded by experimental data, one can use the Barbieri-Giudice criterion to assign a measure of fine-tuning. Originally, in Reference [13], Δ was chosen to be 10, allowing for an order of magnitude cancellation. The choice of a particular value for Δ is obviously very subjective, and after strong experimental constraints from the Large Electron Positron (LEP) and the Large Hadron Collider (LHC), larger choices of Δ became more commonly accepted.

It was also suggested in Reference [3] that the Barbieri-Giudice (BG) measure of Eq. (8.39) can be misleading, as it mixes the concept of sensitivity of a quantity to a given parameter and the notion of naturalness. In essence, the problem is related to the value of Δ that should be considered acceptable because when a quantity is more sensitive to a parameter, one should permit larger values of Δ. Hence, a new measure of fine-tuning was suggested in Reference [3] embodying the idea that 'Observable properties of a system should not be unusually unstable against minute variations of the fundamental parameters'. In essence, the new naturalness measure wants to separate the sensitivity of observables to parameters from the fine-tuning. This point can be understood by examining Eq. (8.33), which shows that changing, e.g., g_0 from, e.g., 0.15 to 0.16 changes Q from 1 GeV to 100 GeV. Hence, Q is very sensitive to g_0 despite the fact that there is no fine-tuning problem. In fact, the BG measure for Q is

$$\Delta_{\mathrm{BG}}(Q, g_0) = \frac{1}{\beta_0 g_0^2}, \tag{8.41}$$

which has no UV sensitivity to a cut-off but does depend strongly on g_0. The solution suggested in Reference [3] is to assign a probability distribution for the parameters of the theory and to calculate the fine-tuning associated with measured values after normalising away the specific sensitivity of physical quantities to parameters of the theory.

Starting from the BG measure in Eq. (8.39), Anderson and Castano (AC) define a new measure as [3]

$$\Delta_{\mathrm{AC}}(\mathcal{O}; a_i) = \frac{\Delta_{\mathrm{BG}}(\mathcal{O}; a_i)}{\langle \Delta_{\mathrm{BG}}(\mathcal{O}; a_i) \rangle}, \tag{8.42}$$

where

$$\frac{1}{\langle \Delta_{\mathrm{BG}}(\mathcal{O}; a_i) \rangle} = \frac{\int da_i a_i f(a_i) \Delta_{\mathrm{BG}}(\mathcal{O}; a_i)^{-1}}{\int da_i f(a_i)} \tag{8.43}$$

and $f(a_i)$ is a probability density. In essence, the new measure takes the BG measure and normalises it to its average over a reasonable range of parameters.

As a simplest option, one can, for instance, take a flat density $f(a_i) = 1$ or some simple form for it. One can then integrate over some minimum and maximum value of a_i, where the chosen values should correspond to a sensible range of variation for this theory parameter, with the chosen probability density. Simple forms of $f(a_i)$ give similar results in terms of the fine-tuning measure $\langle \Delta_{\mathrm{AC}} \rangle$. In particular, the presence or absence of a fine-tuning problem is independent of the choice of $f(a_i)$, and the sensitivity to physical parameters is removed by the normalisation.

As an example, we can consider again the case of the scalar mass with quadratic corrections, Eq. (8.36). The BG measure is given in Eq. (8.40). To compute the AC measure, one needs to evaluate Eq. (8.43). For instance, for $f(g) = 1$, one obtains

$$\langle \Delta_{\mathrm{BG}}(m^2; g_0) \rangle^{-1} = \frac{\int_{g_0^-}^{g_0^+} dg_0 g_0 f(g_0) \Delta_{\mathrm{BG}}(m^2; g_0)^{-1}}{\int_{g_0^-}^{g_0^+} dg_0 g_0 f(g_0)} = \frac{m_0^2 \ln \frac{g_0^+}{g_0^-}}{\Lambda^2 (g_0^{+2} - g_0^{-2})} + \frac{1}{2}, \tag{8.44}$$

$\Phi = 0$, while for $\mu^2 < 0$, the field value $\Phi = 0$ becomes a maximum of the potential, and the minima lie at

$$|\Phi|^2 = -\frac{\mu^2}{2\lambda} = \frac{v^2}{2}, \tag{8.53}$$

with $v = \sqrt{-\mu^2/\lambda}$ the vacuum expectation value (VEV) of Φ. One can now expand the Φ field around the vacuum solutions. In the unitary gauge, one can write

$$\Phi = \frac{1}{\sqrt{2}} \begin{pmatrix} 0 \\ v + H \end{pmatrix}, \tag{8.54}$$

with H a real scalar – i.e., the Higgs field. Up to an irrelevant constant, in this gauge the scalar potential reads

$$V(\Phi) = \frac{1}{2}(2\lambda v^2)H^2 + \lambda v H^3 + \frac{\lambda}{4}H^4, \tag{8.55}$$

which implies that the Higgs field H has a mass

$$m_H = \sqrt{2\lambda}\, v. \tag{8.56}$$

The term involving the covariant derivative of Φ can be also worked out easily. One obtains

$$|D_\mu \Phi|^2 = \frac{1}{2}|\partial_\mu H|^2 + \left[\frac{g^2 v^2}{4} W_\mu^+ W^{-\mu} + \frac{(g^2 + g'^2)\, v^2}{8} Z_\mu Z^\mu\right]\left(1 + \frac{H}{v}\right)^2. \tag{8.57}$$

From this expression, one can read off the values of the W and Z masses:

$$m_W = \frac{gv}{2}, \qquad m_Z = \frac{\sqrt{g^2 + g'^2}\, v}{2}. \tag{8.58}$$

The photon instead remains massless $m_A = 0$. The value of the Higgs VEV can then be obtained from precision measurements of the W and Z mass, from the Fermi constant $G_F \simeq 1.166 \cdot 10^{-5}\,\text{GeV}^{-2}$ using the relation

$$\frac{G_F}{\sqrt{2}} = \frac{g^2}{8m_W^2} = \frac{1}{2v^2}, \tag{8.59}$$

which simply follows from matching the Fermi theory to the full SM. Numerically, one finds

$$v \simeq 246.22\,\text{GeV}. \tag{8.60}$$

The Higgs mass is also known very precisely today [1], to better than $\sim 1\%$ precision, and reads

$$m_H = 125.09 \pm 0.21(\text{stat.}) \pm (\text{syst.})\,\text{GeV}. \tag{8.61}$$

Using the second equality in Eq. (8.53) from the knowledge of the Higgs mass and the VEV, we also know the value of the quartic coupling λ in the SM. One obtains

$$\lambda \simeq 0.13. \tag{8.62}$$

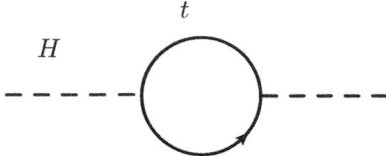

Figure 8.6 A one-loop correction to the Higgs mass coming from a top quark in the loop.

This implies that all interactions of SM particles, fermions, and bosons with the Higgs boson – in particular, also the Higgs interaction with itself through triple and quartic Higgs couplings – are fixed. A broad and detailed experimental program is ongoing at the LHC to test the validity of the Higgs mechanism. At the moment, couplings of the Higgs to fermions of the third generation and to W and Z have been measured to about 10% accuracy and seem to follow the pattern predicted in the SM. Yet couplings to lighter fermions and the Higgs self-coupling remain experimentally largely unconstrained. Better measurements of those quantities are top priorities for the upcoming LHC physics programme. In fact, although at the moment all measurements agree with the SM Higgs mechanism, the Higgs sector suffers from a fine-tuning problem, as explained in the following. Hence, it seems natural to think that the fine-tuning should be removed by the presence of BSM physics, and trying to remove or alleviate the fine-tuning in the EW sector has driven a lot of the theoretical activity in particle physics.

8.4.1 Higher-Order Corrections to the Higgs Mass

The fine-tuning problem in the Higgs sector is related to the fact that the Higgs mass receives very large quantum corrections from particles that couple directly, or even indirectly, to the Higgs boson. This fact was first observed in the context of grand unified theories (GUTs) [28, 57].

Let us consider, for example, the quantum corrections to the Higgs mass coming from a loop of fermions that couples to the Higgs through a Yukawa interaction of the form

$$\mathcal{L} \supset -\lambda_f H \bar{f} f, \tag{8.63}$$

with $\lambda_f = \sqrt{2} m_f / v$ and m_f the mass of the fermion. The diagram illustrated in Figure 8.6 gives rise to a correction to the Higgs mass of the form

$$\Delta m_H^2 = N_c i \lambda_f^2 \int \frac{d^4 l}{(2\pi)^4} \frac{\text{tr}\left[(\slashed{l} + m_t)(\slashed{l} + \slashed{p} + m_t)\right]}{(l^2 - m_t^2)\left((l + p)^2 - m_t^2\right)}, \tag{8.64}$$

which is quadratically divergent in the limit $|l| \to \infty$. The divergent contribution is given by

$$\Delta m_H^2 = N_c \frac{i\lambda_f^2}{16\pi^4} \int d^4 l \frac{4l^2}{l^4} = N_c \frac{i\lambda_f^2}{4\pi^4} \left(-i\pi^2 \Lambda^2\right) = N_c \frac{G_F m_f^2}{\sqrt{2}\pi^2} \Lambda^2, \tag{8.65}$$

Figure 8.7 Two examples of two-loop correction to the Higgs mass coming from a heavy fermion Q that couples only indirectly to the Higgs via the exchange of a SM gauge boson V.

where we used the expression of the divergent integral in Eq. (8.12) and $N_c = 3$ is the colour factor.

Within the SM, the dominant contribution to the Higgs mass comes from the top quark because of its large Yukawa coupling. Also, loops involving massive bosons give rise to divergent corrections, and when one includes all important one-loop corrections, one obtains [55]

$$\frac{\Delta m_H^2}{m_H^2} = \frac{3G_F}{4\sqrt{2}\pi^2}\left(\frac{4m_t^2}{m_H^2} - \frac{2m_W^2}{m_H^2} - \frac{m_Z^2}{m_H^2} - 1\right)\Lambda^2 \simeq \left(\frac{\Lambda}{500\,\text{GeV}}\right)^2. \tag{8.66}$$

The UV cut-off can then be interpreted as the scale at which new physics enters and modifies the high-energy behaviour of the theory. The hierarchy problem of the Higgs mass is now related to the fact that, if in Eq. (8.66) the UV cut-off Λ is taken to be the Planck mass, m_H receives corrections that are more than 15 orders of magnitude larger than the measured value of around 125 GeV. If, on the other hand, new, not-too-heavy particles exist and act as an effective UV cut-off, then this could remove the fine-tuning problem in the Higgs sector.

The interesting thing to note is that in any extension of the SM that introduces one or more heavy states, the Higgs mass squared becomes quadratically sensitive to any new degree of freedom, regardless of whether the new states interact at the tree level with the Higgs. In fact, even if the new particles do not interact directly with the Higgs boson – for instance, because their masses predominantly arise from a mechanism other than EWSB – they will generically affect the Higgs boson mass indirectly through radiative corrections. An example is shown in Figure 8.7, which displays two two-loop diagrams that involve a heavy fermion that couples only indirectly to the Higgs boson through a SM vector boson. The quadratically divergent correction to the Higgs mass squared that arises from the diagram on the right-hand side of Figure 8.7 can be written schematically as follows:

$$\Delta m_H^2 \sim g_1^2 g_2^2 C_1 C_2 \int \frac{d^4 l_1}{(2\pi)^4} \int \frac{d^4 l_2}{(2\pi)^4} \frac{\text{tr}\left[(l_1 + l_2)l_2\right]}{l_1^4 l_2^2 (l_1 + l_2)^2}, \tag{8.67}$$

where the coupling of the intermediate vector boson to the Higgs is called g_1, the coupling of the vector boson to the heavy fermion is denoted by g_2, and C_1 and C_2 are the corresponding Casimir factors. The preceding expression simplifies to

$$\Delta m_H^2 \sim \frac{g_1^2}{16\pi^4} \frac{g_2^2}{16\pi^4} C_1 C_2 \int d^4 l_1 \int d^4 l_2 \frac{1}{l_1^4 (l_1 + l_2)^2}$$

$$\sim \frac{g_1^2}{16\pi^2} \frac{g_2^2}{16\pi^2} C_1 C_2 \Lambda^2 \ln \Lambda, \tag{8.68}$$

where we used Eq. (8.12) and

$$\int d^4 l \frac{1}{l^4} = i \int d\Omega_4 \int_0^\Lambda dl_E \frac{1}{l_E} = i 2\pi^2 \ln \Lambda \tag{8.69}$$

and ignored numerical factors of $\mathcal{O}(1)$ to arrive at the final result in Eq. (8.68).

We see that like the one-loop contribution Eq. (8.65), the two-loop contribution also depends quadratically on the UV cut-off Λ. One finds a similar quadratically divergent contribution from the diagram on the left-hand side of Figure 8.7, but the two contributions do not cancel. This example illustrates that even if a new heavy particle does not interact with the Higgs boson at tree level, as long as the new state has interactions with some of the SM particles, radiative corrections will give rise to a quadratic sensitivity of m_H on the cut-off scale Λ.

Given that the technical naturalness argument is used to explain the lightness of many SM fermions, it is worth examining if an additional symmetry appears within the SM in the limit of $m_H \to 0$. A massless Higgs would imply the absence of the quadratic term in Eq. (8.52). In such a case, the Higgs VEV would be zero, the EW symmetry would be unbroken, and neither the SM gauge bosons nor the fermions would receive a mass from EWSB. In fact, at the classical level, the SM action is conformally symmetric – i.e., invariant under the transformations in Eq. (8.34) and

$$V_\mu \to \kappa^{-1} V_\mu, \qquad \Phi \to \kappa^{-1} \Phi, \tag{8.70}$$

except for the quadratic term in Eq. (8.52). In the preceding transformations, V_μ represents any of the SM gauge fields. It follows that the vanishing of the Higgs mass term $\mu^2 |\Phi|^2$ would increase the symmetry of the SM by making the conformal symmetry exact. At the loop level, there is explicit breaking of the conformal symmetry reflected by the logarithmic running of the gauge and Yukawa couplings. The logarithmic running alone, however, will not generate Λ^2 terms so that in perturbation theory conformal symmetry would protect Δm_H^2 from quadratic corrections and the conventional statement of the EW fine-tuning problem. Yet the fine-tuning issue may reappear if the SM is embedded into a more complex theory, visible at short distances or high energies. Whether conformal invariance can still be used to protect the EW scale then depends on the structure and the dynamics of these more complete formulations.

Before concluding this section, it is also worth commenting on what happens if, instead of regularising the divergent UV behaviour of loop integrals by a cut-off Λ, one considers dimensional regularisation, which is the regularisation procedure usually employed in modern higher-order perturbative calculations. This point is discussed in detail, for instance, in Reference [62]. Dimensional regularisation of UV divergences proceeds by

[5] Arkani-Hamed, N., *et al.* 2002b. 'The Minimal Moose for a Little Higgs'. *Journal of High Energy Physics*, **08**, 021.

[6] Arkani-Hamed, N., Dimopoulos, S., and Dvali, G. R. 1998. 'The Hierarchy Problem and New Dimensions at a Millimeter'. *Physics Letters B*, **429**, 263–272.

[7] Arkani-Hamed, N., Cohen, A. G., and Georgi, H. 2001. 'Electroweak Symmetry Breaking from Dimensional Deconstruction'. *Physics Letters B*, **513**, 232–240.

[8] Athron, P. and Miller, D. J. 2007. 'A New Measure of Fine Tuning'. *Physical Review D*, **76**, 075010.

[9] Aubert, J. J., *et al.* 1974. 'Experimental Observation of a Heavy Particle J'. *Physical Review Letters*, **33**, 1404–1406.

[10] Augustin, J. E., *et al.* 1974. 'Discovery of a Narrow Resonance in e+ e- Annihilation'. *Physical Review Letters*, **33**, 1406–1408.

[11] Baker, C. A., *et al.* 2006. 'An Improved Experimental Limit on the Electric Dipole Moment of the Neutron'. *Physical Review Letters*, **97**, 131801.

[12] Banks, T., Dine, M., and Motl, L. 2001. 'On Anthropic Solutions of the Cosmological Constant Problem'. *Journal of High Energy Physics*, **01**, 031.

[13] Barbieri, R., and Giudice, G. F. 1988. 'Upper Bounds on Supersymmetric Particle Masses'. *Nuclear Physics B*, **306**, 63–76.

[14] Bardeen, W. A. 1995. 'On Naturalness in the Standard Model'. In *Ontake Summer Institute on Particle Physics Ontake Mountain, Japan, August 27–September 2, 1995*.

[15] Barrow, J. D., and Tipler, F. J. 1988. *The Anthropic Cosmological Principle*. Oxford University Press.

[16] Belavin, A. A., *et al.* 1975. 'Pseudoparticle Solutions of the Yang-Mills Equations'. *Physics Letters B*, **59**, 85–87.

[17] Bjorken, J. D. 2001. 'Standard Model Parameters and the Cosmological Constant'. *Physical Review D*, **64**, 085008.

[18] Cabibbo, N. 1963. 'Unitary Symmetry and Leptonic Decays'. *Physical Review Letters*, **10**, 531–533.

[19] Ciafaloni, P., and Strumia, A. 1997. 'Naturalness Upper Bounds on Gauge Mediated Soft Terms'. *Nuclear Physics B*, **494**, 41–53.

[20] Das, T., *et al.* 1967. 'Electromagnetic Mass Difference of Pions'. *Physical Review Letters*, **18**, 759–761.

[21] de Gouvea, A., Hernandez, D., and Tait, T. M. P. 2014. 'Criteria for Natural Hierarchies'. *Physical Review D*, **89**(11), 115005.

[22] Dine, M., Fischler, W., and Srednicki, M. 1981. 'A Simple Solution to the Strong CP Problem with a Harmless Axion'. *Physics Letters B*, **104**, 199–202.

[23] Donoghue, J. F. 2007. 'The Fine-Tuning Problems of Particle Physics and Anthropic Mechanisms'. In Carr, B. (ed.): *Universe or Multiverse*. Cambridge University Press. 231–246.

[24] Farina, M., Perelstein, M., and Rey-Le Lorier, N. 2014. 'Higgs Couplings and Naturalness'. *Physical Review D*, **90**(1), 015014.

[25] Fermi, E. 1934. 'An Attempt of a Theory of Beta Radiation. 1'. *Zeitschrift für Physik*, **88**, 161–177.

[26] Gaillard, M. K., and Lee, B. W. 1974. 'Rare Decay Modes of the K-Mesons in Gauge Theories'. *Physical Review D*, **10**, 897.

[27] Georgi, H., and Kaplan, D. B. 1984. 'Composite Higgs and Custodial SU(2)'. *Physics Letters B*, **145**, 216–220.

[28] Gildener, E. 1976. 'Gauge Symmetry Hierarchies'. *Physical Review D*, **14**, 1667.

[29] Giudice, G. F. 2013. 'Naturalness after LHC8'. *Proceedings of Science*, **EPS-HEP2013**, 163.

[30] Glashow, S. L., Iliopoulos, J., and Maiani, L. 1970. 'Weak Interactions with Lepton-Hadron Symmetry'. *Physical Review D*, **2**, 1285–1292.

[31] Goldberger, W. D., and Wise, M. B. 1999. 'Modulus Stabilization with Bulk Fields'. *Physical Review Letters*, **83**, 4922–4925.

[32] Graham, P. W., Kaplan, D. E., and Rajendran, S. 2015a. 'Cosmological Relaxation of the Electroweak Scale'. *Physical Review Letters*, **115**(22), 221801.

[33] Graham, P. W., *et al.* 2015b. 'Experimental Searches for the Axion and Axion-Like Particles'. *Annual Review of Nuclear and Particle Science*, **65**, 485–514.

[34] Grinbaum, A. 2012. 'Which Fine-Tuning Arguments Are Fine?'. *Foundations of Physics*, **42**, 615–631.

[35] Hogan, C. J. 2000. 'Why the Universe Is Just So'. *Reviews of Modern Physics*, **72**, 1149–1161.

[36] Kallosh, R., and Linde, A. D. 2003. 'M Theory, Cosmological Constant and Anthropic Principle'. *Physical Review D*, **67**, 023510.

[37] Kaplan, D. B., and Georgi, H. 1984. 'SU(2) x U(1) Breaking by Vacuum Misalignment'. *Physics Letters B*, **136**, 183–186.

[38] Kim, J. E. 1979. 'Weak Interaction Singlet and Strong CP Invariance'. *Physical Review Letters*, **43**, 103.

[39] Kobayashi, M., and Maskawa, T. 1973. 'CP Violation in the Renormalizable Theory of Weak Interaction'. *Progress of Theoretical Physics*, **49**, 652–657.

[40] Martel, H., Shapiro, P. R. and Weinberg, S. 1998. 'Likely Values of the Cosmological Constant'. *Astrophysical Journal*, **492**, 29.

[41] Martin, S. P. 1998. *A Supersymmetry Primer*. World Scientific. Pages 1–98.

[42] Murayama, H. 2000. 'Supersymmetry Phenomenology'. *Proceedings, Summer School in Particle Physics: Trieste, Italy, June 21–July 9, 1999*. Pages 296–335.

[43] Olive, K. A. *et al.* 2014. 'Review of Particle Physics'. *Chinese Physics C*, **38**, 090001.

[44] Peccei, R. D., and Quinn, H. R. 1977a. 'Constraints Imposed by CP Conservation in the Presence of Instantons'. *Physical Review D*, **16**, 1791–1797.

[45] Peccei, R. D. and Quinn, H. R. 1977b. 'CP Conservation in the Presence of Instantons'. *Physical Review Letters*, **38**, 1440–1443.

[46] Peskin, M. E., and Schroeder, D. V. 1995. *An Introduction to Quantum Field Theory*. Avalon Publishing.

[47] Randall, L. and Sundrum, R. 1999a. 'A Large Mass Hierarchy from a Small Extra Dimension'. *Physical Review Letters*, **83**, 3370–3373.

[48] Randall, L., and Sundrum, R. 1999b. 'An Alternative to Compactification'. *Physical Review Letters*, **83**, 4690–4693.

[49] Sakharov, A. D. 1967. 'Violation of CP Invariance, c Asymmetry, and Baryon Asymmetry of the Universe'. *Pisma Zh. Eksp. Teor. Fiz.*, **5**, 32–35.

[50] Schmaltz, M. 2004. 'The Simplest Little Higgs'. *Journal of High Energy Physics*, **08**, 056.

[51] Shifman, M. A., Vainshtein, A. I., and Zakharov, V. I. 1980. 'Can Confinement Ensure Natural CP Invariance of Strong Interactions?' *Nuclear Physics B*, **B166**, 493–506.

[52] Streater, R. F., and Wightman, A. S. 1989. *PCT, Spin and Statistics, and All That*. Princeton University Press.

[53] Susskind, L. 1979. 'Dynamics of Spontaneous Symmetry Breaking in the Weinberg-Salam Theory'. *Physical Review D*, **20**, 2619–2625.

interactions are sufficiently strong, the particle will remain in chemical equilibrium, even as the temperature drops somewhat below the mass. If the species is still in chemical equilibrium at temperatures below the mass, the relative abundance of the species (relative to, say, photons) becomes exponentially suppressed by the Boltzmann factor $\exp(-m/T)$. As the Universe further cools below the mass of the species, kinetic equilibrium is maintained, but the particle eventually drops out of chemical equilibrium. This is because, at temperatures much less than the mass of the species, the abundance of χ is so small it is exponentially unlikely for a χ to find another χ with which to annihilate (and it is exponentially unlikely for light particles to have sufficient collisional energy to pair-produce χs in a collision). At this point the species is said to be frozen out (or the abundance frozen in). Since the particle froze out when non-relativistic, it is a *cold* thermal relic. After freeze-out, the abundance of the species only changes because of the dilution in the number density caused by the expansion of the Universe.

So freeze-out (and, hence, the present abundance) of a cold thermal relic is governed by the interplay between the DM-SM interaction strength (set in the realm of particle physics) and the expansion rate of the Universe (gravity). These are very disparate forces that act in concert to determine the present abundance of the DM particle species and, hence, its contribution to Ω.

If one knows the properties of the DM candidate, then it is possible to calculate the cross sections for annihilation and production and find the rate for production and annihilation of DM. The rate of processes keeping the particles in chemical equilibrium should be compared to the expansion rate of the Universe. If the equilibration rates are much greater than the expansion rate, the particle will track its equilibrium chemical abundance. If the equilibration rates are much less than the expansion rate of the Universe, then the particle is frozen out, and its abundance changes only through the expansion of the Universe.

We can do quite a bit better than the preceding qualitative argument. The evolution of n, the number density of a cold thermal relic, is determined by a Boltzmann equation:

$$\dot{n} = -3Hn - \langle \sigma_A v \rangle \left(n^2 - n_{eq}^2 \right) . \tag{9.2}$$

In this equation, the overdot represents the time derivative, H is the expansion rate of the Universe, $\langle \cdots \rangle$ denotes a thermal average, $\sigma_A v$ is the total annihilation cross section times the Møller flux, and n_{eq} is the equilibrium abundance of the species with zero chemical potential. It is easy to understand the physical meaning of the terms on the right-hand side of Eq. (9.2): $-3Hn$ represents the dilution of the number density due to expansion, $-\langle \sigma_A v \rangle n^2$ represents the decrease in the number density due to annihilation, and $\langle \sigma_A v \rangle n_{eq}^2$ is the increase in the number density due to production. The final abundance of the species is calculated assuming the species is in chemical equilibrium at $T > m$ and is kept in chemical equilibrium through production and annihilation processes until the rate for the processes keeping the species in chemical equilibrium drop below the expansion rate of the Universe. The expansion rate is $H = \dot{a}/a$, where $a(t)$ is the Robertson-Walker scale factor, the only dynamical metric degree of freedom in a homogeneous and isotropic Universe. After

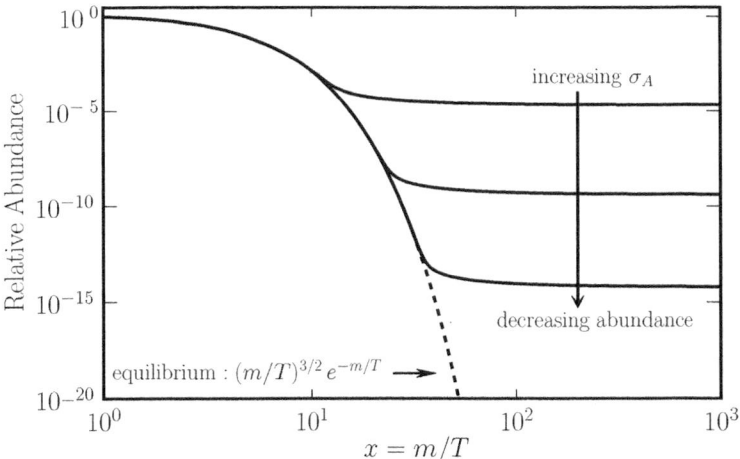

Figure 9.2 Origin of species for a cold thermal relic. Plotted is the relative abundance (say, the abundance relative to photons) as a function of the mass of the species divided by the temperature. The dashed line is the chemical equilibrium abundance. Illustrating the relative abundance, rather than the number density, removes the decrease of the number density due to expansion.

freeze-out, we can ignore interactions that change the number of χ; i.e., set $\langle \sigma_A v \rangle = 0$, and then $\dot{n}/n = -3\dot{a}/a$, which implies $n \propto a^{-3}$. Since a^3 is a volume element, the number density decreases as the volume of the Universe. So, after the species is frozen out, it decreases only due to the expansion of the Universe.

Before giving some quantitative results, it is possible to anticipate how things work. The more strongly the DM particle interacts with SM particles, the longer it will stay in equilibrium, and the lower will be its abundance when eventually it freezes out. This general feature is illustrated in Figure 9.2, where it is indicated that increasing the interaction strength, here characterised by the annihilation cross section σ_A, leads to a smaller abundance today. The more weakly interacting a particle, the larger is its contribution to the present mass density. In cosmology, at least, the weak shall dominate!

There are analytic approximations for the final abundance resulting from the Boltzmann equation. Assuming that $\sigma_A v$ is a constant in the non-relativistic limit (s-wave annihilation), the contribution to Ω from a cold thermal relic is approximately[6]

$$\Omega h^2 \simeq 0.12 \times \frac{10^{-36} \text{ cm}^2}{\sigma_s} \qquad (\langle \sigma_A v \rangle = \sigma_s \text{ in the non-relativistic limit}) . \qquad (9.3)$$

As expected, the result decreases as the cross section increases. It is interesting to note that the result is independent of the mass of the species, or it depends on the mass of the species

---

[6] To be sure, there are many effects that can complicate this simple result: there may be a velocity dependence to the annihilation cross section, there may be resonances in the annihilation process, there is a logarithmic dependence of the mass, there might be asymmetries between particle and antiparticle, etc. The variations on the simple theme have kept many theorists (the author included) busy for many years.

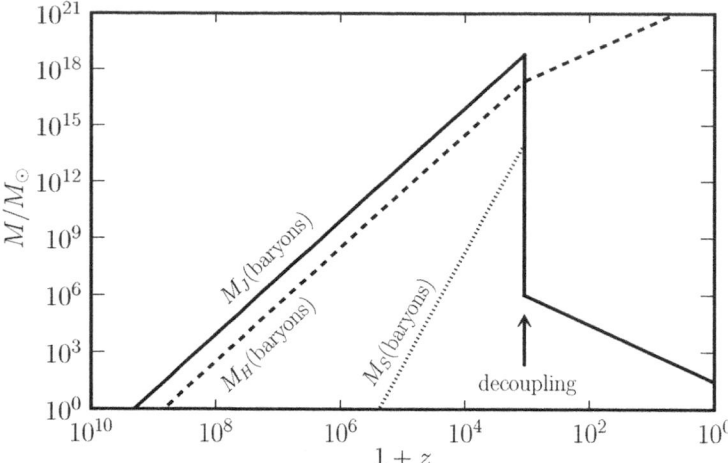

Figure 9.4 The evolution of mass scales associated with baryons as a function of $1 + z = a_0/a(t)$ (a_0 is the present scale factor). The solid curve labelled M_J is the baryonic Jeans mass. Small perturbations grow only for masses larger than the Jeans mass. Before decoupling, pressure support is provided by the coupling of baryons to photons. After decoupling pressure, support is provided by non-relativistic hydrogen atoms, hence the drop in the Jeans mass at decoupling. The dashed curve labelled M_H is the mass in baryons within the Hubble radius. Small perturbations on mass scales larger than the Hubble radius do not grow. Thus, baryonic perturbations only grow after decoupling since the Jeans mass is less than the Hubble-radius mass only after decoupling. The dotted curve labelled M_S is the Silk mass. Collisional (Silk) damping erases perturbations on mass scales less than the Silk mass. This figure is representational. Several crude assumptions have been made, like that the approximation that the Universe becomes matter dominated at decoupling.

Another important effect in a cosmological setting is collisional damping, which erases baryonic perturbations on interesting scales. For most of the history of the early Universe, it is valid to assume that the baryon/photon fluid is a 'perfect' fluid because the baryons and photons are tightly coupled, and the mean-free-paths of baryons and photons are much less than scales of interest.[8] But as decoupling of matter and radiation is approached, the mean-free-path of photons grow as the Universe passes from matter in an ionised state with free electrons and baryons to a state with matter in the form of neutral hydrogen; in other words, the assumption of a perfect fluid breaks down. We are assuming an initial condition of adiabatic (isentropic) perturbations, so the photon and baryon perturbations are correlated. As the photons begin to stream out of over-dense regions, they drag baryons along with them, smoothing the baryonic perturbations. In a cosmological context, this collisional damping is known as Silk damping [33].

---

[8] The mean free path of an electron, $\lambda_e = (n_\gamma \sigma_T)^{-1}$, is much less than the photon mean-free-path, $\lambda_\gamma = (n_e \sigma_T)^{-1}$ because the number density of photons is about 10^{10} that of electrons. Here, σ_T is the Thomson cross section.

The photon mean-free-path in the non-relativistic limit ($T \ll m_e$) is $\lambda_\gamma = (X_e n_e \sigma_T)^{-1}$, where σ_T is the Thomson cross section, X_e, is the ionisation fraction, and n_e is the total electron density (bound and free). In a time t, a photon will have t/λ_γ collisions and undergo a random walk. As it random-walks out of the perturbations, the photons will drag baryons out of the perturbations. The Silk scale reaches a maximum around decoupling; then, after the electrons and protons recombine, the photons decouple from baryons, and photons no longer can drag the baryons out of perturbations. The Silk scale is shown in Figure 9.4. After decoupling, baryon perturbations are damped for mass scales below about $10^{13} M_\odot$.

So now we can put everything together and predict what the growth of structure would be in a dark-matter-less universe where the only matter components are baryons and electrons.

1. For perturbations of a size smaller than the Hubble radius, a Newtonian/Jeans analysis is applicable.
2. In the radiation-dominated (or curvature-dominated or cosmological-constant-dominated) era perturbations do not grow. When the universe becomes matter dominated, perturbations on mass scales greater than the Jeans mass grow as $t^{2/3}$.
3. The Jeans mass becomes smaller than the horizon mass only after recombination.
4. By the time of recombination, collisional (Silk) damping strongly damped perturbations on scales less than about $10^{13} M_\odot$

In the baryon-only matter model, structure formation begins only after decoupling when the baryon Jeans mass drops and only on mass scales larger than the Silk mass scale, about $10^{13} M_\odot$. In this model universe, structure formation proceeds by a top-down fragmentation of perturbations on a mass scale larger than $10^{13} M_\odot$. This does not seem to be the way structure formed in our Universe, suggesting that we do not live in a baryon-only universe.

There is another reason to believe that we do not live in a baryon-only universe. Perturbations do not grow in a radiation-dominated universe; they only grow when the universe becomes matter dominated and then grow as $t^{2/3}$ or linearly in the scale factor a. Observationally, today, Ω in baryons is about 4.4×10^{-2}. (This value is consistent with the expectation from considerations of primordial nucleosynthesis.) The value of Ω in relativistic degrees of freedom (photons and neutrinos) today is about 4.4×10^{-5}, so today $\rho_{\text{radiation}}/\rho_{\text{baryons}} \sim 10^{-3}$. In expansion, $\rho_{\text{radiation}}$ scales as $(1 + z)^4$ and ρ_{baryons} scales as $(1 + z)^3$, so the ratio $\rho_{\text{radiation}}/\rho_{\text{baryons}} \propto (1 + z)$. Since the ratio today is 10^{-3}, the scale factor at the time of equal baryon and radiation density was $1 + z \sim 1,000$. If structure formation does not commence until after decoupling and when the universe becomes matter dominated (which, in a baryon-only model, happens at about the same time), perturbations grew by at most a factor of 10^3 from matter domination until today. In order for cosmological structures to form, the perturbations must grow to become non-linear, so the perturbation amplitude at decoupling must be at least 10^{-3}. This is larger than what we see in cosmic background radiation (CBR) measurements by a factor of at least 10.[9]

---

[9] For isentropic perturbations, the radiation perturbation amplitude is 1/3 as large as the matter perturbation amplitude.

In conclusion, in a baryon-only universe, structure must form top-down from fragmentation of structures of mass larger $10^{13} M_\odot$. Furthermore, for non-linear structures to form today, the amplitude of perturbations must be larger than allowed from CBR measurements.

Clearly structure as observed in our Universe did not form in a baryon-photon-neutrino universe. There must be something more in the mix. That something more is dark matter.

9.4.2 Growth of Dark-Matter Perturbations

Dark matter does not interact with radiation; in fact, it is collisionless. Rather than analyse fluid equations, in principle, one must solve the collisionless Boltzmann equation. In practice, one analyses moments of the collisionless Boltzmann equation. The analysis is similar to the Jeans analysis except the velocity dispersion of dark matter plays the role corresponding to the role played by the sound speed in the fluid equations.

Since dark matter is collisionless, it does not suffer Silk (collisional) damping. However, dark matter perturbations may suffer collisionless damping (or Landau damping). Again, in order to account for collisionless phase mixing, it is necessary to integrate the collisionless Boltzmann equation. However, it is possible to estimate the scale of collisionless damping.

If the collisionless species has a non-zero velocity dispersion with respect to the rest frame of the plasma, they will freely propagate out of over-dense regions and into under-dense regions, smoothing out inhomogeneities. Since we know how the velocity of a freely propagating particle redshifts in expansion, we can calculate the free-streaming distance.

Of course, if the dark matter is dead cold, then the free-streaming distance vanishes and the perturbations are undamped. Note that 'cold' refers to the velocity of the dark matter around the time of decoupling. Cold thermal relics are cold because the mass of the particle species is large compared to the temperature of the species around decoupling. But in general, dark-matter relics are cold because their velocity is small around decoupling. Another example of cold dark matter is the axion. Although axions are expected to have a mass smaller than the temperature of the Universe at decoupling (about 1/3 eV), they are cold because they were produced as a Bose condensate and a have small velocity dispersion.

In the other extreme, if dark matter is 'hot' (has a significant velocity around the time of decoupling), perturbations will suffer collisionless damping; perturbations will be damped on scales less than the free-streaming length. The quintessential example of hot dark matter is a light-mass neutrino. The contribution to Ω for a (two-component) neutrino of mass m_ν is $\Omega_\nu h^2 = m_\nu/91$ eV. So if the neutrino is the dark matter, $\Omega_\nu = 0.24$, and using $h^2 = 1/2$ we find $m_\nu = 11$ eV. We know the neutrino temperature and when it became non-relativistic. Calculating the free-streaming length until the time of matter-radiation equality when the perturbations can grow, we find that the free streaming length corresponds to a present distance of $\lambda_{FS} \simeq 100$ Mpc, containing, on average, a mass of about $2 \times 10^{16} M_\odot$.[10]

---

[10] These numbers come from a more exact calculation [5] of the free-streaming length. The free-streaming length from the more exact calculation is about a factor of 2 larger than the approximate calculation.

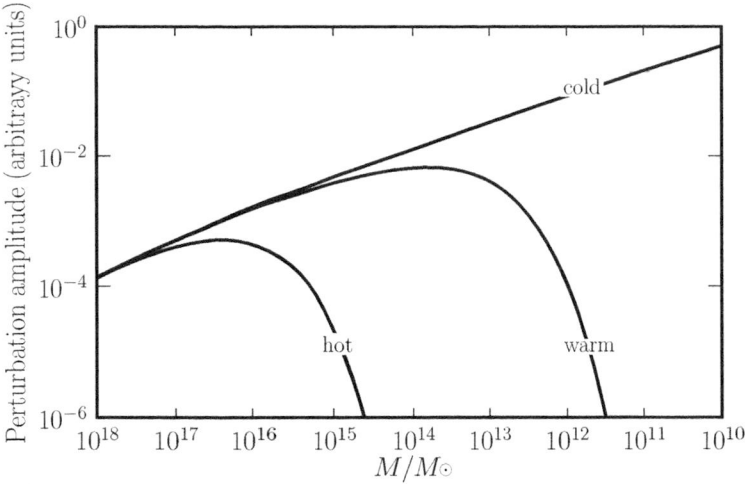

Figure 9.5 The dark matter perturbation spectrum is processed by collisionless damping. Shown are the perturbation spectra at some time after decoupling as a function of the mass in dark matter within the perturbation for three choices for dark matter: *cold*, dark matter that is non-relativistic (zero velocity) at decoupling; *hot*, dark matter that is semi-relativistic at decoupling; and *warm*, dark matter that is somewhere between hot and cold. (For aficionados, the 'perturbation spectrum' is $k^{3/2} |\delta_k|$, and for hot dark matter, I have chosen a neutrino of mass 11 eV, the appropriate mass if that species of neutrino contributes $\Omega = 0.24$.)

The dark matter perturbation amplitude processed by collisionless damping is shown in Figure 9.5. For the figure, it was assumed that dark matter today has a value of $\Omega = 0.24$, $h^2 = 1/2$, and the initial perturbation spectrum is a Harrison-Zeldovich spectrum. The spectrum shown represents the spectrum before the growth of perturbations after the Universe becomes matter dominated. For neutrino hot dark matter, the perturbations are damped for length scales today of about $2 \times 10^{16} M_\odot$. This is much larger than individual galaxy masses, so in a hot dark matter universe, structure forms top down.

Of course, there is the intermediate case of 'warm' dark matter. For warm dark matter, the dark matter is colder than neutrinos but hotter than cold dark matter. A sample warm dark matter spectrum is also shown in Figure 9.5.

Numerical simulations of structure formation highly disfavour hot dark matter. It seems that dark matter must be cold, or if it is warm, it must be very close to the spectrum for cold dark matter down to mass scales as small as astrophysically important for structure formation.

A universe with cold dark matter has another advantage over a baryon-only universe. We know from Big Bang nucleosynthesis and astronomical observations that the baryon density contributes only $\Omega_B = 4.4 \times 10^{-2}$. As mentioned before, this means that a baryon-only universe became matter dominated around $z \sim 1,000$. If we include dark matter, the total dark matter Ω plus the baryon Ω is about 0.3, or about an order of magnitude larger

Part IV
Fine-Tuning for Life

Eddington's Amazing Insight

Decades before we knew the energy source that powered the stars, how did famed astronomer Sir Arthur Eddington conclude that stars are inevitable? Eddington made his reputation by being one of the first to understand the significance of Einstein's theory of general relativity, published in 1915. Perhaps as a Quaker and conscientious objector, he had more time on his hands. But he realised that a key prediction, that gravity bends light, could be tested during the next total eclipse of the Sun by studying the deviations in positions of stars close to the edge of the Sun that otherwise were totally obscured by sunlight. This test occurred in 1919, when Eddington secured the possibility of the leadership of a solar eclipse expedition to Principe to substitute for the military service he refused. Viewed from this island off the west coast of Africa, totality would endure more than six minutes, one of the longest total eclipses of the century, and allowing ample time to photograph the stars visible near the position of the Sun. The measurement was a success, revealing the displacement predicted by Einstein's theory of the bending of light by gravity, twice that expected according to Newtonian gravitation.

Eddington became famous overnight, and even more so did Einstein. Eddington's major achievements, however, were in theoretical astrophysics, where he pioneered our modern understanding of stars. He reasoned that a star is a giant ball of gas supported by gravity. Here is what he wrote in 1926 (with very slight amendment):

We can imagine a physicist on a cloud-bound planet who has never heard tell of the stars calculating the ratio of radiation to gas pressure for a series of globes of gas of various sizes, stating say, with a globe of mass 10 gm., then 100 gm., 1000 gm. and so on, so that her nth globe contains 10^n gm. Regarded as a tussle between gas pressure and radiation pressure, the contest is overwhelmingly one-sided except between Nos. 33–35, where we may expect something interesting to happen. What 'happens' is the stars. We draw aside the veil of cloud beneath which our physicist has been working and have her look up at the sky. There she will find a thousand million globes of gas nearly all of mass between her 33rd and 35th globes, that is to say, between 1/2 and 50 times the sun's mass.

Eddington realised that the pressure of radiation is highly destabilising. He had studied giant globes of gas supported by the balance between their own gravity and the interior pressure of the gas. But as one turned up the mass, and consequently the gravity, the centre became so hot that the pressure of radiation exceeded that of the gas. And this was enough to blow the globe apart. And if the mass was too small, the centre of the globe was so cold that it could not resist gravity, so he reasoned. That is how he deduced, from pure thought, the mass range of the stars.

In the century that followed, astronomers measured the masses of many stars, typically by using the orbits in binary systems, and confirmed Eddington's reasoning. Most stars are similar in mass to the Sun, more typically half or a third the mass of the Sun.

10.3 Stellar Mass-Scales

Let us begin with how stars are made, by fragmentation of molecular clouds. What determines the scale of the first gravitationally bound clumps of gas, the building blocks of stars?

My analysis will be analytic and order of magnitude, designed to bring out the underlying physical scalings. All examples involve fundamental constants and illustrate the role of fine-tuning.

10.3.1 The Minimum Protostellar Fragment Mass

I first derive the collapse time of a collapsing cold gas sphere of either (1) uniform density or (2) non-uniform density. For simplicity, I take an isothermal sphere density profile $\rho(r) = v_s^2/(2\pi G r^2)$, where the sound speed is expressed in terms of the temperature $v_s^2 = kT/m_p$, and turbulence and magnetic support are neglected.

Some simple scaling relations between fragment mass M, radius R, and free-fall time t_{ff} are $R = v_s t_{ff} = M^{1/3}\rho^{-1/3} = GM v_s^{-2}$, and $\rho = M/R^3 = v_s^6 M^{-2} G^{-3}$. The luminosity of an optically thick fragment satisfies

$$L_{rad} = \sigma 4\pi T^4 R^2 = \sigma 4\pi T^4 G^2 M^2 v_s^{-4}. \tag{10.1}$$

The accretion rate onto a protostellar core that is surrounded by such a sphere controls the rate of gravitational energy release. The gravitational energy release from contraction is $L_g = GM^2 R^{-1} t_{dyn}^{-1} = GM^2 R^{-2} v_s = v_s^5 G^{-1}$. Hence, equating $L_{rad} = L_g$, $\sigma 4\pi T^4 G^2 M^2 v_s^{-4} = v_s^5 G^{-1}$ or $M = V_s^{9/2} T^{-2} G^{-3/2} (\sigma 4\pi)^{-1/2} = \text{const} T^{1/4}$. To convert to fundamental units, I use $\sigma = (2\pi^5/15)k^4 h^{-3} c^{-2}$ and write T dimensionlessly as $kT/m_p c^2$ to obtain

$$\frac{M}{m_p} = \alpha_g^{-3/2}\alpha^{1/2}\left(\frac{kT}{m_p c^2}\right)^{1/4}. \tag{10.2}$$

I also write T in units of Rydbergs, $1\text{Ry} = \alpha^2 m_e c^2/2$ to get a crude estimate of the cooling levels and energy radiated by atomic or molecular excitations. Dimensionless constants to a small power can fix this. Then one has

$$M = \alpha_g^{-3/2}\alpha\left(\frac{m_e}{m_p}\right)^{1/4}\left(\frac{kT}{1\text{Ry}}\right)^{1/4}. \tag{10.3}$$

This opacity-limited minimum fragment scale is around $0.01 M_\odot$ [30, 38] (see Figure 10.1).

Note that the definition of 'gravitational coupling constant', which explicitly connects the gravitational and electromagnetic couplings between two protons, is $\alpha_g = Gm_p^2/\hbar c = 6\times 10^{-39}$. One can also write this as $\alpha_g = Gm_p^2/\hbar c = (m_p/m_{pl})^2$, where the Planck mass is $\sqrt{\hbar c/G}$. We shall see that the characteristic mass of a star is $\alpha_g^{-3/2} = (m_{pl}/m_p)^3 = 2\times 10^{57}$. Note that the ratio of the attractive gravitational force to the repulsive electromagnetic force between two protons is $\alpha^{-1}\alpha_g$ (see also Chapter 2).

Attainable temperatures are 10 K in nearby molecular cloud cores and 1,000 K in primordial clouds with only trace amounts of H_2 as coolants. The sound speed and accretion rate vary considerably over this temperature range. However, one always finds that the minimum mass limited by opacity is around $M \sim 0.01 M_\odot$. Remarkably, detailed numerical

10.3.4 The Fate of Stars Like Our Sun: White Dwarfs

We can calculate from first principles the mass of a white dwarf. The Sun will end up as a
a white dwarf, a compact star whose nuclear fuel supply is exhausted and is supported by
degeneracy pressure of the electrons. This phenomenon occurs at sufficiently high density,
when nuclear fuel resources are exhausted and there is no thermal pressure support. The
Pauli exclusion principle comes into play, forbidding fermions from occupying the same
quantum state, and manifests itself as a quantum pressure.

For a white dwarf, there are two regimes:

1. If degenerate but non-relativistic, $p_e = n\Delta p^2/m_e = n^{5/3}\hbar^2/m_e$, where $\Delta p = \hbar n^{1/3}$,
   $n_e = \Delta x^{-3}$, and $(\Delta x)(\Delta p) = \hbar$. This yields $GM/R = p_e/\rho \propto n^{5/3}$.
2. If degenerate and relativistic, then $p_e = n_e c\Delta p = \hbar c n^{4/3}$, with $GM/R = p_e/\rho = \hbar c(M/R^3)^{1/3}/m_p^{4/3}$. This leads to the white dwarf mass $M_{WD} = \left(\hbar c/G^{3/2}m_p^{-3}\right)$.
   Rewriting this expression, the white dwarf mass is

$$M_{Chandra} \approx \alpha_g^{-3/2} m_p. \tag{10.14}$$

In summary, two dimensionless fine-structure constants – combining quantum, electromagnetic, and gravitational forces – determine the future of our sun, the mass of the white
dwarf to which it will collapse. The energy release en route is manifested by the formation
of an ultraluminous red giant, followed by the expulsion of a planetary nebula, leaving the
white-dwarf remnant in its centre. Planetary nebulae are among the most beautiful objects
photographed in the sky.

10.3.5 The Most Massive Stars

Let us next see what determines the upper limit on the mass of a star. One of the greatest
battles of all time occurs when the force of gravity faces up to the electromagnetic force.
Gravity is an incredibly weak force. But it adds up. For stars, its main opponent, electromagnetism, provides pressure and sums over positive and negative charges. So while
the repulsive electromagnetic force between a pair of protons is stronger than the attractive
gravitational force by about 40 powers of 10, if we consider enough atoms with protons and
electrons whose charges cancel, the forces can balance each other. When we turn on gravity,
many of the atoms are crushed into bare nuclei by the immense pressure that itself opposes
the force of gravity. The magic number for an equilibrium between electromagnetic and
gravitational force is about the mass of the Sun, give or take Eddington's factor of 10 or
100 – which, in a nutshell, is why we are here.

One can show that the upper limit on the mass of a hydrogen-burning star is $m_{max} \approx \alpha_g^{-5/3}\alpha^{2/3}m_p$. For a massive star, we use opacity $\tau = \kappa\rho R$, where $\kappa = \sigma_T/m_p$ and σ_T is
the Thomson scattering cross section. The luminosity is given by

$$L = (\kappa\rho)^{-1}\nabla(\sigma 4\pi T^4 R^2) \approx \sigma 4\pi T^4 R^2/\tau = v_s^5/G. \tag{10.15}$$

Then the electron scattering opacity can be expressed as

$$\tau_{es} = \frac{\sigma_T}{m_p} \frac{M}{R^2} = \pi \left(\frac{e^2}{m_e c^2}\right)^2 \frac{1}{m_p} \frac{1}{M} \left(\frac{kT}{Gm_p}\right)^2 = \alpha^2 \alpha_g^{-2} \frac{m_p}{M} \left(\frac{kT}{Gm_p}\right)^2. \tag{10.16}$$

The resulting expression for the maximum stellar mass is

$$M = \alpha_g^{-5/3} \alpha^{7/3} \left(\frac{m_e}{m_p}\right)^{1/6} \left(\frac{\tau}{\tau_{es}}\right)^{1/3}. \tag{10.17}$$

Here, τ is the optical depth expressed in terms of the Rosseland mean opacity. The explicit dependence on opacity is seen to be very weak.

Massive stars are short-lived because of their prolific burning of nuclear fuel. They explode after exhausting their nuclear fuel supply. Their masses are so large that there is no stable end point. Their cores collapse to form neutron stars, objects that are a thousand times more compact than white dwarfs, and supported by the quantum degeneracy pressure of neutrons. Such stars, if of initial mass less than about $25M_\odot$, end up as neutron stars. These are made of the most extreme form of matter with the nuclei so compacted together that the stars are essentially at nuclear density. The maximum mass of a neutron star is calculated to be three solar masses. The parent stars lose matter by driving winds as they evolve and end their lives as neutron stars. The radius of a solar mass neutron star is about 10 kilometres.

So much energy is released in this final collapse that a huge explosion blows off the outer layers of the star. This is a supernova explosion. The light released in this immense explosion, heralding the death throes of a massive star, is about that of the luminosity of an entire galaxy, some tens of billions of suns. The impact of such a violent explosion on the surrounding interstellar gas results in spectacular images of shock waves igniting quiescent gas clouds into glowing ribbons of hot gas that extend over hundreds of light years.

We observe objects well above the maximum mass of compact stars, produced by the death and collapse of massive hydrogen-burning stars. Stars that are initially more than about $25M_\odot$ cannot eject more than 90% of their initial mass. Any compact object that is more massive than $3M_\odot$ must be a black hole. One cannot pack matter any more densely than a neutron star without forming a black hole. The most massive stars collapse into black holes when their supply of nuclear fuel is exhausted. We measure their masses because there often is a close companion star whose atmosphere is heated by the gravity of its black hole companion and which emits prolifically in X-rays. Prior to 2016, the existence of black holes was conjecture and inference, although sound physical reasoning reinforced our belief in the existence of black holes. But the reasoning was indirect. The LIGO gravitational wave detectors provided the ultimate breakthrough via detection of merging black holes [1].

But one question leads to another. What came first, before the star clusters of Population III? Like that of the legendary old lady at a talk, often attributed to nineteenth-century philosopher William James, who had an answer as to what came first. Your cosmology is rubbish, she said – the world is supported by a giant turtle. So what supports the turtle?

It's a second turtle. And what supports that? It's turtles all the way down, she continued. This paradox of infinite regression is more commonly known as the chicken-and-egg problem: which came first? We believe that the smallest galaxies came first. Let us see what determines the mass of a galaxy today.

10.3.6 The First Stars

The most massive stars of all directly collapse into black holes. This is where our search for the first stars leads us. For we expect that the first stars were exclusively massive compared to the Sun. The early Universe was metal-free, consisting just of hydrogen and helium atoms. Now, metals control gas cooling and, hence, gas pressure. Because in today's Universe interstellar clouds are highly contaminated by metals, gas cooling is very effective. This means that the ability to support massive clouds from the compressive effects of gravity is greatly reduced. The clouds fragment into smaller clumps, and it is these that form stars today. The typical star in the Milky Way is about a third the mass of the Sun.

In the early Universe, conditions were different. There was pristine gas, hydrogen and helium. The lack of effective cooling by the metal pollutants meant that fragmentation was relatively ineffective. We refer to the first, zero metallicity, generation of stars as Population III. The gas temperature even in dense clouds is controlled by molecular hydrogen cooling. The typical temperature is determined by the lowest rotational level of H_2 and is around 1,000 K. This means that the sound speed is of order 3 km/s, and, hence, the accretion rate is around $\dot{M} \sim 3 \times 10^{-3} M_\odot \mathrm{yr}^{-1}$. Unimpeded accretion would generate typical first stars of thousands of solar masses. Such masses could accumulate in accreting gas over the typical Kelvin-Helmholtz timescale of a million years before strong energy release in the form of ionising photons occurs.

To better estimate how feedback intervenes to prevent all first stars being thousands of solar masses, consider the relation between mass and luminosity, $L \propto M^3$. Inserting this into the expression for the duration of the Kelvin-Helmholtz phase, one sees that $t_{KH} \propto 1/MR$. Let us assume v_s is controlled by molecular physics and require t_{KH} to be less than t_{acc} for effective feedback above a mass M_{cr}. Then, let's use the protostellar opacity constraint $\kappa \rho R \sim 1$ to set $R \sim (\kappa M)^{1/2}$ to deduce $M_{cr} \propto T^{-1}$, where T is the temperature of the accreting gas. I conclude that feedback is effective today in cold molecular clouds but much less effective by a factor ~ 100 in mass threshold in the primordial stellar case. Hence, the characteristic mass of the first stars is of order $100 M_\odot$.

Numerical simulations, including the effects of feedback on the accreting gas, show that there is considerable mass loss, and the typical mass of a first Population III star ranges from hundreds of solar masses at high redshift ($z > 20$), where there is intense UV radiation, to tens of solar masses at lower redshifts, where the UV background is lower and H_2 cooling is more effective [20].

These massive stars lived fast and furiously, dying in supernova explosions after millions of years. They generated elements like carbon, oxygen, and iron. The explosions polluted their environment. Once enough pollution of heavy elements occurred and enhanced cooling at low temperatures, the typical gas accretion rates onto protostellar fragments were greatly reduced. Once a metallicity $[Z] \gtrsim -4$ was reached, the first generation of exclusively massive stars gave way to formation of normal stars whose typical masses were of order a solar mass. There were fewer more-massive stars, and the mass range of Population II spanned the full range that Eddington had envisaged. Their low metallicity tells us that these are mostly old stars that formed early in the history of our Milky Way galaxy.

The spectra of the most metal-poor halo stars reveal fossil tracers from a previous generation of long-extinct stars of the first generation. The ratios of certain elements to a standard tracer such as iron are useful age indicators since iron is ejected in supernovae throughout the history of the galaxy. If we find a very iron-poor environment, we can be sure it is old. And if, in that environment, there are overabundant traces of unusual elements not normally produced in today's Universe, we can attribute such clues to the ashes from the first generation of short-lived stars, incorporated into the clouds that made all later stars. It as though we have silent witnesses to a long-vanished crime scene.

10.3.7 So What If G Was Different?

Imagine we live in a universe with a very different value of Newton's constant. If G were too small, there would not be time to make stars within the age the Universe. The lifetime of a nuclear-fuelled star is proportional to GM. Since the mass of a star is of order $\alpha_G^{-3/2}$, its lifetime is proportional to $G^{-1/2}$. So if G were smaller by a factor of 10^8, carbon would not be formed, and we would not be here. This argument assumes the age of the Universe has nothing to do with G, as it is dominated by dark energy – that is to say, the energy density associated with the cosmological constant, or some 10 billion years when it becomes dominant (see Chapter 3 for more details). That's the modern view.

Robert Dicke had a somewhat different take on this [10]. In the case of a closed Friedmann universe, considered by many at that time to be the most natural choice of a cosmological model, the age of the universe depends as $G^{-1/2}$, and so stars are inevitably fine-tuned to produce carbon. However, for an Einstein–de Sitter universe, or an open universe, we would need to be at a special epoch. This can be expressed as Dirac's coincidence [11] between the scales of cosmology, involving G and H_0, and quantum physics, involving α and $\hbar$.

10.3.8 Our Lack of Understanding of Star Formation Physics

Here is one recent example to demonstrate the fragility of our current knowledge of star formation. In our Milky Way galaxy, most dense molecular gas, where much of galactic star formation occurs in giant molecular clouds (GMCs), is in the ~ 3 kpc radial distance molecular ring. However, within ~ 0.5 kpc radial distance, the central molecular zone (CMZ) contains massive giant molecular clouds of typical mass $\sim 10^6 M_\odot$ and density

Trace amounts of hydrogen molecules form from residual ionisation and H^- formation, and the enhanced cooling leads fragmentation into stars. This fate can be avoided if there is enough turbulence present, as this can remove the molecules in shocks or via UV radiation from nearby stars and black holes that destroy any molecules in their vicinities.

For a seed black hole, Ly-α cooling is thought to be the prevalent cooling mechanism for black hole feeding. Take v_c to correspond to $m_p v_c^2 = 0.75\text{Ry} = 10.2\text{eV}$, corresponding approximately to the sound speed in a medium where Ly-α cooling ($n = 2$ to $n = 1$) prevails. This is equivalent to $v_{ref} = \alpha c \left(\frac{m_e}{m_p}\right)^{1/2}$. The inferred black hole mass mass is

$$M_{\text{seed}} = \left(\frac{v_c}{v_{ref}}\right)^5 \frac{\alpha^5}{\alpha_g^2} \left(\frac{m_e}{m_p}\right)^{1/2}. \tag{10.35}$$

The ratio of seed black hole to galaxy mass is $\alpha^2 \left(\frac{m_e}{m_p}\right)^{3/2} \sim 10^{-7}$. This will provide the necessary boost for Eddington-limited growth to form supermassive black holes at early epochs.

Astrophysics of SMBH

Most massive galaxies are elliptical galaxies and today are indeed red and dead. All contain massive central black holes. Because of the paucity of interstellar gas, these black holes are not being fed and are not active. Occasionally, there is a merger, most typically with a small galaxy. If this contains gas, some of this will fall into the black hole and drive a new phase of activity.

A massive black hole lurks in the centre of our own galaxy. The Milky Way black hole is located in the constellation of Sagittarius. Studies of orbiting stars allowed a precise mass determination. The stars can be directly resolved in our galactic centre at infrared wavelengths, and their orbital motions and speeds around the black hole have been followed for more than a decade.

Our massive black hole is a bright source of radio waves, one of the brightest in the sky, hence its name Sagittarius A. Despite the radio emission, it is not very active today. The X-ray emission from Sagittarius A is very low, and we infer that it is currently lacking fuel, accreting very little gas from its surroundings. But in the past, the situation was occasionally very different, and Sagittarius A has experienced outbursts of violent activity. About 10 million years ago, a giant explosion occurred that left behind traces in the gamma ray sky. The Fermi gamma ray satellite telescope image has revealed twin bubbles of gamma ray emission hundreds of light years in extent that attest to the violence of this explosion a million years ago.

Quasars are much more vigorous manifestations of black hole activity. Their scaled-up outbursts and X-ray luminosities enable us to detect them at the far end of the Universe. Perhaps the most intriguing result of studies with the largest telescopes show that intense black hole activity and extreme rates of star formation occur in the same objects. They are also driving powerful outflows of gas. This is one of the biggest mysteries of structure formation. Why are these phenomena coinciding at early epochs, especially reaching extreme rates of star formation and mass loss that are rarely seen in the absence of supermassive black holes?

Perhaps black hole feeding, with its immense energy release, is a consequence of star formation, with the stellar debris feeding the black hole. If so, this fails to address the extreme intensity of star formation seen in many of the most distant galaxies. Both AGN activity and star formation bursts may be the collateral damage from gas-accretion events. Gas-rich galaxy mergers can provide the gas supply responsible both for star formation and supermassive black hole fuelling. In a merger, gas-cloud orbits are perturbed, and some clouds are directed into the capture zone of the supermassive black hole to refuel its activity. The increase in gas mass stimulates star formation at the same time.

An alternative pathway has the powerful outflows from the black holes compressing nearby clouds and triggering an intense burst of star formation in the surrounding clouds. This is predicted analytically [42] and in simulations [16]. Of course, the vigorous outflows show that the star formation rate is being quenched. But this may have been preceded by a phase of triggering that induced the jets and winds. Predicted signatures include star formation by positive feedback, leaving an imprint on the kinematics of stars formed via AGN ouflow-triggered star formation. Such kinematic tracers would be both long-lived and differ from the more chaotic kinematics of cloud collision and/or SN-induced star formation in the conventional scenario of gravitationally unstable disc star formation [12].

Observational evidence has recently been found for positive feedback by AGN, despite the expected short duty cycle before negative feedback kicks in. Examples include stacked star-forming galaxies [5] that show evidence of radial trajectories of hypervelocity stars, as found in the Milky Way and interpreted as being induced by Fermi-bubble-type explosions millions of years ago [44], and in AGN-driven outflows [27]. Indications of positive feedback via enhanced star formation rates are also found in green valley, X-ray detected AGN galaxies [26]; in a Seyfert galaxy [8]; along with simultaneous negative and positive feedback in an obscured radio-quiet quasar [7].

SMBH Feedback and the M_{BH}-σ Scaling Relation

Black hole feeding, with its immense energy release, may be a consequence of star formation, with the stellar debris feeding the black hole. It more likely is fed by gas streams similar to those that replenish the gas reservoir and drive star formation. Alternatively, a merger between two galaxies helps shed gas angular momentum and allows gas to pour into the vicinity of the massive central black hole. The paucity of major mergers favours the merger hypothesis only for the most luminous quasars.

The outpouring of energy from the accreting SMBH exerts strong feedback on the surrounding interstellar medium. It is conjectured that there is first a period of positive feedback during which gas clouds are compressed and a burst of star formation is triggered, followed a million years later by a phase of gas-outflow and star formation quenching because of the combined effects of AGN outflows and supernovae from the newly formed massive stars [40]. A key issue is the efficiency of the coupling of the high-energy ejecta with the surrounding ISM. This may involve momentum or energy conservation in order to explain the massive outflows of escaping gas observed from the host galaxy. Nuclear outflows are observed at $\sim 0.1c$, and these drive outflows on galactic scales at typical velocities of thousands of km/s which are responsible for depleting the gas reservoir and quenching star formation [15, 48]. Quenching must occur in order to explain the red and dead nature of early-type galaxies, and SMBHs are the principal culprit.

It is not simply momentum conservation of the SMBH ejecta that drives galactic winds. Reality, as found by simulations and observations, is more complicated. Feedback by massive black holes is controlled by the balance between gravitational attraction and radiation pressure acting on ionised accreting gas. The former is controlled by

$$\frac{L_{\text{Edd}}}{c} = \frac{GMM_{gas}}{r^2} \tag{10.36}$$

and the latter by

$$L_{\text{Edd}} = 4\pi c G M_{\text{BH}} m_p / \sigma_T. \tag{10.37}$$

When blowout occurs at a high enough rate, the gas reservoir is exhausted, and star formation terminates. This should apply for a homogeneous ISM when the $M_{\text{BH}} - \sigma$ relation saturates, and then

$$M_{\text{BH}} = 3 \times 10^9 \left(\frac{\sigma}{300 \text{ km/s}} \right)^4.$$

Mechanical feedback by SMBH is driven both by jets and winds. Jets drive bow shocks that interact much as do winds, geometrical factors aside. The Eddington luminosity and Eddington mass-outflow rate are related by $L_{\text{Edd}} = \eta c^2 \dot{M}_{\text{Edd}}$. We can infer the central wind velocity, since $L_{\text{Edd}}/c = v_w \dot{M}_{\text{Edd}}$ and the outflow speed is $v_w = \eta c \sim 0.1c$ for a nuclear wind, as observed.

There are two limiting cases that correspond to momentum or energy conserving outflows. Momentum balance gives $M_{\text{BH}} = \sigma^4 (\sigma_T/m_p)/(\pi G^2)$. Inclusion of cooling results in failure to eject enough material to quench star formation, failing to attain both the massive galaxy luminosity function and the massive black hole scaling relation.

Energy balance $1/2 M_{\text{outflow}} v_{\text{shock}}^2 = \text{const}$ means that the momentum increases as $P \propto 1/v_w$. This results in enough momentum boost to reduce excessive SMBH feedback and growth, thereby reproducing the normalisation of the Magorrian relation. The predicted scaling is given by

$$M_{\text{BH}} = \alpha \sigma^5 \frac{(\sigma_T/m_p)}{(\pi \eta c G^2)}. \tag{10.38}$$

Naïve application of this expression gives too low a normalisation for the observed relation. These results apply to highly simplified outflows. Cosmological gas accretion is ignored, as is the more realistic ISM structure. In reality, inflow is not quenched in a multiphase ISM. One needs nuclear momentum at a level $\sim 10 L_{\text{Edd}}/c$ [43]. Because the ISM is multiphase, a more realistic treatment is needed, as diffuse gas is ejected more easily, whereas dense gas clumps tend to fall in.

Simulations with cosmological accretion converge on the energy-driving case, albeit needing more effective coupling than single photon radiation scattering as in the preceding naïve formulae. Energy-driven outflows can reach momentum fluxes exceeding $10L_{Edd}/c$ within the innermost 10 kpc of the galaxy. The observed large-scale AGN-driven outflows seem to be energy driven, according to simulations. The momentum of the swept-up shell grows with decreasing shell velocity and, hence, gives effective outflows that shut off star formation and can recover the observed normalisation [6].

Observations favour energy conservation, connecting nuclear outflows from accretion discs around the AGN at $\sim 0.1c$ with kiloparsec-scale galactic-scale molecular outflows at several hundreds of km/s. The more massive the central SMBH is, the stronger the quasar outflow is found to be [32]. The spheroid mass and SMBH mass both grow with time because of the merger and gas accretion history.

There is a large variance in the scaling relations. In particular, while IMBH and SMBH generally lie on the M_{BH}–σ relation, there is more dispersion in the M_{BH}–M_* relation. For early-type massive galaxies, star formation is preferentially quenched in the quasar phase; hence, the SMBH can be overmassive relative to the stellar component. Indeed, a much broader dispersion is found of black hole mass to stellar mass [51] at high redshift relative to the nearby universe, with some extraordinarily massive SMBH outliers [36]. For late-type galaxies, continuing star formation at late times is more important, and black holes can be under-massive. Consequently, while there is expected to be little deviation in the M_{BH}–σ relation, which primarily samples the gravitational potential, there should be a trend for SMBH at high redshift to often be obese and for IMBH, necessarily observed at low redshift, to be anorexic [28], as observed.

10.6 Conclusions

We have made immense progress in star and galaxy formation over the past several decades. Much of this is driven by data from space and terrestrial telescopes–ranging from Spitzer, Hubble, and Herschel to IRAM and ALMA. Much is also driven by numerical simulations at ever-improving resolution. It is possible to understand some of the key scales in terms of the fine-tuning of fundamental parameters. I discussed estimates for, among other characteristic scales, the typical protostellar fragment mass, stellar mass, maximum stellar mass, galaxy mass, maximum galactic mass, supermassive black hole mass, and SMBH seed mass. I showed how feedback may be characterised by ratios of fundamental constants, both for galaxies and supermassive black holes.

The essence of this discussion ultimately is star formation. The truth remains that simulations of star formation, especially in the cosmological context, cannot span the essential dynamical range between galaxy and star formation. Our models for star formation remain phenomenological rather than fundamental, and all predictions for the high-redshift universe should consequently be taken with a grain of salt. Semi-analytical models of galaxy formation have 50 or more parameters. These parameters are set phenomenologically, based on studies of the nearby universe, and one cannot be confident that similar rules for star formation apply at high redshift. Numerical simulations rely on sub-grid modelling of star formation physics. The paradigm is reasonable but, again, is based on experience deduced from nearby regions of star formation. The early Universe differs in being not only gas rich but also AGN and quasars being far more prevalent, given that their duty cycles render them almost permanently in active mode.

Galaxy formation is not well understood in our vicinity, where, for example, only half the baryon content of the Universe is accounted for. There are problems in understanding the cored nature and the abundance of dwarf galaxies, for which, admittedly, there are solutions that, however, require a certain degree of fine-tuning. Similar remarks apply to the prevalence of bulgeless galaxies and of ultra-diffuse galaxies in our vicinity.

The most direct evidence for fine-tuning is usually associated with the adopted strength of gravity. If the gravitational constant is too low, stars do not evolve sufficiently to produce the carbon needed for life. On the larger scales, galaxies provide the backdrop for star formation. Here, the most relevant parameter is the strength of the initial density fluctuations, often attributed to inflation-boosted quantum fluctuations (see Chapter 4). Over a modest range, this determines the abundance of galaxies and their epoch of formation.

If, however, one varies the fluctuation strength by an order of magnitude or more, one runs into serious problems. If the initial fluctuations are much too large, the Universe contains relatively few galaxies but predominantly supermassive black holes. If the fluctuations are much too small, there are no galaxies [47]. The fluctuation strength is a phenomenological parameter that cannot be directly related to fundamental constants; rather, it is a product of inflation and of poorly constrained inflationary physics.

It is not clear, short of a compelling theory of inflation, how one can control the initial conditions of the Universe. Inflation cannot generate isotropy and flatness of space from generic initial conditions. Nor is there a simple measure of initial conditions, although we know that our observed Universe represents a small subset of possible universes. Here, I prefer to focus on the phenomenology of galaxies and stars and to link this phenomenology to the fundamental constants of nature, motivated by the possibility that these may vary between possible universes. That is the cosmic connection that one can develop without straying too far beyond observational constraints.

References

[1] Abbott, B. P., *et al.* 2016. 'Observation of Gravitational Waves from a Binary Black Hole Merger'. *Physical Review Letters*, **116**(6), 061102.

[2] Barnes, A. T., *et al.* 2017. 'Star Formation Rates and Efficiencies in the Galactic Centre'. *Monthly Notices of the Royal Astronomical Society*, **469**(Aug.), 2263–2285.

[3] Birnboim, Y., Federrath, C., and Krumholz, M. 2018. 'Compression of Turbulent Magnetized Gas in Giant Molecular Clouds'. *Monthly Notices of the Royal Astronomical Society*, **473**(Jan.), 2144–2159.

[4] Chabrier, G., Hennebelle, P., and Charlot, S. 2014. 'Variations of the Stellar Initial Mass Function in the Progenitors of Massive Early-Type Galaxies and in Extreme Starburst Environments'. *Astrophysical Journal*, **796**(Dec.), 75.

[5] Cicone, C., Maiolino, R., and Marconi, A. 2016. 'Outflows and Complex Stellar Kinematics in SDSS Star-Forming Galaxies'. *Astronomy and Astrophysics*, **588**(Apr.), A41.

[6] Costa, T., *et al.* 2018. 'Quenching Star Formation with Quasar Outflows Launched by Trapped IR Radiation'. *Monthly Notices of the Royal Astronomical Society*, **479**(Sept.), 2079–2111.

[7] Cresci, G., *et al.* 2015a. 'Blowin' in the Wind: Both "Negative" and "Positive" Feedback in an Obscured High-z Quasar'. *Astrophysical Journal*, **799**(Jan.), 82.

[8] Cresci, G., *et al.* 2015b. 'The MAGNUM Survey: Positive Feedback in the Nuclear Region of NGC 5643 Suggested by MUSE'. *Astronomy and Astrophysics*, **582**(Oct.), A63.

[9] Dekel, A., and Birnboim, Y. 2006. 'Galaxy Bimodality Due to Cold Flows and Shock Heating'. *Monthly Notices of the Royal Astronomical Society*, **368**(May), 2–20.

[10] Dicke, R. H. 1961. 'Dirac's Cosmology and Mach's Principle'. *Nature*, **192**(Nov.), 440–441.

[11] Dirac, P. A. M. 1937. 'The Cosmological Constants'. *Nature*, **139**(Feb.), 323.

[12] Dugan, Z., Gaibler, V., and Silk, J. 2017. 'Feedback by AGN Jets and Wide-Angle Winds on a Galactic Scale'. *Astrophysical Journal*, **844**(July), 37.

[13] Faber, S. M., and Jackson, R. E. 1976. 'Velocity Dispersions and Mass-to-Light Ratios for Elliptical Galaxies'. *Astrophysical Journal*, **204**(Mar.), 668–683.

[14] Ferrarese, L., and Merritt, D. 2000. 'A Fundamental Relation between Supermassive Black Holes and Their Host Galaxies'. *Astrophysical Journal Letters*, **539**(Aug.), L9–L12.

[15] Feruglio, C., *et al.* 2015. 'The Multi-Phase Winds of Markarian 231: From the Hot, Nuclear, Ultra-Fast Wind to the Galaxy-Scale, Molecular Outflow'. *Astronomy and Astrophysics*, **583**(Nov.), A99.

[16] Gaibler, V., *et al.* 2012. 'Jet-Induced Star Formation in Gas-Rich Galaxies'. *Monthly Notices of the Royal Astronomical Society*, **425**(Sept.), 438–449.

[17] Gebhardt, K., *et al.* 2000. 'A Relationship between Nuclear Black Hole Mass and Galaxy Velocity Dispersion'. *Astrophysical Journal Letters*, **539**(Aug.), L13–L16.

[18] Gnat, O., and Sternberg, A. 2007. 'Time-Dependent Ionization in Radiatively Cooling Gas'. *Astrophysical Journal Supplements*, **168**(Feb.), 213–230.

[19] Hennebelle, P., and Chabrier, G. 2008. 'Analytical Theory for the Initial Mass Function: CO Clumps and Prestellar Cores'. *Astrophysical Journal*, **684**(Sept.), 395–410.

[20] Hirano, S., *et al.* 2015. 'Primordial Star Formation under the Influence of Far Ultraviolet Radiation: 1540 Cosmological Haloes and the Stellar Mass Distribution'. *Monthly Notices of the Royal Astronomical Society*, **448**(Mar.), 568–587.

[21] Kennicutt, Jr., R. C. 1989. 'The Star Formation Law in Galactic Disks'. *Astrophysical Journal*, **344**(Sept.), 685–703.

[22] Krumholz, M. R. 2011. 'On the Origin of Stellar Masses'. *Astrophysical Journal*, **743**(Dec.), 110.

[23] Larson, R. B. 1981. 'Turbulence and Star Formation in Molecular Clouds'. *Monthly Notices of the Royal Astronomical Society*, **194**(Mar.), 809–826.

[24] Leroy, A. K., *et al.* 2015. 'ALMA Reveals the Molecular Medium Fueling the Nearest Nuclear Starburst'. *Astrophysical Journal*, **801**(Mar.), 25.

at least intriguing: (1) the lack of super-Earths and (2) the lack of very close-in planets. Let's briefly discuss each one of these characteristics.

11.2 Lack of Super-Earths

Super-Earths are planets with masses typically between 1 and 10 Earth masses. They are very common in exoplanetary systems. In fact, more than half of the observed Sun-like stars in the solar neighbourhood are orbited by one or more super-Earths with periods of days to months (e.g., [12, 23, 55, 109]). Furthermore, systems observed to contain a super-Earth usually have more than one. Most recently, observations of the star Trappist-1 found that it hosts at least seven Earth-size planets. Their orbits, six of which form a near-resonant chain (with orbital periods of 1.51, 2.42, 4.04, 6.06, 9.1, and 12.35 days), suggest that these planets formed farther from their host star and then migrated inwards (e.g., [146]). If super-Earths form via mergers of inwardly migrating cores, then having more than one super-Earth is theoretically expected (e.g., [39]).

The fact that the solar system does not contain any close-in super-Earths does set it somewhat apart from most observed exoplanetary systems. Before I examine potential *reasons* for this lack of super-Earths, it is interesting to contemplate whether not containing super-Earths (by a planetary system) can in any way be related to the emergence of life in such a system. It is interesting to note that the presence of super-Earths may affect the formation process and properties of terrestrial planets. In particular, Izidoro *et al.* [80] found that if a super-Earth migrates slowly enough through the habitable zone of its host star (the range of orbital radii that allows for liquid water to exist on a rocky planetary surface), then terrestrial planets subsequently forming in this region would be rich in volatiles – far from being an Earth twin. A super-Earth at a very small orbital radius could also (in principle, at least) disturb the dynamical stability of a terrestrial planet in the habitable zone. In contrast, the terrestrial planets in the solar system are expected to remain dynamically stable until the Sun expands to become a red giant (e.g., [86]). Finally, a super-Earth relatively close to the orbital distance of a terrestrial planet could significantly affect the rate of asteroid impacts on such a planet (see Section 11.5.3 for an extensive discussion).

11.3 Lack of Close-In Planets or Debris

The second element that appears to distinguish the solar system from most observed exoplanetary systems is the lack of very close-in planets (or, for that matter, any type of mass). Specifically, while Mercury is at 0.39 AU from the Sun, other exoplanetary systems harbour planets much closer to their host star [15]. In particular, most systems observed to have three or more planets contain a planet with a semimajor axis smaller than that of Mercury.

Again, it is not clear to what extent (if at all) the absence of close-in planets is related to the Earth's habitability. In the next section, I examine the conditions that are necessary for the formation of super-Earths and their potential implications for the two characteristics that make the solar system somewhat special.

11.4 On the Formation of Super-Earths

Super-Earths either form in situ with no significant migration through the protoplanetary disc or they form outside the snow line (the distance from the central star where the temperature is sufficiently low for water to solidify; [88]) and then migrate inwards. Chiang and Laughlin [35] used observations of super-Earths with orbital periods of $P \lesssim 100$ days to construct a minimum-mass extrasolar nebula (MMEN) with a surface density of the gas disc of (R is the distance from the star)

$$\Sigma_{\text{MMEN}} = 9,900 \left(\frac{R}{1 \text{ AU}} \right)^{-1.6} \text{ g cm}^{-2}. \tag{11.5}$$

Equation (11.5) gives a somewhat higher value for the surface density than the minimum mass solar nebula (MMSN, which gives 1,700 g cm^{-2} at $R = 1$ AU, required to form the planets in our solar system [70, 155]). However, it is not clear whether the MMSN is applicable at distances smaller than 0.4 AU (inside Mercury's orbit). Hansen and Murray [67, 68] found that forming super-Earths in situ required having about 50–100 $M_{\oplus}$ of rocky material interior to 1 AU. Other researchers suggested that super-Earths form farther out in the disc (where solid material is more readily available) and migrate inwards (e.g., [39, 78, 147]).

Martin and Livio [103] showed that in fully turbulent disc models (i.e., discs in which the magnetorotational instability generates viscosity throughout) planets that are close to their host stars cannot form in situ since the mass of the disc interior to $R = 1$ AU is too low (Figure 11.5). Martin and Livio [103] have also shown, however, that in fully turbulent discs super-Earths can (in principle, at least) form farther out, followed by inward migration.

Still, it is generally believed that protoplanetary discs are not fully turbulent. Rather, they most likely contain an unionised region of low or no turbulence, known as a 'dead zone' (e.g., [58, 141]). The dead zone blocks the accretion and allows material to accumulate until it becomes self-gravitating. (Note though that it has been suggested that the Hall effect can 'revive' the dead zone under certain conditions, e.g., [89]).

Martin and Livio [103] considered two prescriptions for determining the surface density in the dead zone. In the first, they assumed that the disc surface layers are ionised by external sources to a maximum surface density depth of $\Sigma_{\text{crit}}/2$ (see discussion in Section 11.4.1). In the second, they assumed that the surface density is determined via a critical magnetic Reynolds number (e.g., [52, 69]), $Re_{\text{M,crit}}$, such that the zone is 'dead' if $Re_M < Re_{\text{M,crit}}$. By following the time-dependent evolution of the disc, they showed that at early times the disc undergoes FU Orionis-type outbursts (see also [5, 165]). These outbursts occur as the extra heating by self-gravity triggers the magnetorotational instability within the dead zone (when the infall accretion rate is still high). At later times, as the accretion rate decreases, there are no further outbursts (Figure 11.6). Planets that survive must form after the cessation of outbursts; otherwise, they are likely to be swept into the star during accretion episodes. Through time-dependent numerical simulations, Martin and Livio [103] have demonstrated that depending on the dead-zone parameters, the disc surface density and the mass inside of 1 AU can build up to several times that of the MMEN exists, the eccentricities and inclinations of the asteroids are damped by tidal interactionsx with the protoplanetary discs (e.g., [7, 85, 153]). The lifetime of the gas disc is typically a few million years [64], after which it is being dispersed by photoevaporation (see discussion in [4]). After the gas disc is removed, gravitational perturbations clear asteroids at many resonance locations (e.g., [37, 61, 112]). This clearing creates potential Earth (and other terrestrial planets) impactors. In the solar system, Jupiter and Saturn, which are the largest planets, are the main drivers of the dynamical evolution of the asteroid belt, even though the effects that Jupiter had on the collision rate with the Earth is debated [75, 76]. Similarly, giant planets can be the driving force of evolution (if an asteroid belt exists) in exoplanetary systems.

Resonances that play a major role in the asteroid belt dynamics are *mean-motion* resonances and *secular* resonances. In a mean-motion resonance, the ratio of the orbital periods of two objects is a ratio of two integers. Several of the known mean-motion resonances in the solar system, such as the Kirkwood gaps [43, 115, 128], are found within the asteroid belt. Secular resonances arise when the apsidal precession rate of two objects orbiting a common central object are close to each other (e.g., [56, 162]). From our perspective here, the most important secular resonance in the solar system is the v_6 resonance ([19, 113] and references therein). It relates the apsidal precessions of the asteroids and Saturn, and it sets the inner boundary (at about 2 AU) of the solar system's asteroid belt. Each resonance has a certain libration width in semimajor axis over which it is effective. Asteroids that fall within a libration width undergo perturbations that cause their eccentricities to increase to the point where they are either ejected from the system or they collide with a planet or the central star. Regions in which libration widths overlap are dynamically chaotic regions [125], and almost all asteroids are cleared from such regions. The outer edge of the asteroid belt in the solar system (at about 3.3 AU) is determined by the overlapping of Jupiter's resonances.

The fact that many of the exoplanetary systems contain a super-Earth (e.g., [82]; Figure 11.10) raises the additional question of how the presence of a super-Earth in the solar system might have affected the rate of asteroid impacts on Earth. Smallwood *et al.* [143] studied in particular multi-planet systems in which there had been no process (e.g., migration) that could have destroyed the asteroid belt or could have prevented the formation of terrestrial planets. More specifically, they investigated how the architecture of systems such as the solar system affects the rate of asteroid impacts on Earth (or a similar terrestrial planet).

The Numerical Method: N-body Simulations

Smallwood *et al.* [143] used the hybrid symplectic integrator in the orbital dynamics package MERCURY to model the structure of the asteroid belt and the rate of impacts on Earth (see [30] for details of the package). The motions of Jupiter, Saturn, a super-Earth, Earth, and a distribution of asteroids all orbiting a central object were simulated. The asteroids were assumed to gravitationally interact with the star and the planets but not with each other. The evolution of the orbit of each asteroid was followed for ten million years.

Figure 11.10 Planet masses and semimajor axes of observed exoplanets (from [65]). The area between the dashed lines contains the range of super-Earth masses used in the simulations described in Section 11.5.3. The transparent grey box highlights the observed super-Earths with a semi-major axis corresponding to the inner solar system.

In the solar system, there are more than 10,000 asteroids with high-accuracy measurements of their semimajor axes. The mean of those values is $\langle a \rangle = 2.74 \pm 0.616$ AU. The mean eccentricity is $\langle e \rangle = 0.148 \pm 0.086$, and the mean inclination $\langle i \rangle = 8.58° \pm 6.62°$ (see description in [124]). At the same time, the precise initial distribution of the asteroids within the belt (immediately following the dispersal of the protoplanetary gas disc) is not accurately known. In their simulations, Smallwood *et al.* [143] therefore assumed a uniform distribution (see also [87]), with the semimajor axis of each asteroid given by

$$a_i = (a_{max} - a_{min})\,\xi_r + a_{min}, \tag{11.8}$$

where $a_{min} = 1.558$ AU is the inner boundary of the distribution, $a_{max} = 4.138$ AU is the outer boundary, and ξ_r is a randomly generated number between 0 and 1. The inner and outer boundaries were based on the structure of the solar system, with a_{min} being three Hill radii (the region in which a body's gravity dominates) beyond the semimajor axis of Mars, and a_{max} being three Hill radii inside Jupiter's orbit. The Hill radius is given by

$$R_H = a_p \left(\frac{M_p}{3\,M_s} \right)^{1/3}, \tag{11.9}$$

where a_p is the planet's semimajor axis, M_p is its mass, and M_s is the mass of the star. No asteroids are likely to be located within three Hill radii because this is the planet's gravitational reach (e.g., [31, 60, 123]).

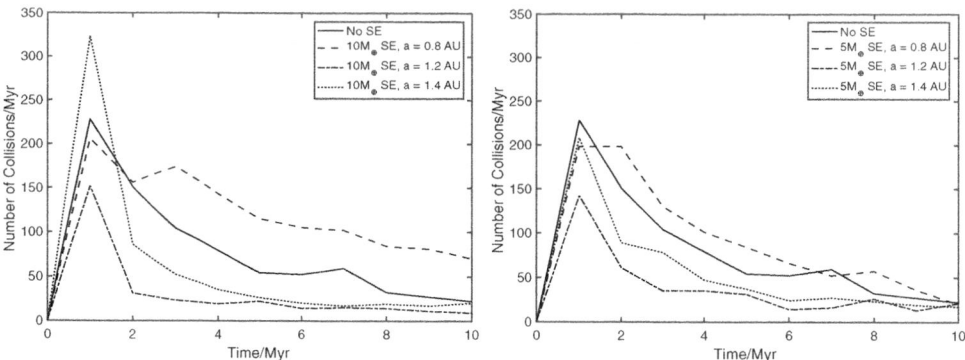

Figure 11.20 The collision rate with Earth per million years. The left panel involves a 10 $M_\oplus$ super-Earth, and the right panel a 5 $M_\oplus$ super-Earth.

1.4 AU, it creates a chaotic zone which is produced by the overlapping of the libration widths of the super-Earth's 6:5, 4:3, and 7:5 mean-motion resonances. In turn, this chaotic zone clears a large number of asteroids from the inner parts of the asteroid belt. The chaotic region can be clearly seen in Figure 11.19.

The effects of the mass of the super-Earth on the rate of asteroid impacts on Earth is shown in Figure 11.20. What is depicted is the number of impacts per million years for a super-Earth with a mass of $10M_\oplus$ (left panel) and a mass of $5M_\oplus$ (right panel), for various semimajor axis values. The initial spike in the impact rate is due to the fact that the number of asteroids in the belt is larger at the beginning of the simulations than at their end. The fact that a $10M_\oplus$ super-Earth at 1.4 AU produces the highest impact rate at 1 Myr is due to the chaotic region created by the super-Earth. The rate, however, rapidly declines as asteroids are cleared out. Over the 10 Myr duration covered by the simulation, a super-Earth located at 0.8 AU produces the highest rates, both for a $10M_\oplus$ super-Earth and a $5M_\oplus$ one.

As I have noted earlier, the ν_6 resonance is a major contributor to the rate of asteroid impacts on Earth. In fact, the majority of the asteroids colliding with the Earth originate from the location of the ν_6 resonance in the asteroid belt. Specifically, in the simulations of Smallwood *et al.* [143] without a super-Earth, the total number of asteroid impacts on Earth produced by secular resonances was about 2.5 times higher than that produced by mean-motion resonances. *If* asteroid impacts were indeed important for the emergence and/or evolution of life on Earth, then the ν_6 resonance may have played a significant role in our planet's habitability. The ν_6 resonance involves both Saturn and Jupiter. Basically, Jupiter increases the precession frequency of the asteroids so that they fall into a resonance with the apsidal precession rate of Saturn. In Figure 11.21 (taken from [143]), I show the precession rate of a test particle as a function of orbital separation. The solid horizontal line marks the eigenfrequency of Saturn (found using a generalised form of secular perturbation theory; e.g., [124]). The top-left panel of Figure 11.21 represents our solar system since it includes

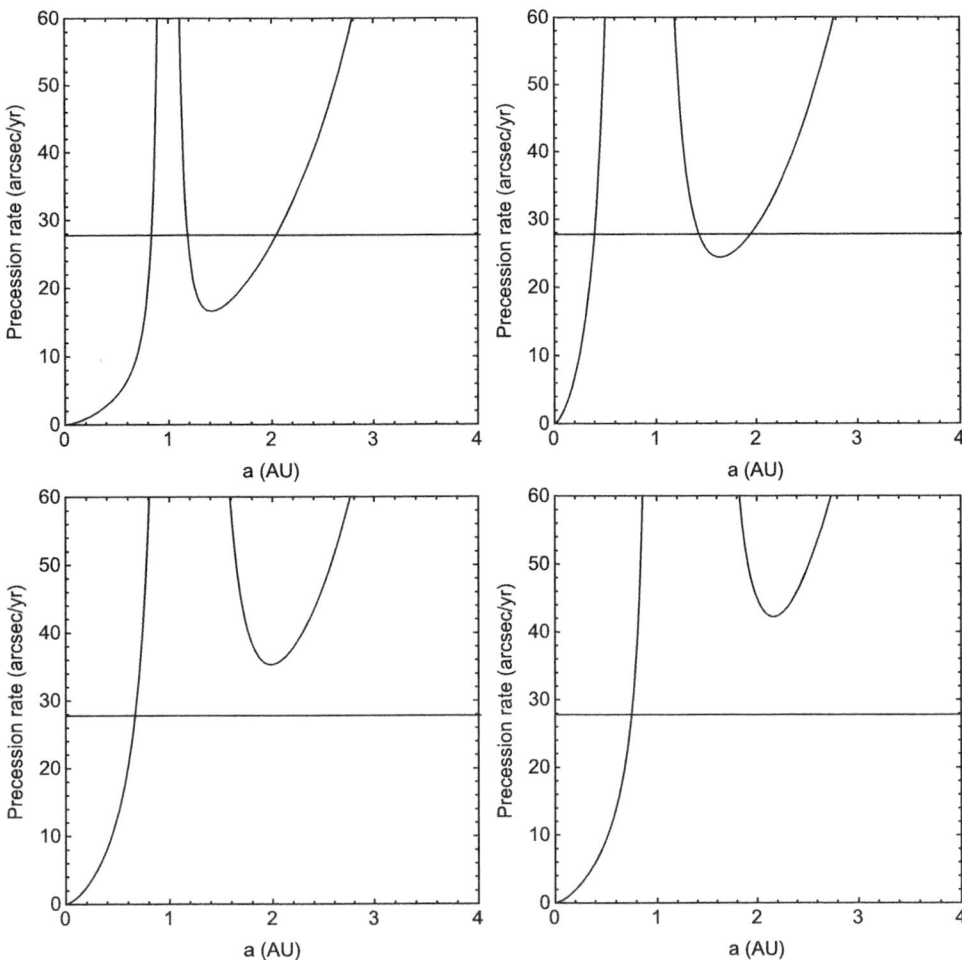

Figure 11.21 The precession rate of a test particle as a function of semimajor. The horizontal line represents the g_i eigenfrequency of Saturn. The intersection of the precession rate with the eigenfrequency denotes the location of a secular resonance. Top-left: a system with no super-Earth. Top-right: a $10M_\oplus$ super-Earth at $a = 0.8$ AU. Bottom-left: a $10M_\oplus$ super-Earth at $a = 1.2$ AU. Bottom-right: a $10M_\oplus$ super-Earth at $a = 1.4$ AU.

Earth, Jupiter, Saturn, and the asteroid belt. The intersection of the particle's precession rate with Saturn's eigenfrequency represents the location of the ν_6 resonance, at about 2 AU.

The introduction of a super-Earth can change the asteroid precession rate so as to either enhance or altogether remove the resonance with Saturn, depending on the super-Earth's location. For example, when the super-Earth has a semimajor axis of 0.8 AU, the precession rate (of the test particle) is close to Saturn's eigenfrequency for semimajor axis values in the range 1.5–2.0 AU, which enhances the ν_6 resonance (top-right panel in Figure 11.21). On the other hand, when the super-Earth is exterior to Earth's orbit, the ν_6 resonance is

11.5.4 The CO Snow Line

Introduction

The most abundant volatiles in a protoplanetary disc are H_2O, CO, and CO_2. A snow line marks the location in the disc where the midplane temperature is sufficiently low so that a volatile condenses out of the gas phase and becomes solid. The snow line of each volatile has its own radial distance from the central star, with the water snow line being the closest to the star. As I have noted earlier, giant planets are expected to form outside the water snow line because the density of solids there is higher (e.g., [130, 135]). In general, the composition of planets and their atmospheres is largely determined by the location of their formation relative to the snow lines [127]. While the water snow line is found at a temperature of about 170 K (see Section 11.5.1), the CO snow line occurs at $T_{CO,snow} = 17$ K [126]. Comets from the solar system's Kuiper belt show different amounts of CO, suggesting that they formed close to the CO snow line [2]. The Kuiper belt is thought to have formed in the region extending from about 27 AU (from the Sun) to about 35 AU [90]. Since the CO snow line would have been in this region at the time of planetesimal formation, it could mark the transition from the planet-forming zone to the dwarf planet/small icy body–forming zone.

While it is very difficult to detect the water snow line in exosolar systems (because of its relative proximity to the host star), the CO snow line presents an easier target because it is farther away. The best observed extrasolar snow line is in the disc around TW Hya. This star has a mass of $0.8M_\odot$ and an age of less than 10 Myr [72]. Qi *et al.* [133] detected in the star's vicinity the reactive ion N_2H^+, which is only present when CO is frozen out. These authors determined the radial distance of the CO snow line to be 28–31 AU, very similar to our own solar system.

Since CO ice is needed to form methanol – a building block of more complex organic molecules – we need to understand the evolution of the CO snow line because that evolution plays a crucial role in the origin of the prebiotic molecules that had led to the emergence of life on Earth [156]. Recall that comets had intensely bombarded the young Earth, apparently delivering those ingredients that were necessary for life. I should note that debris discs – the equivalent of the solar system's Kuiper belt – are extremely common in exosolar systems and, hence, the solar system is not special in that sense. Similarly, systems that have giant planets that are the equivalent of Jupiter, may be expected to have Oort clouds.

Martin and Livio [102] studied the evolution of the CO snow line using two protoplanetary disc configurations: (1) a fully turbulent disc model and (2) the more likely model for a protoplanetary disc – a disc with a dead zone (low-turbulence region).

A Turbulent Disc Model

In fully turbulent discs, material is assumed to orbit at radius R with a Keplerian angular velocity $\Omega = \sqrt{GM_*/R^3}$, where M_* is the star's mass (e.g., [131]). The effective viscosity in a disc in which turbulence is driven by the magnetorotational instability (MRI) is usually parameterised in the form [139]

$$\nu = \alpha \frac{c_s^2}{\Omega},\qquad(11.10)$$

where α is the viscosity parameter and c_s is the sound speed at midplane. In a steady state, mass conservation produces a surface density of

$$\Sigma = \frac{\dot{M}}{3\pi\nu},\qquad(11.11)$$

where $\dot{M}$ is the infall (onto the disc) accretion rate. The surface temperature, T_e, in such a disc is given by energy conservation

$$\sigma T_e^4 = \frac{9}{8}\frac{\dot{M}}{3\pi}\Omega^2 + \sigma T_{irr}^4,\qquad(11.12)$$

where T_{irr} is the irradiation (by the central star) temperature (e.g., [24]), given by

$$T_{irr} = \left(\frac{2}{3\pi}\right)^{1/4}\left(\frac{R}{R_*}\right)^{-3/4} T_*.\qquad(11.13)$$

Here, R_* and T_* are the star's radius and temperature [36]. The midplane temperature of the disc is related to its surface temperature through $T_c^4 = \tau T_e^4$, where τ is the optical depth given by

$$\tau = \frac{3}{8}\kappa\frac{\Sigma}{2},\qquad(11.14)$$

and the opacity is $\kappa = aT_c^b$.

While the precise values of a and b do not have a strong effect on the inferred disc's temperature, values appropriate for the low temperatures in the vicinity of the CO snow line are those obtained from absorption by dust, $a = 0.053$, $b = 0.74$ (e.g., [164]). Martin and Livio [102] solved for the CO snow line radius by equating the central temperature T_c to $T_{CO,snow}$. Their results for $R_{CO,snow}$ as a function of the accretion rate are presented by the short-dashed curve in Figure 11.24, where the assumed parameters were $M_* = 1M_\odot$, $R_* = 3R_\odot$, $T_* = 4{,}000$ K, $T_{CO,snow} = 17$ K, and $\alpha = 0.01$. Since the accretion rate drops in time, time is the implicit coordinate in this figure. The value of the viscosity parameter α is rather uncertain [84], consequently the figure also presents the results (by the long-dashed curve) obtained for $\alpha = 10^{-4}$.

The figure demonstrates that the observed location of the CO snow line in our solar system (at 27–35 AU), and in the solar system analogue TW Hydra (at 28–31 AU), *cannot* be explained by a fully turbulent disc model. Specifically, in this model, the CO snow line moves in too close to the central star during the low-accretion-rate phase towards the end of the disc's lifetime.

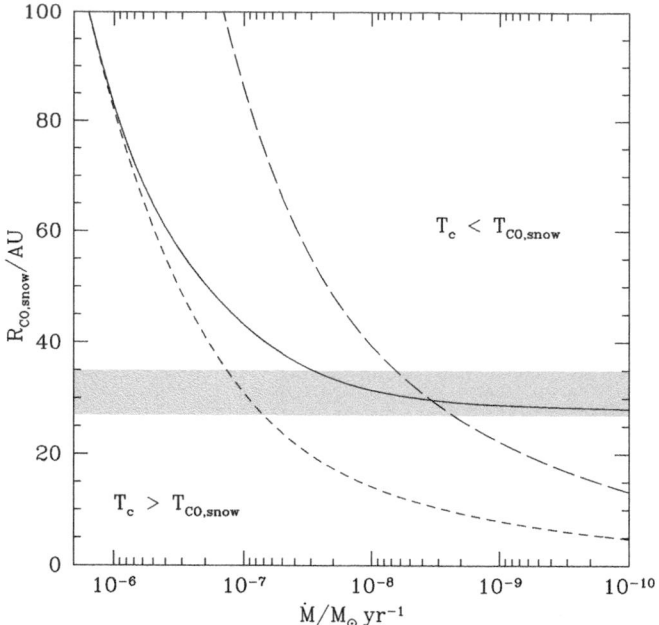

Figure 11.24 Evolution of the CO snow line as a function of the accretion rate in a steady-state disc. The dashed lines show a fully MRI turbulent disc with a viscosity parameter of 0.01 (short-dashed) and 0.0001 (long-dashed). The solid line represents a disc with a self-gravitating dead zone. The shaded region indicates the location of the CO snow line in our solar system at the time of planetesimal formation.

I should note that since irradiation is the dominant heating source (over viscous heating) on the scale of tens of AU, we can ignore the viscous heating term in Eq. (11.12), to find an approximate analytic solution that represents very closely the dashed lines in Figure 11.24 for accretion rates of up to about 10^{-8} $M_\odot$ yr^{-1}. This analytic approximation is given by

$$
R_{CO,\,snow}^{MRI} \simeq 13.2 \left(\frac{\alpha}{0.01}\right)^{-2/9} \left(\frac{M_*}{M_\odot}\right)^{1/9} \left(\frac{\dot{M}}{10^{-8} M_\odot \ yr^{-1}}\right)^{2/9}
$$
$$
\left(\frac{T_{CO,\,snow}}{17 \ K}\right)^{-0.95} \left(\frac{R_*}{3R_\odot}\right)^{2/3} \left(\frac{T_*}{4000 \ K}\right)^{8/9} \ AU.
$$

(11.15)

Since the fully turbulent disc model fails to reproduce the observed CO snow line, Martin and Livio [102] also calculated time-dependent disc models with a dead zone to follow the evolution of the CO snow line.

A Disc with a Dead Zone

A dead zone is formed when the ionisation fraction is not high enough for the MRI to drive turbulence. Typically, the hot part of the disc (close to the central star), where the midplane temperature is higher than some critical value, $T_c \simeq 800$ K [149], is thermally ionised (and, therefore, MRI turbulent.) Farther away from the central star, either cosmic rays or the X-ray flux from the star are the dominant sources of ionisation [62]. These sources can only penetrate the surface layers, down to a surface density of Σ_{crit}. Where $\Sigma > \Sigma_{\text{crit}}$, a dead zone exists at the midplane, with a surface density of $\Sigma_g = \Sigma - \Sigma_{\text{crit}}$. The precise value of Σ_{crit} depends on the ionising source, being around $\Sigma_{\text{crit}} \simeq 200$ g cm^{-2} if cosmic rays dominate [57] but much lower if X-rays dominate [108]. In the very outer parts of the disc (in terms of distance from the central star), where $\Sigma < \Sigma_{\text{crit}}$, the external ionisation sources can penetrate all the way to the midplane, and that part is again MRI active.

In the dead-zone (low-viscosity) region, material can accumulate to the point where the disc becomes self-gravitating. This is expected to occur when the Toomre [148] parameter $Q = c_s \Omega / \pi G \Sigma$ drops below a critical value Q_{crit}. Martin and Livio [102] took $Q_{\text{crit}} = 2$. This drives gravitational turbulence, with an effective viscosity parameter

$$\nu = \alpha_g \frac{c_s^2}{\Omega}. \tag{11.16}$$

Martin and Livio [102] adopted the functional form $\alpha_g = \alpha \exp(-Q^4)$ [104, 165].

They considered a model in which a molecular cloud collapses onto a disc, and they took the initial accretion rate to be $2 \times 10^{-6} M_\odot$ yr^{-1} and assumed that the accretion rate decreases exponentially on a timescale of 10^5 years. The initial surface density of the disc was taken to be that of a turbulent steady disc around a $1 M_\odot$ star, with an accretion rate of $2 \times 10^{-6} M_\odot$ yr^{-1}. The disc was modelled on a radial grid of 200 points evenly distributed in log R from $R = 1$ AU to $R = 200$ AU. The infalling material was added at $R = 195$ AU. To allow for a comparison, Martin and Livio [102] modelled one disc to be fully turbulent and the other to contain a dead zone, with $\Sigma_{\text{crit}} = 10$ g cm^{-2} (corresponding to ionisation by X-rays).

The disc was found to be gravo-magneto unstable for high accretion rates, which caused unsteady accretion onto the central star (for a similar behaviour, see also [5, 104]). The evolution of the CO snow line as a function of time is shown in Figure 11.25. The dashed line represents the fully turbulent disc, and the solid line, represents the disc with a dead zone. At late times, the model with the dead zone has a snow line radius that is considerably larger than that obtained for a turbulent disc and one which is consistent with the observations of the solar system and of TW Hya. The small (and brief) repetitive increases in the snow line radius at early times are caused by FU Orionis-type outbursts. Basically, with a dead zone, the small amount of self-gravity heats the more massive disc with a dead zone, and this causes the CO snow line radius to move outward, as required by observations. For low accretion rates, Martin and Livio [102] were able to find an analytic solution for the CO snow line radius in a disc with a dead zone. This is given by

$$R_{\text{CO, snow}}^{\text{dead}} \simeq 29.3 \left(\frac{M_*}{M_\odot} \right)^{1/9} \left(\frac{R_*}{3 R_\odot} \right)^{2/3} \left(\frac{T_*}{4,000 \text{ K}} \right)^{8/9} \left(\frac{T_{\text{CO, snow}}}{17 \text{ K}} \right)^{-0.61} \text{ AU.} \tag{11.17}$$

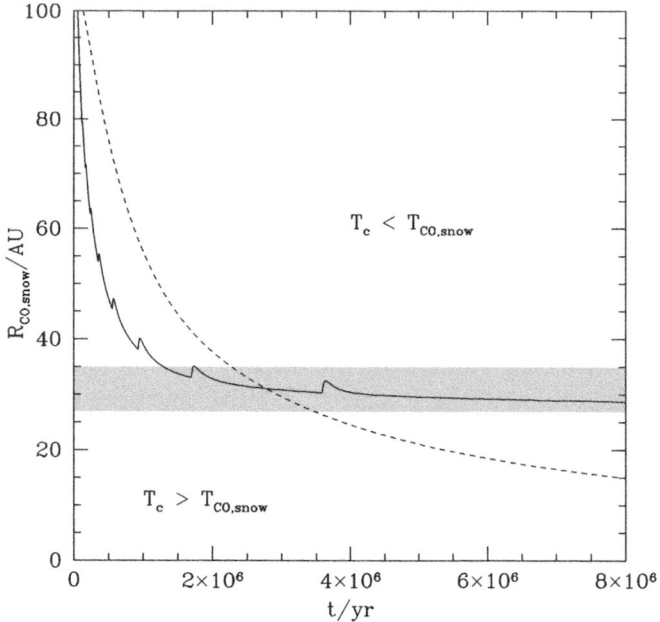

Figure 11.25 Evolution of the CO snow line in a time-dependent disc with an exponentially decreasing infall accretion rate. The dashed line shows a fully MRI turbulent disc, and the solid line, a disc with a dead zone. The shaded region indicates the location of the CO snow line in our solar system at the time of planetesimal formation.

This approximate solution agrees very well with the solid line in Figure 11.24, where the CO snow line radius is shown as a function of the accretion rate.

I should note that the CO snow line in the disc around the Herbig Ae star HD 163296 has been found to lie at a radius of 155 AU [107, 132]. For the observed parameters of this system: $M_* = 2.3 M_\odot$, $R_* = 2 R_\odot$, $T_* = 9,333$ K, and $\dot{M} = 7.6 \times 10^{-8} M_\odot$ yr^{-1}, the model of Martin and Livio [102] with a dead zone would predict a snow line radius of only 62 AU (a fully turbulent disc would give the even much smaller 37 AU). Disc flaring in this system could potentially account for this discrepancy, since approximations for the temperatures of flared discs would result in CO snow line radii larger than 100 AU (e.g., [36]).

In general, the analytic solution for the radius of the CO snow line could prove useful in determining the composition (and thereby maybe the habitability) of exosolar planets.

11.5.5 Conclusion of Sections 11.1–11.5

An examination of the physical properties of our solar system reveals that it is not extremely unusual when those are compared to the characteristics of the other observed exoplanetary systems. Still, there is no doubt that a few of the solar system's parameters have made it conducive to the emergence and evolution of life. For example, low eccentricity planets (as observed in the solar system) have a more stable temperature throughout the entire orbit, which may make them more likely to harbour life [158]. Planetary systems with a low mean eccentricity are also more likely to have a long-term dynamical stability.

The age of the solar system (about 4.5 billion years) may also be favourable for the emergence of complex life (which on Earth took some three billion years). However, I should note that since the current age of the Sun is about half of its total lifetime and about half the age of the Milky Way's disc, one can expect that roughly half of the stars in the disc are even older than the Sun. It is still the case, though, that if life *can* emerge and evolve around low-mass stars (that are much more numerous and live much longer), then one could expect complex life to be much more abundant in the future (about a trillion years from now; see, e.g., Chapter 12). If that is indeed the case, then the appearance of complex life on Earth could be regarded as very early. More recent work suggests, however, that planets around most M-dwarfs may experience serious atmosphere erosion and are therefore not likely to harbour life [94].

As I have already noted, the existence of terrestrial planets in the habitable zone around their host star is quite common [18, 46].

The metallicity of the host star (and of the protoplanetary disc) also does not appear too special in the solar system. The metallicity plays a role in determining the structure of the planetary system that forms [22, 151]. The metallicity was found to be correlated with the probability for the star to have a giant planet orbiting it [51, 134, 144]. However, the correlation is much less clear for lower-mass planets [21, 110]. While planets with radii smaller than four Earth radii are observed around stars with a wide range of metallicities, the average metallicity of stars hosting planets with $R_p < 1.7 R_\oplus$ is close to solar [22]. It is unclear, therefore, whether having a near-solar metallicity is somehow related to forming Earth-size planets.

The variability of the Sun has also been compared to the activity of the stars in the Kepler sample, with rather ambiguous conclusions. While Basri and collaborators [10, 11] found the Sun to be quite typical, with only a quarter to a third of the sample being more active, McQuillan and his collaborators [111] found the Sun to be relatively quiet, with 60% of the stars being more active. The difference in the conclusions stems primarily from the fact that fainter stars were included in McQuillan *et al.* [111] (and also from some differences in defining the activity level of the Sun). Since 90% of M-dwarfs are more active than the Sun, the inclusion of lower-mass stars makes the Sun quiet in comparison. Compared to other Sun-like stars, however, the Sun is quite typical.

As I have noted throughout Section 11.5, the existence of a compact asteroid belt may be conducive to initiating life. Current observations do not allow us to determine categorically how common asteroid belts are. However, the list of unresolved debris disc candidates now contains hundreds of examples [34], and of these, about two-thirds are better modelled by a two-component (rather than a single) dust disc arising from two separate belts [83]. The two-belt configuration of our solar system may therefore be quite ubiquitous.

In Sections 11.2 and 11.3, I identified the lack of super-Earths and the lack of planets interior to Mercury's orbit as perhaps the two main characteristics that make the solar of the cosmological constant Λ (in the context of a ΛCDM cosmology). It is interesting to note, in this respect, that using cosmological N-body simulations, Piran *et al.* [129] concluded that we find ourselves in a favourable spot in the cosmological phase space, in that the exposure of the solar system to GRBs is minimised while the number of hydrogen-burning stars (around which complex life can in principle evolve) is maximised.

The bottom line is clear: Carter's argument should not be taken as a discouragement from searching for other intelligent civilisations in the Milky Way. With the realisation that punctuated equilibrium could characterise the evolution of life even on the galactic scale, even the possibility of $\tau_\ell \ll \tau_*$ cannot be convincingly rejected, since life can independently appear several times and then become extinct through catastrophic events.

Irrespective of Carter's argument, however, there are other reasons to suspect that we are not the only intelligent species to have ever existed in the Universe as a whole. For example, Frank and Sullivan [54] used the exoplanets statistics obtained by the Kepler Space Observatory to evaluate the probability that humanity is the only technological civilisation to have ever existed. They showed that for that to be true, the probability that a habitable-zone planet develops a technological species must be smaller than 10^{-24}, essentially the reciprocal of the expected number of rocky, habitable-zone planets in the observable Universe. So, unless the evolution to technology is truly extraordinarily improbable, chances are that other such civilisations existed at some point in the Universe's lifetime.

To conclude this entire chapter, I have not identified any physical parameters which convincingly demonstrate that life on Earth, or even intelligent life, is unique, either in the Universe as a whole or even in the Milky Way galaxy. Ongoing space mission such as TESS (launched in 2019) and upcoming missions such as JWST and WFIRST (to be launched in 2021 and the mid-2020s, respectively), as well as ground-based telescopes (such as a next-generation Extremely Large Telescope) will detect and start to characterise the atmospheres of super-Earth and Earth-like planets, in the search for biosignatures. Chances are that even if we do not detect extrasolar life in the next two to three decades, we will at least be able to place some meaningful limits on how rare life that dominates the planetary surface chemistry (so that it significantly alters the atmosphere) really is.

Acknowledgements

Most of the work presented in this chapter was done in collaboration with Rebecca Martin. I am also grateful to Jim Pringle, Andrew King, Joe Silk, Jeremy Smallwood, Stephen Lepp, Martin Beer, Steve Lubow, Phil Armitage, Arik Kopelman, and Lev Yungelson, who contributed to various parts of the work.

References

[1] Abe, Y., *et al.* 2000. 'Water in the Early Earth'. In Canup, R. M., and Righter, K. (eds.) *Origin of the Earth and Moon.* University of Arizona Press. Pages 413–433.
[2] A'Hearn, M. F., *et al.* 2012. Cometary Volatiles and the Origin of Comets. *Astrophysical Journal*, **758**(Oct.), 29.

[3] Alvarez, L. W., *et al.* 1980. 'Extraterrestrial Cause for the Cretaceous-Tertiary Extinction'. *Science*, **208**(June), 1095–1108.

[4] Armitage, P. J. 2013. *Astrophysics of Planet Formation* Cambridge University Press.

[5] Armitage, P. J., Livio, M., and Pringle, J. E. 2001. 'Episodic Accretion in Magnetically Layered Protoplanetary Discs'. *Monthly Notices of the Royal Astronomical Society*, **324**(June), 705–711.

[6] Armitage, P. J., *et al.* 2002. 'Predictions for the Frequency and Orbital Radii of Massive Extrasolar Planets'. *Monthly Notices of the Royal Astronomical Society*, **334**(July), 248–256.

[7] Artymowicz, P. 1993. 'Disk-Satellite Interaction via Density Waves and the Eccentricity Evolution of Bodies Embedded in Disks'. *Astrophysical Journal*, **419**(Dec.), 166.

[8] Backman, D., *et al.* 2009. 'Epsilon Eridani's Planetary Debris Disk: Structure and Dynamics Based on Spitzer and Caltech Submillimeter Observatory Observations'. *Astrophysical Journal*, **690**(Jan.), 1522–1538.

[9] Barrow, J. D., and Tipler, F. J. 1986. *The Anthropic Cosmological Principle*. Oxford University Press.

[10] Basri, G., *et al.* 2010. 'Photometric Variability in Kepler Target Stars: The Sun Among Stars: A First Look'. *Astrophysical Journal Letters*, **713**(Apr.), L155–L159.

[11] Basri, G., Walkowicz, L. M., and Reiners, A. 2013. 'Comparison of Kepler Photometric Variability with the Sun on Different Timescales'. *Astrophysical Journal*, **769**(May), 37.

[12] Batalha, N. M., *et al.* 2013. 'Planetary Candidates Observed by Kepler. III. Analysis of the First 16 Months of Data'. *Astrophysical Journal Supplements*, **204**(Feb.), 24.

[13] Batalha, N. M. 2014. 'Exploring Exoplanet Populations with NASA's Kepler Mission'. *Proceedings of the National Academy of Science*, **111**(Sept.), 12647–12654.

[14] Batygin, K., and Brown, M. E. 2016. 'Evidence for a Distant Giant Planet in the Solar System'. *Astronomical Journal*, **151**(Feb.), 22.

[15] Batygin, K., and Laughlin, G. 2015. 'Jupiters Decisive Role in the Inner Solar System's Early Evolution'. *Proceedings of the National Academy of Sciences*.

[16] Berkner, L. V. 1952. 'Signposts to Future Ionospheric Research'. In Katz, L. and Gerson, N.C. (eds.), *Proceedings of the Conference on Ionospheric Physics*. Page 13.

[17] Berkner, L. V., and Marshall, L. C. 1965. 'On the Origin and Rise of Oxygen Concentration in the Earth's Atmosphere.' *Journal of Atmospheric Sciences*, **22**(May), 225–261.

[18] Bonfils, X., *et al.* 2013. 'The HARPS Search for Southern Extra-solar Planets. XXXI. The M-dwarf Sample'. *Astronomy and Astrophysics*, **549**(Jan.), A109.

[19] Bottke, W. F., *et al.* 2002. 'Debiased Orbital and Absolute Magnitude Distribution of the Near-Earth Objects'. *Icarus*, **156**(Apr.), 399–433.

[20] Box, G. E. P., and Cox, D. R. 1964. 'An Analysis of Transformations'. *Journal of the Royal Statistical Society. Series B (Methodological)*, **26**(2), 211–252.

[21] Buchhave, L. A., *et al.* 2012. 'An Abundance of Small Exoplanets around Stars with a Wide Range of Metallicities'. *Nature*, **486**(June), 375–377.

[22] Buchhave, L. A., *et al.* 2014. 'Three Regimes of Extrasolar Planet Radius Inferred from Host Star Metallicities'. *Nature*, **509**(May), 593–595.

[23] Burke, C. J., *et al.* 2015. 'Terrestrial Planet Occurrence Rates for the Kepler GK Dwarf Sample'. *Astrophysical Journal*, **809**(Aug.), 8.

[143] Smallwood, J. L., *et al.* 2018. 'Asteroid Impacts on Terrestrial Planets: The Effects of Super-Earths and the Role of the ν_6 Resonance'. *Monthly Notices of the Royal Astronomical Society*, **473**(Jan.), 295–305.

[144] Sousa, S. G., *et al.* 2011. 'Spectroscopic Stellar Parameters for 582 FGK Stars in the HARPS Volume-Limited Sample: Revising the Metallicity-Planet Correlation'. *Astronomy and Astrophysics*, **533**(Sept.), A141.

[145] Stecher, T. P. 1970. 'Stellar Spectrophotometry from a Pointed Rocket'. *Astrophysical Journal*, **159**(Feb.), 543.

[146] Tamayo, D., Rein, H., Petrovich, C., and Murray, N. 2017. 'Convergent Migration Renders TRAPPIST-1 Long-Lived'. *Astrophysical Journal*, **840**(May), L19.

[147] Terquem, C., and Papaloizou, J. C. B. 2007. 'Migration and the Formation of Systems of Hot Super-Earths and Neptunes'. *Astrophysical Journal*, **654**, 1110–1120.

[148] Toomre, A. 1964. 'On the Gravitational Stability of a Disk of Stars'. *Astrophysical Journal*, **139**(May), 1217–1238.

[149] Umebayashi, T., and Nakano, T. 1988. 'Chapter 13: Ionization State and Magnetic Fields in the Solar Nebula'. *Progress of Theoretical Physics Supplement*, **96**, 151–160.

[150] Walsh, K. J., *et al.* 2011. 'A Low Mass for Mars from Jupiter's Early Gas-Driven Migration'. *Nature*, **475**(July), 206–209.

[151] Wang, J., and Fischer, D. A. 2015. 'Revealing a Universal Planet-Metallicity Correlation for Planets of Different Sizes around Solar-Type Stars'. *Astronomical Journal*, **149**(Jan.), 14.

[152] Ward, P., and Brownlee, D. 2000. *Rare Earth : Why Complex Life Is Uncommon in the Universe*. Copernicus.

[153] Ward, W. R. 1989. 'On the Rapid Formation of Giant Planet Cores'. *Astrophysical Journal Letters*, **345**(Oct.), L99–L102.

[154] Watanabe, K. 1958. 'Ultraviolet Absorption Processes in the Upper Atmosphere'. *Advances in Geophysics*, **5**, 153–221.

[155] Weidenschilling, S. J. 1977. 'The Distribution of Mass in the Planetary System and Solar Nebula'. *Astrophysics and Space Science*, **51**(Sept.), 153–158.

[156] Wickramasinghe, J., Wickramasinghe, C., and Napier, W. 2010. *Comets and the Origin of Life*. World Scientific Publishing Co.

[157] Willbold, M., Elliott, T., and Moorbath, S. 2011. 'The Tungsten Isotopic Composition of the Earth's Mantle before the Terminal Bombardment'. *Nature*, **477**(Sept.), 195–198.

[158] Williams, D. M., and Pollard, D. 2002. 'Earth-Like Worlds on Eccentric Orbits: Excursions beyond the Habitable Zone'. *International Journal of Astrobiology*, **1**(Jan.), 61–69.

[159] Wilson, P. A. 1994. 'Carter on Anthropic Principle Predictions'. *The British Journal for the Philosophy of Science*, **45**(1), 241–253.

[160] Winn, J. N. and Fabrycky, D. C. 2015. 'The Occurrence and Architecture of Exoplanetary Systems'. *Annual Review of Astronomy and Astrophysics*, **53**(Aug.), 409–447.

[161] Wright, J. T., *et al.* 2011. 'The Exoplanet Orbit Database'. *Publications of the Astronomical Society of the Pacific*, **123**(Apr.), 412.

[162] Yoshikawa, M. 1987. 'A Simple Analytical Model for the Secular Resonance nu6 in the Asteroidal Belt'. *Celestial Mechanics*, **40**(Sept.), 233–272.

[163] Zakamska, N. L., Pan, M., and Ford, E. B. 2011. 'Observational Biases in Determining Extrasolar Planet Eccentricities in Single-Planet Systems'. *Monthly Notices of the Royal Astronomical Society*, **410**, 1895.

[164] Zhu, Z., Hartmann, L., and Gammie, C. 2009. 'Nonsteady Accretion in Protostars'. *Astrophysical Journal*, **694**(Apr.), 1045–1055.

[165] Zhu, Z., Hartmann, L., and Gammie, C. 2010. 'Long-Term Evolution of Protostellar and Protoplanetary Disks: II. Layered Accretion with Infall'. *Astrophysical Journal*, **713**(Apr.), 1143–1158.

12

On the Temporal Habitability of Our Universe

ABRAHAM LOEB

Abstract

Is life most likely to emerge at the present cosmic time near a star like the Sun? We consider the habitability of the Universe throughout cosmic history and conservatively restrict our attention to the context of 'life as we know it' and the standard cosmological model, ΛCDM. The habitable cosmic epoch started shortly after the first stars formed, about 30 Myr after the Big Bang, and will end about 10 Tyr from now, when all stars will die. We review the formation history of habitable planets and find that unless habitability around low-mass stars is suppressed or appears preferentially early, life is most likely to exist near $\sim 0.1 M_\odot$ stars 10 trillion years from now. Spectroscopic searches for biosignatures in the atmospheres of transiting Earth-mass planets around low-mass stars will determine whether present-day life is indeed premature or typical from a cosmic perspective.

12.1 Introduction

The known forms of terrestrial life involve carbon-based chemistry in liquid water [130, 131]. In the cosmological context, life could not have started earlier than 10 Myr after the Big Bang ($z \gtrsim 140$) since the entire Universe was bathed in a thermal radiation background above the boiling temperature of liquid water. Later on, however, the Universe cooled to a *habitable epoch* at a comfortable temperature of 273–373 K between 10 and 17 Myr after the Big Bang [159].

The phase diagram of water allows it to be liquid only under external pressure in an atmosphere which can be confined gravitationally on the surface of a planet. To keep the atmosphere bound against evaporation requires strong surface gravity of a rocky planet with a mass comparable to or above that of the Earth [229, 276].

The emergence of 'life as we know it' requires stars for two reasons. Stars are needed to produce the heavy elements (carbon, oxygen, and so on, up to iron) out of which rocky planets and the molecules of life are made. Stars also provide a heat source for powering the chemistry of life on the surface of their planets. Each star is surrounded by a habitable zone where the surface temperature of a planet allows liquid water to exist. The approximate distance of the habitable zone, r_{HZ}, is obtained by equating the heating rate per unit area from the stellar luminosity, L, to the cooling rate per unit area at a surface temperature of $T_{HZ} \sim 300$ K, namely $(L/4\pi r_{HZ}^2) \sim \sigma T_{HZ}^4$, where σ is the Stefan-Boltzman constant [130, 131].

Starting from the vicinity of particular stars, life could potentially spread. This process, so-called panspermia, could be mediated by the transfer of rocks between planetary systems [1]. The 'astronauts' on such rocks could be microscopic animals such as *tardigrades*, which are known to be resilient to extreme vacuum, dehydration, and exposure to radiation that characterise space travel. In 2007, dehydrated tardigrades were taken into a low Earth orbit for 10 days. After returning to Earth, most of them revived after rehydration, and many produced viable embryos [87]. Life which arose via spreading will exhibit more clustering than life which arose spontaneously, and so the existence of panspermia can be detected statistically through excess spatial correlations of life-bearing environments. Future searches for biosignatures in the atmospheres of exoplanets could test for panspermia: a smoking-gun signature would be the detection of large regions in the Milky Way where life saturates its environment, interspersed with voids where life is very uncommon. In principle, detection of as few as several tens of biologically active exoplanets could yield a highly significant detection of panspermia [151]. Once life emerges on the surface of a planet, it is difficult to extinguish it completely through astrophysical events (such as quasar activity, supernovae, gamma ray bursts, or asteroid impacts) other than the death of the host star. Life is known to be resilient to extreme environments and could be protected from harmful radiation or heat if it resides underground or on the deep ocean floor [236].

Panspermia is not limited to galactic scales and could extend over cosmological distances. Some stars and their planets are ejected from their birth galaxies at a speed approaching the speed of light, through a gravitational slingshot from pairs of supermassive black holes which are formed during galaxy mergers [104, 161]. The resulting population of relativistic stars roaming through intergalactic space could potentially transfer life between galaxies separated by vast distances across the Universe.

The spread of life could be enhanced artificially through the use of spacecrafts by advanced civilisations. Our own civilisation is currently starting to develop the technology needed to visit the nearest stars with a travel time of decades through the propulsion of lightweight sails to a fraction of the speed of light by a powerful laser.[1] The existence of advanced civilisations could be revealed through the detection of industrial pollution in the atmospheres of planets [152], the detection of powerful beacons of light used for propulsion [103] and communications [164], or through artificial lights [163]. The search for signatures of advanced civilisations is the richest interdisciplinary frontier, offering interfaces between astronomy and other disciplines, such as biology (astrobiology), chemistry (astrochemistry), statistics (astro-statistics), or engineering (astroengineering). Moreover, the prospects for communication with aliens could open new disciplines on the interface with linguistics (astro-linguistics), psychology (astro-psychology), sociology

---

[1] www.breakthroughinitiatives.org/Concept/3.

temperature inside the dark matter halo, then it could fragment into dense clumps and make stars.

At redshifts $z \gtrsim 140$, Compton cooling on the CMB is effective on a timescale comparable to the age of the Universe, given the residual fraction of free electrons left over from cosmological recombination (see section 2.2 in Reference [160] and also Reference [215]). The thermal coupling to the CMB tends to bring the gas temperature to T_{cmb}, which at $z \sim 140$ is similar to the temperature floor associated with molecular hydrogen cooling [108, 113, 244]. In order for virialised gas in a dark matter halo to cool, condense, and fragment into stars, the halo virial temperature T_{vir} has to exceed $T_{min} \approx 300$ K, corresponding to T_{cmb} at $(1 + z) \sim 110$. This implies a halo mass in excess of $M_{min} = 10^4 M_\odot$, corresponding to a baryonic mass $M_{b, min} = 1.5 \times 10^3 M_\odot$, a circular virial velocity $V_{c, min} = 2.6$ km s^{-1}, and a virial radius $r_{vir, min} = 6.3$ pc (see section 3.3 in Reference [160]). This value of M_{min} is close to the minimum halo mass to assemble baryons at that redshift (see section 3.2.1 in Reference [160] and figure 2 of Reference [250]).

The corresponding number of star-forming halos on our past light cone is given by [193],

$$N = \int_{(1+z)=100}^{(1+z)=137} n(M > M_{min}, z') \frac{dV}{dz'} dz', \tag{12.1}$$

where $n(M > M_{min})$ is the co-moving number density of halos with a mass $M > M_{min}$ [233], and $dV = 4\pi r^2 dr$ is the co-moving volume element with $dr = c dt/a(t)$. Here, $a(t) = (1 + z)^{-1}$ is the cosmological scale factor at time t, and $r(z) = c \int_0^z dz'/H(z')$ is the co-moving distance. The Hubble parameter for a flat Universe is

$$H(z) \equiv (\dot{a}/a) = H_0 \sqrt{\Omega_m (1 + z)^3 + \Omega_r (1 + z)^4 + \Omega_\Lambda}, \tag{12.2}$$

with Ω_m, Ω_r and Ω_Λ being the present-day density parameters of matter, radiation, and vacuum, respectively. The total number of halos that existed at $(1 + z) \sim 100$ within our entire Hubble volume (not restricted to the light cone), $N_{tot} \equiv n(M > M_{min}, z = 99) \times (4\pi/3)(3c/H_0)^3$, is larger than N by a factor of $\sim 10^3$.

For the standard cosmological parameters [211], we find that the first star-forming halos on our past light cone reached its maximum turnaround radius[2] (with a density contrast of 5.6) at $z \sim 112$ and collapsed (with an average density contrast of 178) at $z \sim 71$. Within the entire Hubble volume, a turnaround at $z \sim 122$ resulted in the first collapse at $z \sim 77$. This result includes the delay by $\Delta z \sim 5.3$ expected from the streaming motion of baryons relative to the dark matter [82].

The preceding calculation implies that rocky planets could have formed within our Hubble volume by $(1 + z) \sim 78$ but not by $(1 + z) \sim 110$ if the initial density perturbations were perfectly Gaussian. However, the host halos of the first planets are extremely rare, representing just $\sim 2 \times 10^{-17}$ of the cosmic matter inventory. Since they lie ~ 8.5 standard deviations (σ) away on the exponential tail of the Gaussian probability distribution of initial density perturbations, $P(\delta)$, their abundance could have been significantly enhanced by

---

[2] In the spherical collapse model, the turnaround time is half the collapse time.

primordial non-Gaussianity [166, 172, 189] if the decline of $P(\delta)$ at high values of δ/σ is shallower than exponential. The needed level of deviation from Gaussianity is not ruled out by existing data sets [212]. Non-Gaussianity below the current limits is expected in generic models of cosmic inflation [173] that are commonly used to explain the initial density perturbations in the Universe.

12.2.3 Section Summary and Implications

In the discussion thus far, we highlighted a new regime of habitability made possible for ~ 6.6 Myr by the uniform CMB radiation at redshifts $(1 + z) = 100 - 137$, just when the first generation of star-forming halos (with a virial mass $\gtrsim 10^4 M_\odot$) turned around in the standard cosmological model with Gaussian initial conditions. Deviations from Gaussianity in the far (8.5σ) tail of the probability distribution of initial density perturbations, already at these redshifts, could have led to the birth of massive stars, whose heavy elements triggered the formation of rocky planets with liquid water on their surfaces.[3]

Thermal gradients are needed for life. These can be supplied by geological variations on the surface of rocky planets. Examples for sources of free energy are geothermal energy powered by the planet's gravitational binding energy at formation and radioactive energy from unstable elements produced by the earliest supernova. These internal heat sources (in addition to possible heating by a nearby star) may have kept planets warm even without the CMB, extending the habitable epoch from $z \sim 100$ to later times. The lower CMB temperature at late times may have allowed ice to form on objects that delivered water to a planet's surface and helped to maintain the cold trap of water in the planet's stratosphere. Planets could have kept a blanket of molecular hydrogen that maintained their warmth [208, 241], allowing life to persist on internally warmed planets at late cosmic times. If life persisted at $z \lesssim 100$, it could have been transported to newly formed objects through panspermia. Under the assumption that interstellar panspermia is plausible, the redshift of $z \sim 100$ can be regarded as the earliest cosmic epoch after which life was possible in our Universe.

Finally, we note that an increase in the initial amplitude of density perturbations on the mass-scale of $10^4 M_\odot$ by a modest factor of $1.4 \times [(1 + z)/110]$ would have enabled star formation within the Hubble volume at redshifts $(1 + z) > 110$ even for perfectly Gaussian initial conditions.

12.3 CEMP Stars: Possible Hosts to Carbon Planets in the Early Universe

12.3.1 Section Background

The questions of when, where, and how the first planetary systems actually formed in cosmic history remain crucial to our understanding of structure formation and the emergence

---

[3] The dynamical time of galaxies is shorter than $\sim 1/\sqrt{200} = 7\%$ of the age of the Universe at any redshift since their average density contrast is $\gtrsim 200$. After the first stars formed, the subsequent delay in producing heavy elements from the first supernovae could have been as short as a few Myr. The supernova ejecta could have produced high-metallicity islands that were not fully mixed with the surrounding primordial gas, leading to efficient formation of rocky planets within them.

of life in the early Universe [159]. The short-lived, metal-free, massive first-generation stars ultimately exploded as supernovae (SNe) and enriched the interstellar medium (ISM) with the heavy elements fused in their cores. The enrichment of gas with metals that had otherwise been absent in the early Universe enabled the formation of the first low-mass stars and perhaps marked the point at which star systems could begin to form planets [42, 57, 89]. In the core accretion model of planet formation (e.g., [123, 201]), elements heavier than hydrogen and helium are necessary not only to form the dust grains that are the building blocks of planetary cores but also to extend the lifetime of the protostellar disc long enough to allow the dust grains to grow via merging and accretion to form planetesimals [80, 126, 138, 271].

In the past four decades, a broad search has been launched for low-mass Population II stars in the form of extremely metal-poor sources within the halo of the Galaxy. The HK survey [23], the Hamburg/ESO Survey [55, 268], the Sloan Digital Sky Survey (SDSS; [274]), and the SEGUE survey [270] have all significantly enhanced the sample of metal-poor stars with [Fe/H] < −2.0. Although these iron-poor stars are often referred to in the literature as 'metal-poor' stars, it is critical to note that [Fe/H] does not necessarily reflect a stellar atmosphere's total metal content. The equivalence between 'metal-poor' and 'iron-poor' appears to fall away for stars with [Fe/H] < −3.0 since many of these stars exhibit large overabundances of elements such as C, N, and O; the total mass fractions, Z, of the elements heavier than He are therefore not much lower than the solar value in these iron-poor stars.

Carbon-enhanced metal-poor (CEMP) stars comprise one such chemically anomalous class of stars, with carbon-to-iron ratios [C/Fe] ≥ 0.7 (as defined in [11, 45, 196]). The fraction of sources that fall into this category increases from ∼15%–20% for stars with [Fe/H] < −2.0, to 30% for [Fe/H] < −3.0, to ∼75% for [Fe/H] < −4.0 [22, 88, 196]. Furthermore, the degree of carbon enhancement in CEMP stars has been shown to notably increase as a function of decreasing metallicity, rising from [C/Fe] ∼ 1.0 at [Fe/H] = −1.5 to [C/Fe] ∼ 1.7 at [Fe/H] = −2.7. [45]. Given the significant frequency and level of carbon excess in this subset of metal-poor Population II stars, the formation of carbon planets around CEMP stars in the early Universe presents itself as an intriguing possibility.

From a theoretical standpoint, the potential existence of carbon exoplanets, consisting of carbides and graphite instead of Earth-like silicates, has been suggested by Reference [143]. Using the various elemental abundances measured in planet-hosting stars, subsequent works have sought to predict the corresponding variety of terrestrial exoplanet compositions expected to exist [34, 47, 48]. Assuming that the stellar abundances are similar to those of the original circumstellar disc, related simulations yield planets with a whole range of compositions, including some that are almost exclusively C and SiC; these occur in discs with C/O > 0.8, favourable conditions for carbon condensation [144]. Observationally, there have also been indications of planets with carbon-rich atmospheres – e.g., WASP-12b [170] and carbon-rich interiors – e.g., 55 Cancri e [171].

In this section, we explore the possibility of carbon planet formation around the iron-deficient but carbon-rich subset of low-mass stars, mainly CEMP stars. Standard definitions of elemental abundances and ratios are adopted. For element X, the logarithmic absolute abundance is defined as the number of atoms of element X per 10^{12} hydrogen atoms, $\log \epsilon(X) = \log_{10}(N_X/N_Y) + 12.0$. For elements X and Y, the logarithmic abundance ratio relative to the solar ratio is defined as $[X/Y] = \log_{10}(N_X/N_Y) - \log_{10}(N_X/N_Y)_\odot$. The solar abundance set is that of Reference [14], with a solar metallicity $Z_\odot = 0.0134$.

12.3.2 Star-Forming Environment of CEMP Stars

A great deal of effort has been directed in the literature towards theoretically understanding, the origin of the most metal-poor stars and, in particular, the large fraction that is carbon-rich. These efforts have been further perturbed by the fact that CEMP stars do not form a homogenous group but, rather, can be further subdivided into two main populations [22]: carbon-rich stars that show an excess of heavy neutron-capture elements (CEMP-s, CEMP-r, and CEMP-r/s), and carbon-rich stars with a normal pattern of the heavy elements (CEMP-no). In the following sections, we focus on stars with $[Fe/H] \leq -3.0$, which have been shown to fall almost exclusively in the CEMP-no subset [10].

A number of theoretical scenarios have been proposed to explain the observed elemental abundances of these stars, though there is no universally accepted hypothesis. The most extensively studied mechanism to explain the origin of CEMP-no stars is the mixing and fallback model, where a 'faint' Population III SN explodes but, due to a relatively low explosion energy, only ejects its outer layers, rich in lighter elements (up to magnesium); its innermost layers, rich in iron and heavier elements, fall back onto the remnant and are not recycled in the ISM [253, 254]. This potential link between primeval SNe and CEMP-no stars is supported by recent studies which demonstrate that the observed ratio of carbon-enriched to carbon-normal stars with $[Fe/H] < -3.0$ is accurately reproduced if SNe were the main source of metal enrichment in the early Universe [60, 65]. Furthermore, the observed abundance patterns of CEMP-no stars have been found to be generally well matched by the nucleosynthetic yields of primordial faint SNe [35, 121, 122, 125, 133, 177, 178, 248, 254, 272]. These findings suggest that most of the CEMP-no stars were probably born out of gas enriched by massive, first-generation stars that ended their lives as Type II SNe with low levels of mixing and a high degree of fallback.

Under such circumstances, the gas clouds which collapse and fragment to form these CEMP-no stars and their protostellar discs may contain significant amounts of carbon dust grains. Observationally, dust formation in SNe ejecta has been inferred from isotopic anomalies in meteorites where graphite, SiC, and Si_3N_4 dust grains have been identified as SNe condensates [278]. Furthermore, in situ dust formation has been unambiguously detected in the expanding ejecta of SNe such as SN 1987A [119, 167] and SN 1999em [79]. The existence of cold dust has also been verified in the supernova remnant of Cassiopeia A by SCUBA's recent submillimetre observations, and a few solar masses worth of dust is estimated to have condensed in the ejecta [75].

Theoretical calculations of dust formation in primordial core-collapsing SNe have demonstrated the condensation of a variety of grain species, starting with carbon, in the where $\sigma_d = \pi a^2$ is the cross section of the dust grain with radius a and $c_s = \sqrt{k_B T(r)/\mu m_H}$ is the isothermal sound speed with m_H being the mass of a hydrogen atom and $\mu = 1.36$ being the mean molecular weight of the gas (including the contribution of helium). $\Omega_K = \sqrt{GM_*/r^3}$ is the Keplerian velocity of the disc at a distance r from the central star of mass M_*, which we take to be $M_* = 0.8\,M_\odot$ as representative of the low masses associated with CEMP-no stars [53, 88]. The disc is assumed to be in hydrostatic equilibrium with a density given by

$$\rho(z,r) = \frac{\Sigma(r)}{h\sqrt{2\pi}} \exp\left(-\frac{z^2}{2h^2}\right), \tag{12.5}$$

where the disc scale height is $h = c_s/\Omega_k$. For the disc surface density $\Sigma(r)$ and temperature $T(r)$ profiles, we adopt the radial power-law distributions fitted to (sub-)millimetre observations of circumstellar discs around young stellar objects [6–8],

$$T(r) = 200\text{ K}\left(\frac{r}{1\text{ AU}}\right)^{-0.6} \tag{12.6}$$

$$\Sigma(r) = 10^3\text{ g/cm}^2\left(\frac{r}{1\text{ AU}}\right)^{-0.9}. \tag{12.7}$$

Although these relations were observationally inferred from discs with solar-like abundances, we choose to rely on them for our purposes given the lack of corresponding measurements for discs around stars with different abundance patterns.

The rate of grain growth, dm/dt, is determined by the rate at which grains, subject to small-scale Brownian motion, collide and stick together as they drift towards the disc midplane through a sea of smaller solid particles. If coagulation results from every collision, then the mass growth rate of a particle is effectively the amount of solid material in the volume swept out the particle's geometric cross section,

$$\frac{dm}{dt} = f_{dg}\rho\sigma_d\left(v_{rel} + \frac{dz}{dt}\right) \tag{12.8}$$

where dz/dt is the dust-settling velocity given by Equation (12.3) and

$$v_{rel} = \sqrt{\frac{8k_B T(m_1 + m_2)}{\pi m_1 m_2}} \approx \sqrt{\frac{8k_B T}{\pi m}} \tag{12.9}$$

is the relative velocity in the Brownian motion regime between grains with masses $m_1 = m_2 = m$. To calculate the dust-to-gas mass ratio in the disc f_{dg}, we follow the approach in Reference [124] and relate two expressions for the mass fraction of C: (1) the fraction of carbon in the dust, $f_{dg}M_{C,dust}/M_{dust}$, where M_{dust} is the total dust mass and $M_{C,dust}$ is the carbon dust mass ; and (2) the fraction of carbon in the gas, $\mu_C n_C/\mu n_H$, where μ_C is the molecular weight of carbon ($\sim 12m_p$) and n_C and n_H are the carbon and hydrogen number densities, respectively.

We then assume that a fraction f_{cond} (referred to from now on as the carbon condensation efficiency) of all the carbon present in the gas cloud is locked up in dust, such that

$$f_{cond} \frac{\mu_C n_C}{\mu n_H} = f_{dg} \frac{M_{C,dust}}{M_{dust}}. \tag{12.10}$$

Since faint Population III SNe are believed to have polluted the birth clouds of CEMP-no stars, and the only grain species that forms in non-negligible amounts in these ejecta is amorphous carbon [177, 178], we set $M_{dust} = M_{C,dust}$. Rewriting Eq. (12.9) in terms of abundances relative to the Sun, we obtain

$$f_{dg} = f_{cond} \frac{\mu_C}{\mu} 10^{[C/H]+\log \epsilon(C)_\odot - 12}, \tag{12.11}$$

where $\log \epsilon(C)_\odot = 8.43 \pm 0.05$ [14] is the solar carbon abundance.

For a specified metallicity [C/H] and radial distance r from the central star, we can then estimate the time it takes for dust grains to settle in the disc by integrating Eqs. (3.2) and (3.6) from an initial height of $z(t = 0) = 4h$ with an initial dust grain mass of $m(t = 0) = 4\pi a_{init}^3 \rho_d/3$. The specific weight of dust is set to $\rho_d = 2.28$ g cm^{-3}, reflecting the material density of carbon grains expected to dominate the circumstellar discs of CEMP-no stars. The initial grain size a_{init} is varied between 0.01 and 1 μm to reflect the range of characteristic radii of carbon grains found when modelling CEMP-no star abundance patterns [177]. Comparing the resulting dust-settling timescale to the disc lifetime given by Eq. 12.1 for the specified metallicity, we can then determine whether there is enough time for carbon dust grains to settle in the midplane of the disc and there undergo runaway planetesimal formation before the disc is dissipated by photoevaporation. For the purposes of this simple model, we neglected possible turbulence in the disc which may counteract the effects of vertical settling, propelling particles to higher altitudes and thus preventing them from fully settling into the disc midplane [12]. We have also not accounted for the effects of radial drift, which may result in the evaporation of solid material in the hot inner regions of the circumstellar disc.

As the dust-settling timescale is dependent on the disc surface density $\Sigma(r)$ and temperature $T(r)$, we find that for a given metallicity, [C/H], there is a maximum distance r_{max} from the central star out to which planetesimal formation is possible. At larger distances from the host star, the dust-settling timescale exceeds the disc lifetime, and so carbon planets with semimajor axes $r > r_{max}$ are not expected to form. A plot of the maximum semimajor axis expected for planet formation around a CEMP-no star as a function of the carbon abundance relative to the Sun [C/H] is shown in Figure 12.1 for carbon condensation efficiencies ranging between $f_{cond} = 0.1$ and 1. As discovered in Reference [127], where the critical iron abundance for terrestrial planet formation is considered as a function of the distance from the host star, we find a linear relation between [C/H] and r_{max},

$$[C/H] = \log \left(\frac{r_{max}}{1 \text{ AU}} \right) - \alpha, \tag{12.12}$$

where $\alpha = 1.3, 1.7$, and 1.9 for $f_{cond} = 0.1, 0.5$, and 1, respectively, assuming an initial grain size of $a_{init} = 0.1\mu$m. These values for α change by less than 1% for smaller initial

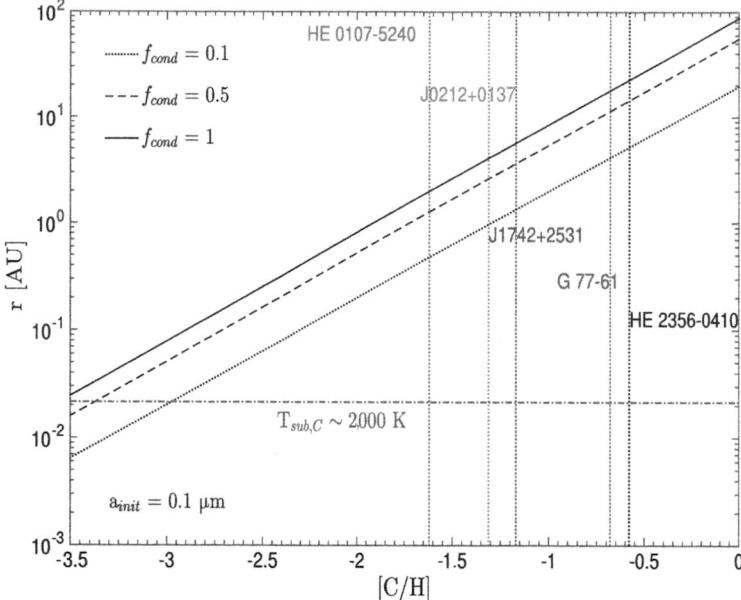

Figure 12.1 The maximum distance r_{max} from the host star out to which planetesimal formation is possible as a function of the star's metallicity, expressed as the carbon abundance relative to that of the Sun, [C/H]. The dotted, dashed, and solid black curves correspond to the results obtained assuming carbon condensation efficiencies of 10%, 50%, and 100%, respectively, and an initial grain size of $a_{init} = 0.1$ µm. The grey dash-dotted curve corresponds to the distance at which the disc temperature approaches the sublimation temperature of carbon dust grains, $T_{sub,C} \sim 2{,}000$ K; the formation of carbon planetesimals will therefore be suppressed at distances that fall below this line, $r \lesssim 0.02$ AU. The coloured vertical lines represent various observed CEMP stars with measured carbon abundances, [C/H].

grain sizes, $a_{init} = 0.01$ µm, and by no more than 5% for larger initial grain sizes $a_{init} = 1$ µm; given this weak dependence on a_{init}, we only show our results for a single initial grain size of $a_{init} = 0.1$ µm. The distance from the host star at which the temperature of the disc approaches the sublimation temperature of carbon dust, $T_{sub,C} \sim 2{,}000$ K [135], is depicted as well (dash-dotted grey curve). At distances closer to the central star than $r \simeq 0.02$ AU, temperatures well exceed the sublimation temperature of carbon grains; grain growth and subsequent carbon planetesimal formation are therefore quenched in this inner region.

Figure 12.1 shows lines representing various observed CEMP stars with measured carbon abundances – mainly HE 0107-5240 [53, 54], SDSS J0212+0137 [35], SDSS J1742+2531 [35], G 77-61 [24, 62, 213], and HE 2356-0410 [195, 223]. These stars all have iron abundances (relative to solar) [Fe/H] < −3.0, carbon abundances (relative to solar) [C/Fe] > 2.0, and carbon-to-oxygen ratios C/O > 1. This latter criteria maximises the abundance of solid carbon available for planet formation in the circumstellar discs by

Table 12.1 *Basic data^a for CEMP stars considered in this section.*

Star	$\log g^b$	[Fe/H]	[C/Fe]	C/O^c	Source
HE 0107-5240	2.2	−5.44	3.82	14.1	[54, 58]
SDSS J0212+0137	4.0	−3.57	2.26	2.6	[35]
SDSS J1742+2531	4.0	−4.77	3.60	2.2	[35]
G 77-61	5.1	−4.03	3.35	12.0	[24, 213]
HE 2356-0410^d	2.65	−3.19	2.61	>14.1	[223]

^{a} Abundances based on one-dimensional LTE model-atmosphere analyses
^{b} Logarithm of the gravitational acceleration at the surface of stars expressed in cm s^{-2}
^{c} C/O = $N_C/N_O = 10^{[C/O]+\log \epsilon(C)_\odot - \log \epsilon(O)_\odot}$
^{d} CS 22957-027

optimising carbon grain growth both in stratified SNe ejecta and later, in the collapsing molecular birth clouds of these stars. It also advances the possibility of carbon planet formation by ensuring that planet formation proceeds by a carbon-rich condensation sequence in the protoplanetary disc. SDSS J0212+0137 and HE 2356-0410 have both been classified as CEMP-no stars, with measured barium abundances [Ba/Fe] < 0 (as defined in [22]); the other three stars are barium indeterminate with only high upper limits on [Ba/Fe], but are believed to belong to the CEMP-no subclass given their light-element abundance patterns. The carbon abundance, [C/H], dominates the total metal content of the stellar atmosphere in these five CEMP objects, contributing more than 60% of the total metallicity in these stars. A summary of the relevant properties of the CEMP stars considered in this analysis can be found in Table 12.1. We find that carbon planets may be orbiting iron-deficient stars with carbon abundances [C/H] $\sim$ −0.6, such as HE 2356-0410, as far out as $\sim$ 20 AU from their host star in the case where $f_{cond} = 1$. Planets forming around stars with less carbon enhancement – i.e., HE 0107-5240 with [C/H] $\sim$ −1.6 are expected to have more compact orbits, with semimajor axes $r < 2$ AU. If the carbon condensation efficiency is only 10%, the expected orbits grow even more compact, with maximum semimajor axes of $\sim$ 5 and 0.5 AU, respectively.

12.3.4 Mass-Radius Relationship for Carbon Planets

Next we present the relationship between the mass and radius of carbon planets that we have shown may theoretically form around CEMP-no stars. These mass-radius relations have already been derived in the literature for a wide range of rocky and icy exoplanet compositions [86, 148, 230, 256, 277]. Here, we follow the approach of Reference [277] and solve the three canonical equations of internal structure for solid planets:

1. Mass conservation

$$\frac{dm(r)}{dr} = 4\pi r^2 \rho(r) \tag{12.13}$$

2. Hydrostatic equilbrium

$$\frac{dP(r)}{dr} = -\frac{Gm(r)\rho(r)}{r^2} \tag{12.14}$$

3. The equation of state (EOS)

$$P(r) = f\left(\rho(r), T(r)\right) \tag{12.15}$$

$m(r)$ is the mass contained within radius r, $P(r)$ is the pressure, $\rho(r)$ is the density of the spherical planet, and f is the unique equation of state (EOS) of the material of interest – in this case, carbon.

Carbon grains in circumstellar discs most likely experience many shock events during planetesimal formation which may result in the modification of their structure. The coagulation of dust into clumps, the fragmentation of the disc into clusters of dust clumps, the merging of these clusters into ~ 1 km planetesimals, the collision of planetesimals during the accretion of meteorite parent bodies, and the subsequent collision of the parent bodies after their formation all induce strong shock waves that are expected to chemically and physically alter the materials. Subject to these high temperatures and pressures, the amorphous carbon grains polluting the protoplanetary discs around CEMP stars are expected to undergo graphitisation and may even crystallise into diamond [202, 243, 246]. In our calculations, the equation of state at low pressures, $P \leq 14$ GPa, is set to the third-order finite strain Birch-Murnagham EOS (BME; [30]) for graphite,

$$P = \frac{3}{2}K_0\left(\eta^{7/3} - \eta^{5/3}\right)\left[1 + \frac{3}{4}\left(K_0' - 4\right)\left(\eta^{2/3} - 1\right)\right], \tag{12.16}$$

where $\eta = \rho/\rho_0$ is the compression ratio with respect to the ambient density, ρ_0, K_0 is the bulk modulus of the material, and K_0' is the pressure derivative. Empirical fits to experimental data yield a BME EOS of graphite ($\rho_0 = 2.25$ g cm^{-3}) with parameters $K_0 = 33.8$ GPa and $K_0' = 8.9$ [109]. At 14 GPa, we incorporate the phase transition from graphite to diamond [109, 192] and adopt the Vinet EOS [262, 263],

$$P = 3K_0\eta^{2/3}\left(1 - \eta^{-1/3}\right)\exp\left[\frac{3}{2}\left(K_0' - 1\right)\left(1 - \eta^{-1/3}\right)\right] \tag{12.17}$$

with $K_0 = 444.5$ GPa and $K_0' = 4.18$ empirically fit for diamond, $\rho_0 = 3.51$ g cm^{-3} [66]. (As pointed out in Reference [230], the BME EOS is not fit to be extrapolated to high pressures since it is derived by expanding the elastic potential energy as a function of pressure keeping only the lowest-order terms.) Finally, at pressures $P \gtrsim 1,300$ GPa, where electron degeneracy becomes increasingly important, we use the Thomas-Fermi-Dirac (TFD) theoretical EOS ([227]; equations (40)–(49)), which intersects the diamond EOS at $P \sim 1,300$ GPa. Given that the full temperature-dependent carbon EOSs are either undetermined or dubious at best, all three EOSs adopted in this section are room-temperature EOSs for the sake of practical simplification.

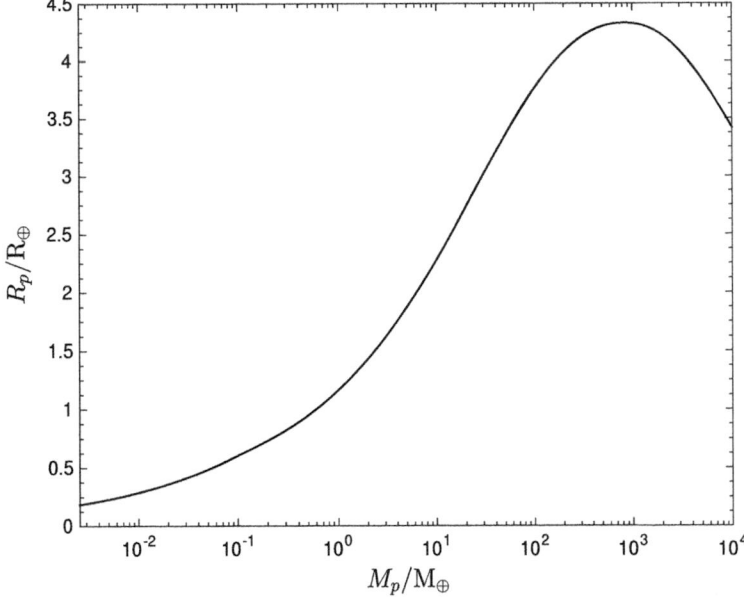

Figure 12.2 Mass-radius relation for solid homogenous, pure carbon planet.

Using a fourth-order Runge-Kutta scheme, we solve the system of equations simultaneously, numerically integrating Eqs. (12.13) and (12.14) begining at the planet's centre with the inner boundary conditions $M(r=0)=0$ and $P(r=0)=P_{\text{central}}$, where P_{central} is the central pressure. The outer boundary condition $P(r=R_p)=0$ then defines the planetary radius R_p and total planetary mass $M_p=m(r=R_p)$. Integrating these equations for a range of P_{central}, with the appropriate EOS, $P=P(\rho)$, to close the system of equations, yields the mass-radius relationship for a given composition. We show this mass-radius relation for a purely solid carbon planet in Figure 12.2. We find that for masses $M_p \lesssim 800 \, M_\oplus$, gravitational forces are small compared with electrostatic Coulomb forces in hydrostatic equilibrium, and so the planet's radius increases with increasing mass, $R_p \propto M_p^{1/3}$. However, at larger masses, the electrons are pressure ionised, and the resulting degeneracy pressure becomes significant, causing the planet radius to become constant and even decrease for increasing mass, $R_p \propto M_p^{-1/3}$ [117]. Planets which fall within the mass range $500 \lesssim M_p \lesssim 1{,}300 \, M_\oplus$, where the competing effects of Coulomb forces and electron degeneracy pressure cancel each other out, are expected to be approximately the same size, with $R_p \simeq 4.3 \, R_\oplus$, the maximum radius of a solid carbon planet. (In the case of gas giants, the planet radius can increase due to accretion of hydrogen and helium.)

Although the mass-radius relation illustrated in Figure 12.2 may alone not be enough to confidently distinguish a carbon planet from a water or silicate planet, the unique spectral features in the atmospheres of these carbon planets may provide the needed fingerprints. At high temperatures ($T \gtrsim 1{,}000$ K), the absorption spectra of massive ($M \sim 10$–$60 \, M_\oplus$)

carbon planets are expected to be dominated by CO, in contrast with the H_2O-dominated spectra of hot massive planets with solar-composition atmospheres [143]. The atmospheres of low-mass ($M \lesssim 10$ $M_\oplus$) carbon planets are also expected to be differentiable from their solar-composition counterparts due to their abundance of CO and CH_4 and lack of oxygen-rich gases like CO_2, O_2, and O_3 [143]. Furthermore, carbon planets of all masses at low temperatures are expected to accommodate hydrocarbon synthesis in their atmospheres; stable long-chain hydrocarbons are, therefore, another signature feature that can help observers distinguish the atmospheres of cold carbon planets and more confidently determine the bulk composition of a detected planet [143].

The detection of theoretically proposed carbon planets around CEMP stars will provide us with significant clues regarding how early planet formation may have started in the Universe. While direct detection of these extrasolar planets remains difficult given the low luminosity of most planets, techniques such as the transit method are often employed to indirectly spot exoplanets and determine physical parameters of the planetary system. When a planet 'transits' in front of its host star, it partially occludes the star and causes its observed brightness to drop by a minute amount. If the host star is observed during one of these transits, the resulting dip in its measured light curve can yield information regarding the relevant sizes of the star and the planet, the orbital semimajor axis, and the orbital inclination, among other characterising properties.

12.3.5 Section Summary and Implications

We explored the possibility of carbon planet formation around the iron-deficient, carbon-rich subset of low-mass stars known as CEMP stars. The observed abundance patterns of CEMP-no stars suggest that these stellar objects were probably born out of gas enriched by massive first-generation stars that ended their lives as Type II SNe with low levels of mixing and a high degree of fallback. The formation of dust grains in the ejecta of these primordial core-collapsing SNe progenitors has been observationally confirmed and theoretically studied. In particular, amorphous carbon is the only grain species found to condense and form in non-negligible amounts in SN explosion models that are tailored to reproduce the abundance patterns measured in CEMP-no stars. Under such circumstances, the gas clouds which collapse and fragment to form CEMP-no stars and their protoplanetary discs may contain significant amounts of carbon dust grains imported from SNe ejecta. The enrichment of solid carbon in the protoplanetary discs of CEMP stars may then be further enhanced by Fischer-Tropsch-type reactions and carbon-rich condensation sequences, where the latter occurs specifically in nebular gas with C/O $\gtrsim 1$.

For a given metallicity [C/H] of the host CEMP star, the maximum distance out to which planetesimal formation is possible can then be determined by comparing the dust-settling timescale in the protostellar disc to the expected disc lifetime. Assuming that disc dissipation is driven by a metallicity-dependent photoevaporation rate, we find a linear relation between [C/H] and the maximum semimajor axis of a carbon planet orbiting its host CEMP star. Very carbon-rich CEMP stars, such as G 77-61 and HE 2356-0410 with [C/H] $\simeq$ -0.7--0.6, can host carbon planets with semimajor axes as large ~ 20 AU for 100% carbon condensation efficiencies; this maximum orbital distance reduces to ~ 5 AU when the condensation efficiency drops by an order of magnitude. In the case of the observed CEMP-no stars HE 0107-5240, SDSS J0212+0137, and SDSS J1742+2531, where the carbon abundances are in the range [C/H] $\simeq$ -1.6--1.2, we expect more compact orbits, with maximum orbital distances $r_{\max} \simeq 2$, 4, and 6 AU, respectively, for $f_{cond} = 1$ and $r_{\max} \simeq 0.5 - 1$ AU for $f_{cond} = 0.1$.

While the shallow transit depths of Earth-mass carbon planets around HE 0107-5240 and HE 2356-0410 may evade detection, current and future space-based transit surveys promise to achieve the precision levels ($\sim 0.001\%$) necessary to detect planetary systems around CEMP stars such as SDSS J0212+0137, SDSS J1742+2531, and G 77-61. If gas giant (Jupiter-scale) planets form around CEMP stars, their transits would be much easier to detect than rocky planets. However, they are not likely to host life as we know it. There are a number of ongoing, planned, and proposed space missions committed to this cause, including CoRot (COnvection ROtation and planetary Transits), Kepler, PLATO (PLAnetary Transits and Oscillations of stars), TESS (Transiting Exoplanet Survey Satellite), and ASTrO (All Sky Transit Observer), which are expected to achieve precisions as low as 20–30 ppm (parts per million) [26, 267].

Short orbital periods and long transit durations are also key ingredients in boosting the probability of transit detection by observers. G 77-61 is not an optimal candidate in these respects since, given its large carbon abundance ([C/Fe] ~ 3.4), carbon planets may form out to very large distances and take up to a century to complete an orbit around the star for $f_{cond} = 1$ ($P_{\max} \sim 10$ years for 10% carbon condensation efficiency). The small stellar radius, $R_* \sim 0.5$ $R_\odot$, also reduces chances of spotting the transit since the resulting transit duration is only ~ 30 hours at most. Carbon planets around larger CEMP stars with an equally carbon-rich protoplanetary disc, such as HE 2356-0410 ($R_* \sim 7$ $R_\odot$), have a better chance of being spotted, with transit durations lasting up to ~ 3 weeks. The CEMP-stars SDSS J0212+0137, and SDSS J1742+2531 are expected to host carbon planets with much shorter orbits, $P_{\max} \sim 16$ years for 100% condensation efficiency ($P_{\max} \sim 1$ year for $f_{cond} = 0.1$), and transit durations that last as long as ~ 60 hours. If the ability to measure transit depths improves to a precision of 1 ppm, then potential carbon planets around HE 0107-5240 are the most likely to be spotted (among the group of CEMP-no stars considered here), transiting across the host star at least once every ~ 5 months (10% condensation efficiency) with a transit duration of 6 days.

While our calculations place upper bounds on the distance from the host star out to which carbon planets can form, we note that orbital migration may alter a planet's location in the circumstellar disc. As implied by the existence of 'hot Jupiters', it is possible for a protoplanet that forms at radius r to migrate inward through gravitational interactions with other protoplanets, resonant interactions with planetesimals with more compact orbits, or tidal interactions with gas in the surrounding disc [201]. Since Figure 12.1 only plots $r_{\max}$, temperatures T in the range 250–350 K, H_2O may be abundant, comparable to or higher than that found in diffuse galactic clouds, provided that the FUV intensity to density ratio is smaller than a critical value.

12.4.2 Model Ingredients

We calculate the abundance of gas-phase H_2O for low-metallicity gas parcels that are exposed to external FUV radiation and cosmic-ray and/or X-ray fluxes. Given our chemical network, we solve the steady-state rate equations using our dedicated Newton-based solver and obtain x_{H_2O} as function of T and the FUV intensity to density ratio.

We adopt a 10^5 K diluted black-body spectrum, representative of radiation produced by massive Population III stars. The photon density in the 6–13.6 eV interval, is $n_\gamma \equiv n_{\gamma,0} I_{UV}$, where $n_{\gamma,0} = 6.5 \times 10^{-3}$ photons cm^{-3} is the photon density in the interstellar radiation field [70], and I_{UV} is the 'normalised intensity'. Thus, $I_{UV} = 1$ corresponds to the FUV intensity in the Draine ISRF.

Cosmic-ray and/or X-ray ionisation drive the ion-molecule chemical network. We assume an ionisation rate per hydrogen nucleon ζ (s^{-1}). In the galaxy, [63] and References [120] showed that ζ lies within the relatively narrow range 10^{-15}–10^{-16} s^{-1}. We therefore introduce the 'normalised ionisation rate' $\zeta_{-16} \equiv (\zeta/10^{-16}$ s$^{-1})$. The ionisation rate and the FUV intensity are likely correlated, as both should scale with the formation rate of massive OB stars. We thus set $\zeta_{-16} = I_{UV}$ as our fiducial case but also consider the cases $\zeta_{-16} = 10^{-1} I_{UV}$ and $\zeta_{-16} = 10 I_{UV}$.

Dust shielding against the FUV radiation becomes ineffective at low metallicities. However, self-absorption in the H_2 lines may significantly attenuate the destructive Lyman-Werner (11.2–13.6 eV) radiation [71, 240], and high abundances of H_2 may be maintained even at low metallicity [28]. In the models presented here, we assume an H_2 shielding column of at least 5×10^{21} cm^{-2}. (For such conditions, CO is also shielded by the H_2.) The LW flux is then attenuated by a self-shielding factor of $f_{shield} \sim 10^{-8}$, and the H_2 photodissociation rate is only $5.8 \times 10^{-19} I_{UV}$ s^{-1}. With this assumption, H_2 photodissociation by LW photons is negligible compared to cosmic-ray and/or X-ray ionisation as long as $I_{UV} < 85\zeta_{-16}$.

However, even when the Lyman-Werner band is fully blocked, OH and H_2O are photodissociated because their energy thresholds for photodissociation are 6.4 and 6 eV, respectively. For the low metallicities that we consider, photodissociation is generally the dominant removal mechanism for H_2O and OH. We adopt the calculated OH and H_2O photodissociation rates calculated by Reference [28].

We assume thermal and chemical steady states. In the Milky Way, the bulk of the ISM gas is considered to be at approximate thermal equilibrium, set by cooling and heating processes. We discuss the relevant chemical and thermal timescales in Section 12.4.4.

Given the before mentioned assumptions, the steady-state solutions for the species abundances depend on only two parameters, the temperature T and the intensity-to-density ratio I_{UV}/n_4. Here, $n_4 \equiv (n/10^4$ cm$^{-3})$ is the total number density of hydrogen nuclei normalised to typical molecular cloud densities. T and I_{UV}/n_4 form our basic parameter space in our study.

12.4.3 Chemical Model

We consider a chemical network of 503 two-body reactions, between 56 atomic, molecular, and ionic species of H, He, C, O, S, and Si. We assume cosmological elemental helium abundance of 0.1 relative to hydrogen (by number). For the metal elemental abundances, we adopt the photospheric solar abundances, multiplied by a metallicity factor Z' (i.e., $Z' = 1$ is solar metallicity). In our fiducial model, we assume $Z' = 10^{-3}$, but we also explore cases with $Z' = 10^{-2}$ and $Z' = 10^{-4}$. Since our focus here is on the very-low-metallicity regime, where dust grains play a lesser role, we neglect any depletion on dust grains and dust-grain chemistry (except for H_2, as discussed further below). Direct and induced ionisations and dissociations by the cosmic-ray/X-ray field ($\propto \zeta$) are included. For the gas-phase reactions, we adopt the rate coefficients given by the UMIST 2012 database [183].

The formation of heavy molecules relies on molecular hydrogen. We consider two scenarios for H_2 formation: (1) pure gas-phase formation and (2) gas-phase formation plus formation on dust grains. In the gas phase radiative-attachment

$$H + e \rightarrow H^- + \nu \tag{12.18}$$

followed by associative-detachment

$$H^- + H \rightarrow H_2 + e \tag{12.19}$$

is the dominant H_2 formation route.

H_2 formation on the surface of dust grains is considered to be the dominant formation channel in the Milky Way. We adopt a standard rate coefficient [49, 116, 128],

$$R \simeq 3 \times 10^{-17} \, T_2^{1/2} \, Z' \, \text{cm}^3 \, \text{s}^{-1}, \tag{12.20}$$

where $T_2 \equiv (T/100 \text{ K})$. In this expression, we assume that the dust-to-gas ratio is linearly proportional to the metallicity Z'. Thus, in scenario (b) H_2 formation on dust grains dominates even for $Z' = 10^{-3}$. Scenario (a) is the limit where the gas-phase channel dominates, as appropriate for dust-free environments or for superlinear dependence of the dust-to-gas ratio on Z'.

12.4.4 Timescales

The timescale for the system to achieve chemical steady state is dictated by the relatively long H_2 formation timescale. In the gas phase (scenario (1)) it is

$$t_{H_2} = \frac{1}{k_2 \, n \, x_e} \approx 8 \times 10^8 \, \zeta_{-16}^{-1/2} \, n_4^{-1/2} \, T_2^{-1} \text{ yr}, \tag{12.21}$$

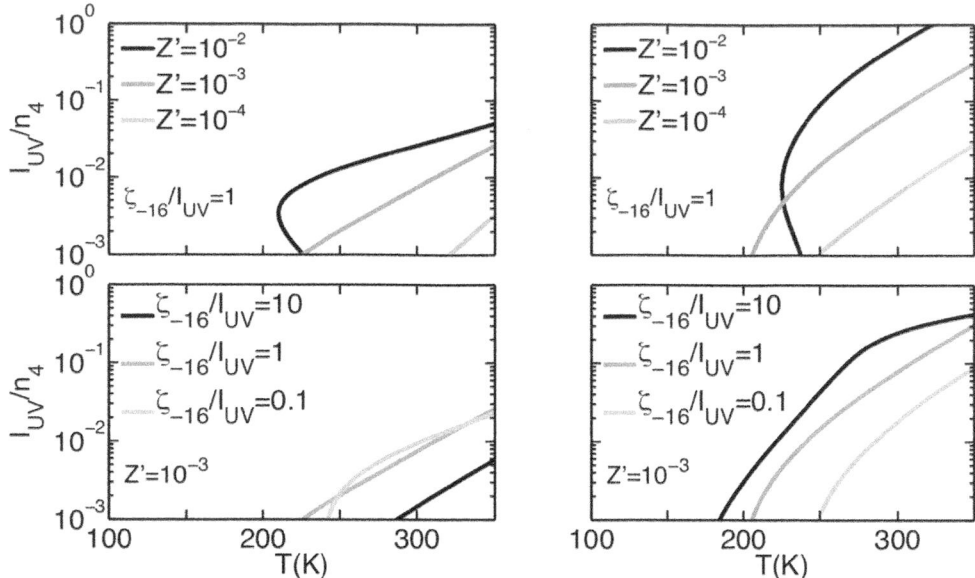

Figure 12.4 The $x_{H_2O} = 10^{-8}$ contour, for variations in Z' (upper panels) and in ζ_{-16}/I_{UV} (lower panels), assuming pure gas-phase chemistry (scenario (a) – left panels) and including H_2 formation on dust grains (scenario (b) – right panels).

12.4.7 Variations in Z' and ζ_{-16}/I_{UV}

In Figure 12.4, we investigate the effects of variations in the value of Z' and the normalisation ζ_{-16}/I_{UV}. The figure shows the $x_{H_2O} = 10^{-8}$ contours for scenarios (a) (left panels) and (b) (right panels). As discussed earlier, H_2O is generally more abundant in scenario (b) because the hydrogen is fully molecular in this case, and, therefore, the 10^{-8} contours are located at higher I_{UV}/n_4 values in both right panels.

The upper panels show the effect of variations in the metallicity value Z' for our fiducial normalisation $\zeta_{-16} = I_{UV}$. In both panels, the oxygen abundance rises, and x_{H_2O} increases with increasing Z'. Thus, at higher Z', the 10^{-8} contours shift to lower T and higher I_{UV}/n_4 and vice versa. An exception is the behaviour of the $Z' = 10^{-2}$ curve, for which the metallicity is already high enough so that reactions with metal species dominate H_2O removal for $I_{UV}/n_4 \lesssim 10^{-2}$. The increase in metallicity then results in a *decrease* of the H_2O abundance, and the 10^{-8} contour shifts to the right. For $I_{UV}/n_4 \gtrsim 10^{-2}$, removal by FUV dominates, and the behaviour is similar to that in the $Z' = 10^{-3}$ and $Z' = 10^{-4}$ cases.

The lower panels show the effects of variations in the ionisation rate normalisation ζ_{-16}/I_{UV} for our fiducial metallicity value of $Z' = 10^{-3}$. First, we consider the pure gas phase formation case (scenario (a) – lower-left panel). For the two cases $\zeta_{-16}/I_{UV} = 1$ and 10^{-1}, the H_2O removal is dominated by FUV photodissociation and, therefore, is independent of ζ. As shown by Reference [28], the H_2O formation rate is also independent of ζ when the H_2 forms in the gas phase. Therefore, x_{H_2O} is essentially independent of ζ, and the contours overlap. For the high ionisation rate $\zeta_{-16}/I_{UV} = 10$, the proton abundance becomes high, and H_2O reactions with H^+ dominate H_2O removal. In this limit, x_{H_2O} decreases with ζ, and the 10^{-8} contour moves down.

When H_2 forms on dust (scenario (b)–lower-right panel), the H_2O formation rate via the ion-molecule sequence is proportional to the H^+ and H_3^+ abundances, which rise with ζ. Since the gas is molecular, the proton fraction is low, and the removal is always dominated by FUV photodissociations (independent of ζ). Therefore, in this case, x_{H_2O} *increases* with ζ_{-16}/I_{UV}, and the curves shift up and to the left, towards lower T and higher I_{UV}/n_4.

12.4.8 Section Summary and Implications

We have demonstrated that the H_2O gas phase abundance may remain high even at very low metallicities of $Z' \sim 10^{-3}$. The onset of the efficient neutral-neutral formation sequence at $T \sim 300$ K may compensate for the low metallicity and form H_2O in abundance, similar to that found in diffuse clouds within the Milky Way.

We have considered two scenarios for H_2 formation, representing two limiting cases: one in which H_2 is formed in pure gas phase (scenario (a)), and one in which H_2 forms both in gase phase and on dust grains, assuming that the dust abundance scales linearly with Z' (scenario (b)). Recent studies by References [84, 91, 112] suggest that the dust abundance might decrease faster than then linearly with decreasing Z'. As shown by Reference [28], for $Z' = 10^{-3}$ and dust abundance that scales as Z'^{β} with $\beta \geq 2$, H_2 formation is dominated by gas phase formation channel. Therefore, our scenario (a) is also applicable for models in which dust grains are present, with an abundance that scales superlinearly with Z'. For both scenarios (a) and (b), we have found that the neutral-neutral formation channel yields $x_{H_2O} \gtrsim 10^{-8}$, provided that I_{UV}/n_4 is smaller than a critical value. For the first scenario, we have found that this critical value is $(I_{UV}/n_4)_{crit} = 2 \times 10^2$. For the second scenario, $(I_{UV}/n_4)_{crit} = 3 \times 10^{-1}$.

In our analysis, we have assumed that the system had reached a chemical steady state. For initially atomic (or ionised) gas, this assumption offers the best conditions for the formation of molecules. However, chemical steady state might not always be achieved within a cloud lifetime or even within the Hubble time. The timescale to achieve chemical steady state (from an initially dissociated state) is dictated by the H_2 formation process and is generally long at low metallicities. For $Z' = 10^{-3}$ and our fiducial parameters, the timescales for both scenarios are of order of a few 10^8 years (see, e.g., Reference [27]) and are comparable to the age of the Universe at redshifts of ~ 10. The generically high water abundances we find for warm conditions and low metallicities will be maintained in dynamically evolving systems, so long as they remain H_2 shielded.

Our results might have interesting implications for the question of how early could have life originated in the Universe [159]. Our study addresses the first step of H_2O formation in early enriched, molecular gas clouds. If such a cloud is to collapse and form a protoplanetary disc, some of the H_2O may make its way to the surfaces of forming planets [258].

However, the question of to what extent the H_2O molecules that were formed in the initial molecular clouds are preserved all the way through the process of planet formation is beyond the scope of this section.

12.5 An Observational Test for the Anthropic Origin of the Cosmological Constant

12.5.1 Section Background

The distance to Type Ia supernovae [142, 206] and the statistics of the cosmic microwave background anisotropies [211] provide conclusive evidence for a finite vacuum energy density of $\rho_V = 4$ keV cm^{-3} in the present-day Universe. This value is a few times larger than the mean cosmic density of matter today. The expected exponential expansion of the Universe in the future (for a time-independent vacuum density) will halt the growth of all bound systems such as galaxies and groups of galaxies including the nearby Virgo cluster [44, 76, 190]. It will also redshift all extragalactic sources out of detectability (except for the merger remnant of the Milky Way and the Andromeda galaxies to which we are bound) – marking the end of extragalactic astronomy, as soon as the Universe ages by another factor of ten [157].

The observed vacuum density is smaller by tens of orders of magnitude than any plausible zero-point scale of the Standard Model of particle physics. Weinberg [264] first suggested that such a situation could arise in a theory that allows the cosmological constant to be a free parameter. On a scale much bigger than the observable Universe, one could then find regions in which the value of ρ_V is very different. However, if one selects those regions that give life to observers, then one would find a rather limited range of ρ_V values near its observed magnitude, since observers are most likely to appear in galaxies as massive as the Milky Way galaxy which assembled at the last moment before the cosmological constant started to dominate our Universe. Vilenkin [261] showed that this so-called 'anthropic argument' [18] can be used to calculate the probability distribution of vacuum densities with testable predictions. This notion [77, 92, 93, 180, 244, 245, 265] gained popularity when it was realised that string theory predicts the existence of an extremely large number [38, 95, 129, 175], perhaps as large as $\sim 10^{100}$–10^{500} [13], of possible vacuum states. The resulting landscape of string vacua [242] in the 'multiverse,' encompassing a volume of space far greater than our own inflationary patch, made the anthropic argument appealing to particle physicists and cosmologists alike [214, 245, 266].

The time is therefore ripe to examine the prospects for an experimental test of the anthropic argument. Any such test should be welcomed by proponents of the anthropic argument, since it would elevate the idea to the status of a falsifiable physical theory. At the same time, the test should also be welcomed by opponents of anthropic reasoning, since such a test would provide an opportunity to diminish the predictive power of the anthropic proposal and suppress discussions about it in the scientific literature.

Is it possible to dispute the anthropic argument without visiting regions of space that extend far beyond the inflationary patch of our observable Universe? The answer is *yes* if one can demonstrate that life could have emerged in our Universe even if the cosmological constant would have had values that are much larger than observed. In this section, we propose a set of astronomical observations that could critically examine this issue. We make use of the fact that dwarf galaxies formed in our Universe at redshifts as high as $z \sim 10$ when the mean matter density was larger by a factor of $\sim 10^3$ than it is today[4] [160]. If habitable planets emerged within these dwarf galaxies or their descendents (such as old globular clusters which might be the tidally truncated relics of early galaxies [184, 187]), then life would have been possible in a universe with a value of ρ_V that is a thousand times bigger than observed.

12.5.2 Prior Probability Distribution of Vacuum Densities

On the Planck scale of a quantum field theory which is unified with gravity (such as string theory), the vacuum energy densities under discussion represent extremely small deviations around $\rho_V = 0$. Assuming that the prior probability distribution of vacuum densities, $\mathcal{P}_*(\rho_V)$, is not divergent at $\rho_V = 0$ (since $\rho_V = 0$ is not favoured by any existing theory), it is natural to expand it in a Taylor series and keep only the leading term. Thus, in our range of interest of ρ_V values [92, 93, 265],

$$\mathcal{P}_*(\rho_V) \approx constant. \tag{12.27}$$

This implies that the probability of measuring a value equal to or smaller than the observed value of ρ_V is $\sim 10^{-3}$ if habitable planets could have formed in a Universe with a value of ρ_V that is a thousand times bigger than observed.

Numerical simulations indicate that our Universe would cease to make new bound systems in the near future [44, 76, 190]. A universe in which ρ_V is a thousand times larger would, therefore, make dwarf galaxies until $z \sim 10$ when the matter density was a thousand times larger than today. The question of whether planets can form within these dwarf galaxies can be examined observationally, as we discuss next. It is important to note that once a dwarf galaxy forms, it has an arbitrarily long time to convert its gas into stars and planets because its internal evolution is decoupled from the global expansion of the universe (as long as outflows do not carry material out of its gravitational pull).

12.5.3 Extragalactic Planet Searches

Gravitational microlensing is the most effective search method for planets beyond our galaxy. The planet introduces a short-term distortion to the otherwise smooth light curve produced by its parent star as that star focuses the light from a background star which happens to lie behind it [176, 203]. In an extensive search for planetary microlensing

---

[4] We note that although the cosmological constant started to dominate the mass density of our Universe at $z \sim 0.4$, its impact on the formation of bound objects became noticeable only at $z \sim 0$ or later [44, 76, 190]. For the purposes of our discussion, we therefore compare the matter density at $z \sim 10$ to that today. Coincidentally, the Milky Way galaxy formed before ρ_V dominated but it could have also formed later.

signatures, a number of collaborations named PLANET [3], μFUN [273], and RoboNET are performing follow-up observations on microlensing events which are routinely detected by the groups MOA [33] and OGLE [251]. Many 'planetary' events have been reported, including a planet of a mass of $\sim$ 5 Earth masses at a projected separation of 2.6 AU from a 0.2 $M_\odot$ M-dwarf star in the microlensing event OGLE-2005-BLG-390Lb [20] and a planet of 13 Earth masses at a projected separation of 2.3 AU from its parent star in the event OGLE-2005-BLG-169 towards the galactic bulge – in which the background star was magnified by the unusually high factor of $\sim$ 800 [102]. Based on the statistics of these events and the search parameters, one can infer strong conclusions about the abundance of planets of various masses and orbital separations in the surveyed star population [32, 94, 102]. The technique can be easily extended to lenses outside our galaxy and out to the Andromenda galaxy (M31) using the method of pixel lensing [15, 56, 61]. For the anthropic experiment, we are particularly interested in applying this search technique to lensing of background Milky Way stars by old stars in foreground globular clusters (which may be the tidally truncated relics of $z \sim 10$ galaxies), or to lensing of background M31 stars by foreground globular clusters [118] or dwarf galaxies such as Andromeda VIII [188]. In addition, self-lensing events in which foreground stars of a dwarf galaxy lens background stars of the same galaxy are of particular interest. Such self-lensing events were observed in the form of caustic-crossing binary lens events in the Large Magellanic Cloud (LMC) and the Small Magellanic Cloud (SMC) [67]. In the observed cases, there is enough information to ascertain that the most likely lens location is in the Magellanic Clouds themselves. Yet each caustic-crossing event represents a much larger number of binary lens events from the same lens population; the majority of these may be indistinguishable from point-lens events. It is therefore possible that some of the known single-star LMC lensing events are due to self-lensing [67], as hinted at by their geometric distribution [105, 225].

As mentioned earlier, another method for finding extragalactic planets involves transit events in which the planet passes in front of its parent star and causes a slight temporary dimming of the star. Spectral modelling of the parent star allows to constrain both the size and abundance statistics of the transiting planets [52, 205]. Existing surveys reach distance scales of several kpc [174, 255] with some successful detections [37, 136, 252]. So far, a Hubble Space Telescope search for transiting Jupiters in the globular cluster 47 Tucanae resulted in no detections [96] (although a pulsar planet was discovered later by a different technique in the low-metallicity globular cluster Messier 4 [234], potentially indicating early planet formation). A future space telescope (beyond the TESS,[5] Kepler,[6] and COROT[7] missions which focus on nearby stars) or a large-aperture ground-based facility (such as the Giant Magellan Telescope [GMT],[8] the Thirty-Meter Telescope [TMT],[9] or the

[5] https://tess.gsfc.nasa.gov.
[6] http://kepler.nasa.gov/.
[7] http://smsc.cnes.fr/COROT/.
[8] www.gmto.org/.
[9] www.tmt.org/.

European Extremely Large Telescope [EELT][10]) could extend the transit search technique to planets at yet larger distances (but see Reference [205]). Existing searches [52] identified the need for a high signal-to-noise spectroscopy as a follow-up technique for confirming real transits out of many false events. Such follow-ups would become more challenging at large distances, making the microlensing technique more practical.

12.5.4 Observations of Dwarf Galaxies at High Redshifts

Our goal is to study stellar systems in the local Universe which are the likely descendents of the early population of $z \sim 10$ galaxies [221]. In order to refine this selection, it would be desirable to measure the characteristic size, mass, metallicity, and star formation histories of $z \sim 10$ galaxies (see Reference [160] for a review on their theoretically expected properties). As already mentioned, it is possible that the oldest globular clusters are descendents of the first galaxies [220].

Recently, a large number of faint early galaxies, born less than a billion years after the Big Bang, have been discovered (see, e.g., [134, 219]). These include starburst galaxies with star formation rates in excess of ~ 0.1 $M_\odot$ yr^{-1} and dark matter halos [239] of $\sim 10^{9-11}$ $M_\odot$ [39, 40, 218, 219, 228] at $z \sim 5$–10. Luminous Lyα emitters are routinely identified through continuum dropout and narrow-band imaging techniques [39, 40]. In order to study fainter sources which were potentially responsible for reionisation, spectroscopic searches have been undertaken near the critical curves of lensing galaxy clusters [134, 228], where gravitational magnification enhances the flux sensitivity. Because of the foreground emission and opacity of the Earth's atmosphere, it is difficult to measure spectral features other than the Lyα emission line from these feeble galaxies from ground-based telescopes.

In one example, gravitational lensing by the massive galaxy cluster A2218 allowed the detection of stellar system at $z = 5.6$ with an estimated mass of $\sim 10^6$ $M_\odot$ in stars [228]. Detection of additional low-mass systems could potentially reveal whether globular clusters formed at these high redshifts. Such a detection would be feasible with the James Webb Space Telescope.[11] Existing designs for future large-aperture (> 20 m) infrared telescopes (such as the GMT, TMT, and EELT mentioned earlier) would also enable researchers to measure the spectra of galaxies at $z \sim 10$ and infer their properties.

Based the characteristics of high-z galaxies, one would be able to identify present-day systems (such as dwarf galaxies or globular clusters) that are their likely descendents [69, 269] and search for planets within them. Since the lifetime of massive stars that explode as core-collapse supernovae is two orders of magnitude shorter than the age of the Universe at $z \sim 10$, it is possible that some of these systems would be enriched to a high metallicity despite their old age. For example, the cores of quasar host galaxies are known to possess super-solar metallicities at $z \gtrsim 6$ [68].

---

[10] www.eso.org/sci/facilities/eelt.
[11] www.jwst.nasa.gov/.

12.5.5 Section Summary and Implications

In future decades, it would be technologically feasible to search for microlensing or transit events in local dwarf galaxies or old globular clusters and to check whether planets exist in these environments. Complementary observations of early dwarf galaxies at redshifts $z \sim 10$ can be used to identify nearby galaxies or globular clusters that are their likely descendents. If planets are found in local galaxies that resemble their counterparts at $z \sim 10$, then the precise version of the anthropic argument [77, 92, 180, 261, 264] would be weakened considerably, since planets could have formed in our Universe even if the cosmological constant, ρ_V, was three orders of magnitude larger. For a flat probability distribution at these ρ_V values (which represents infinitesimal deviations from $\rho_V = 0$ relative to the Planck scale), this would imply that the probability for us to reside in a region where ρ_V obtains its observed value is lower than $\sim 10^{-3}$. The precise version of the anthropic argument [77, 92, 180, 261, 264] could then be ruled out at a confidence level of $\sim 99.9\%$, which is a satisfactory measure for an experimental test. The envisioned experiment resonates with two of the most active frontiers in astrophysics, namely the search for planets and the study of high-redshift galaxies, and if performed, it would have many side benefits to conventional astrophysics.

We note that in the hypothetical universe with a large cosmological constant, life need not form at $z \sim 10$ (merely 400 million years after the Big Bang) but rather any time later. Billions of years after a dwarf galaxy had formed, a typical astronomer within it would see the host galaxy surrounded by a void which is dominated by the cosmological constant. Of course, the volume density of life in such a case is smaller than in our Universe.

An additional factor that enters the likelihood function of ρ_V values involves the conversion efficiency of baryons into observers in the Universe. A universe in which observers only reside in galaxies that were made at $z \sim 10$ might be less effective at making observers. The fraction of baryons that have assembled into star-forming galaxies above the hydrogen cooling threshold by $z \sim 10$ is estimated to be $\sim 10\%$ [160], comparable to the final fraction of baryons that condensed into stars in the present-day Universe [90]. It is possible that more stars formed in smaller systems down to the Jeans mass of $\sim 10^{4-5} \, M_\odot$ through molecular hydrogen cooling [41]. Although today most baryons reside in a warm-hot medium of $\sim 2 \times 10^6$ K that cannot condense into stars [50, 64], most of the cosmic gas at $z \sim 10$ was sufficiently cold to fragment into stars as long as it could have cooled below the virial temperature of its host halos [160]. The star formation efficiency can be inferred [69] from dynamical measurements of the star and dark matter masses in local dwarfs or globulars that resemble their counterparts at $z \sim 10$. If only a small portion of the cosmic baryon fraction (Ω_b/Ω_m) in dwarf galaxies is converted into stars, then the probability of obtaining habitable planets would be reduced accordingly. Other physical factors, such as metallicty, may also play an important role [156]. Preliminary evidence indicates that planet formation favours environments which are abundant in heavy elements [83], although notable exceptions exist [234].

Unfortunately, it is not possible to infer the planet production efficiency for an alternative universe purely based on observations of our Universe. In our Universe, most of the baryons which were assembled into galaxies by $z \sim 10$ are later incorporated into bigger galaxies. The vast majority of the $z \sim 10$ galaxies are consumed through hierarchical mergers to make bigger galaxies; isolated descendents of $z \sim 10$ galaxies are rare among low-redshift galaxies. At any given redshift below 10, it would be difficult to separate observationally the level of planet formation in our Universe from the level that would have occurred otherwise in smaller galaxies if these were not consumed by bigger galaxies within a universe with a large vacuum density, ρ_V. In order to figure out the planet production efficiency for a large ρ_V, one must adopt a strategy that mixes observations with theory. Suppose we observe today the planet production efficiency in the descendents of $z \sim 10$ galaxies. One could then use numerical simulations to calculate the abundance that these galaxies would have had today if ρ_V was $\sim 10^3$ times bigger than its observed value. This approach implicitly takes into account the possibility that planets may form relatively late (after ~ 10 Gyr) within these isolated descendents, irrespective of the value of ρ_V. The late time properties of gravitationally bound systems are expected to be independent of the value of ρ_V.

In our discussion, we assumed that as long as rocky planets can form at orbital radii that allow liquid water to exist on their surface (the so-called *habitable zone* [130]), life would develop over billions of years and eventually mature in intelligence. Without a better understanding of the origin of intelligent life, it is difficult to assess the physical conditions that are required for intelligence to emerge beyond the minimal requirements stated earlier. If life forms early, then civilisations might have more time to evolve to advanced levels. On the other hand, life may be disrupted more easily in early galaxies because of their higher density (making the likelihood of stellar encounters higher) [92, 245], and so it would be useful to determine the environmental density observationally. In the more distant future, it might be possible to supplement the study proposed here by the more adventurous search for radio signals from intelligent civilisations beyond the boundaries of our galaxy. Such a search would bring an extra benefit. If the anthropic argument turns out to be wrong and intelligent civilisations are common in nearby dwarf galaxies, then the older, more advanced civilisations among them might broadcast an explanation for why the cosmological constant has its observed value.

12.6 The Relative Likelihood of Life as a Function of Cosmic Time

12.6.1 Section Background

Currently, we only know of life on Earth. The Sun formed ~ 4.6 Gyr ago and has a lifetime comparable to the current age of the Universe. But the lowest-mass stars near the hydrogen burning threshold at $0.08\, M_\odot$ could live a thousand times longer, up to 10 trillion years [146, 224]. Given that habitable planets may have existed in the distant past and might exist in the distant future, it is natural to ask, *what is the relative probability for the emergence of life as a function of cosmic time?* In this section, we answer this question conservatively by restricting our attention to the context of 'life as we know it' and the standard cosmological model (ΛCDM).[12] Note that since the probability distribution is normalised to have a unit integral, it only compares the relative importance of different epochs for the emergence of life but does not calibrate the overall likelihood for life in the Universe. This feature makes our results robust to uncertainties in normalisation constants associated with the likelihood for life on habitable planets.

Next, we express the relative likelihood for the appearance of life as a function of cosmic time in terms of the star formation history of the Universe, the stellar mass function, the lifetime of stars as a function of their mass, and the probability of Earth-mass planets in the habitable zone of these stars. We define this likelihood within a fixed co-moving volume which contains a fixed number of baryons. In predicting the future, we rely on an extrapolation of star formation rate until the current gas reservoir of galaxies is depleted.

12.6.2 Formalism

Master Equation

We wish to calculate the probability $dP(t)/dt$ for life to form on habitable planets per unit time within a fixed co-moving volume of the Universe [153]. This probability distribution should span the time interval between the formation time of the first stars and the maximum lifetime of all stars that were ever made (~ 10 Tyr).

The probability $dP(t)/dt$ involves a convolution of the star formation rate per co-moving volume, $\dot{\rho}_*(t')$, with the temporal window function, $g(t - t', m)$, due to the finite lifetime of stars of different masses, m, and the likelihood, $\eta_{\text{Earth}}(m)$, of forming an Earth-mass rocky planet in the habitable zone (HZ) of stars of different masses, given the mass distribution of stars, $\xi(m)$, times the probability, $p(\text{life}|\text{HZ})$, of actually having life on a habitable planet. With all these ingredients, the relative probability per unit time for life within a fixed co-moving volume can be written in terms of the double integral,

$$\frac{dP}{dt}(t) = \frac{1}{N} \int_0^t dt' \int_{m_{\min}}^{m_{\max}} dm' \xi(m') \dot{\rho}_*(t', m') \eta_{\text{Earth}}(m') p(\text{life}|\text{HZ}) g(t - t', m'), \quad (12.28)$$

where the prefactor $1/N$ assures that the probability distribution is normalised to a unit integral over all times. The window function, $g(t - t', m)$, determines whether a habitable planet that formed at time t' is still within a habitable zone at time t. This function is non-zero within the lifetime $\tau_*(m)$ of each star, namely $g(t - t', m) = 1$ if $0 < (t - t') < \tau_*(m)$, and zero otherwise. The quantities $m_{\min}$ and $m_{\max}$ represent the minimum and maximum masses of viable host stars for habitable planets, respectively. Next, we provide more details on each of the various components of the preceding master equation.

---

[12] We address this question from the perspective of an observer in a single co-moving Hubble volume formed after the end of inflation. As such, we do not consider issues of self-location in the multiverse nor of the measure on eternally inflating regions of space-time. We note, however, that any observers in a post-inflationary bubble will, by necessity of the eternal inflationary process, only be able to determine the age of their own bubble. We therefore restrict our attention to the question of the probability distribution of life in the history of our own inflationary bubble.

Stellar Mass Range

Life requires the existence of liquid water on the surface of Earth-mass planets during the main stage lifetime of their host star. These requirements place a lower bound on the lifetime of the host star and, thus, an upper bound on its mass.

There are several proxies for the minimum time needed for life to emerge. Certainly, the star must live long enough for the planet to form, a process which took $\sim$ 40 Myr for Earth [185]. Moreover, once the planet has formed, sufficient cooling must follow to allow the condensation of water on the planet's surface. The recent discovery of the earliest crystals, zircons, suggests that these were formed during the Archean era, as much as 160 Myr after the planet formed [257]. Thus, we arrive at a conservative minimum of 200 Myr before life could form; any star living less than this time could not host life on an Earth-like planet. At the other end of the scale, we find that the earliest evidence for life on Earth comes from around 800 Myr after the formation of the planet [185], yielding an upper bound on the minimum lifetime of the host star. For the relevant mass range of massive stars, the lifetime, τ_*, scales with stellar mass, m, roughly as $(\tau_*/\tau_\odot) = (m/M_\odot)^{-3}$, where $\tau_\odot \approx 10^{10}$ yr. Thus, we find that the maximum mass of a star capable of hosting life ($m_{\max}$) is in the range 2.3–3.7 $M_\odot$. Due to their short lifetimes and low abundances, high-mass stars do not provide a significant contribution to the probability distribution, $dP(t)/dt$, and so the exact choice of the upper mass cut-off in the preceding range is unimportant. The lowest-mass stars above the hydrogen burning threshold have a mass $m = 0.08\ M_\odot$.

Time Range

The stars resulted in a second generation of stars, enriched by heavy elements, merely a few Myr later. The theoretical expectation that the second generation stars should have hosted planetary systems can be tested observationally by searching for planets around metal-poor stars in the halo of the Milky Way galaxy [182].

Star formation is expected to exhaust the cold gas in galaxies as soon as the Universe ages by a factor of a few (based on the ratio between the current reservoir of cold gas in galaxies [90] and the current star formation rate), but low-mass stars would survive long after that. The lowest-mass stars near the hydrogen burning limit of 0.08 $M_\odot$ have a lifetime of order 10 trillion years [146]. The probability $dP(t)/dt$ is expected to vanish beyond that time.

Initial Mass Function

The initial mass function (IMF) of stars $\xi(m)$ is proportional to the probability that a star in the mass range between m and $m + dm$ is formed. We adopt the empirically calibrated, Chabrier functional form [51], which follows a log-normal form for masses under a solar mass, and a power-law above a solar mass, as follows:

$$\xi(m) \propto \begin{cases} \left(\dfrac{m}{M_\odot}\right)^{-2.3} & m > 1\ M_\odot \\ a \exp\left(-\dfrac{\ln(m/m_c)^2}{2\sigma^2}\right)\dfrac{M_\odot}{m} & m \leq 1\ M_\odot \end{cases}, \qquad (12.29)$$

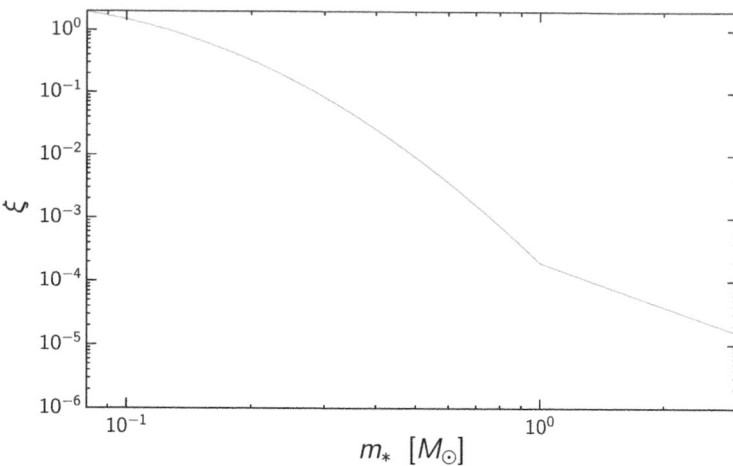

Figure 12.5 The Chabrier [51] mass function of stars, $\xi(m_*)$, plotted with a normalisation integral of unity.

where $a = 790$, $\sigma = 0.69$, and $m_c = 0.08\ M_\odot$. This IMF is plotted as a probability distribution normalised to a unit integral in Figure 12.5.

For simplicity, we ignore the evolution of the IMF with cosmic time and its dependence on galactic environment (e.g., galaxy type or metallicity [59]), as well as the uncertain dependence of the likelihood for habitable planets around these stars on metallicity [182].

Stellar Lifetime

The lifetime of stars, τ_*, as a function of their mass, m, can be modelled through a piece-wise power-law form. For $m < 0.25\ M_\odot$, we follow Reference [146]. For $0.75\ M_\odot < m < 3\ M_\odot$, we adopt a scaling with an average power-law index of -2.5 and the proper normalisation for the Sun [226]. Finally, we interpolate in the range between 0.25 and 0.75 $M_\odot$ by fitting a power-law form there and enforcing continuity. In summary, we adopt,

$$
\tau_*(m) = \begin{cases}
1.0 \times 10^{10}\ \text{yr} \left(\dfrac{m}{M_\odot} \right)^{-2.5} & 0.75\ M_\odot < m < 3\ M_\odot \\[2mm]
7.6 \times 10^{9}\ \text{yr} \left(\dfrac{m}{M_\odot} \right)^{-3.5} & 0.25\ M_\odot < m \leq 0.75\ M_\odot \\[2mm]
5.3 \times 10^{10}\ \text{yr} \left(\dfrac{m}{M_\odot} \right)^{-2.1} & 0.08\ M_\odot \leq m \leq 0.25\ M_\odot
\end{cases} \tag{12.30}
$$

This dependence is depicted in Figure 12.6.

Star Formation Rate

We adopt an empirical fit to the star formation rate per co-moving volume as a function of redshift, z [169],

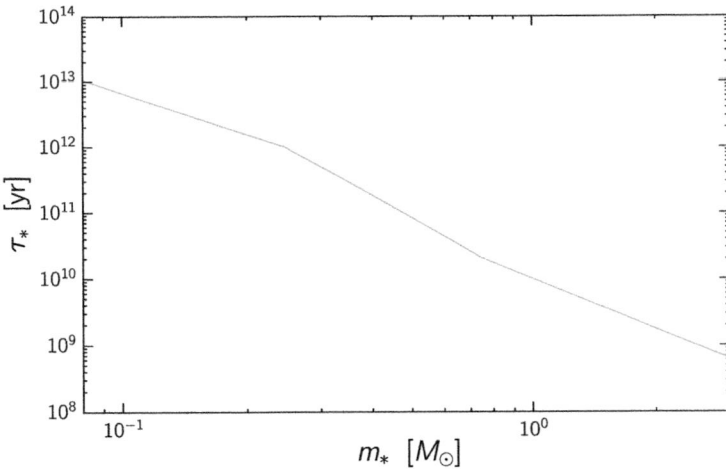

Figure 12.6 Stellar lifetime (τ_*) as a function of mass.

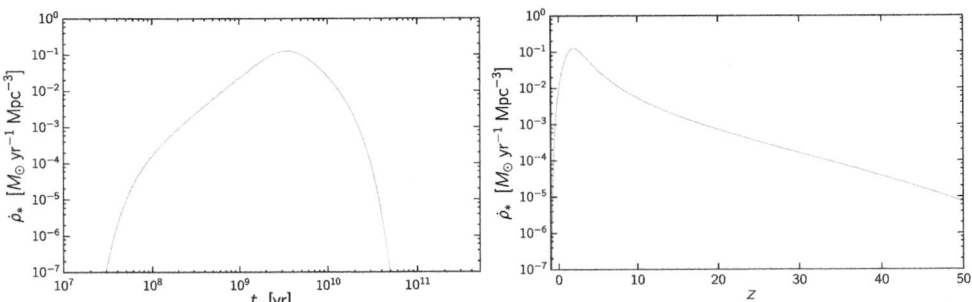

Figure 12.7 Star formation rate, $\dot{\rho}_*$, as a function of the cosmic time t (left panel) and redshift z (right panel), based on an extrapolation of a fitting function to existing data [169]. Although this extrapolation has inherent uncertainties, it provides a reasonable benchmark for theoretical expectations.

$$\dot{\rho}_*(z) = 0.015 \frac{(1+z)^{2.7}}{1 + [(1+z)/2.9]^{5.6}} \, M_\odot \mathrm{yr}^{-1} \mathrm{Mpc}^{-3}, \qquad (12.31)$$

and truncate the extrapolation to early times at the expected formation time of the first stars [159]. We extrapolate the cosmic star formation history to the future or equivalently negative redshifts $-1 \leq z < 0$ (see, e.g., ref. [16]) and find that the co-moving star formation rate drops to less than 10^{-5} of the current rate at 56 Gyr into the future. We cut off the star formation at roughly the ratio between the current reservoir mass of cold gas in galaxies [90] and the current star formation rate per co-moving volume. The resulting star formation rate as a function of time and redshift is shown in Figure 12.7.

Probability of Life on a Habitable Planet

The probability for the existence of life around a star of a particular mass m can be expressed in terms of the product between the probability that there is an Earth-mass planet in the star's habitable zone (HZ) and the probability that life emerges on such a planet: $P(\text{life}|m) = P(\text{HZ}|m)P(\text{life}|\text{HZ})$. The first factor, $P(\text{HZ}|m)$, is commonly labelled η_{Earth} in the exoplanet literature [249].

Data from the NASA Kepler mission implies η_{Earth} values in the range of $6.4^{+3.4}_{-1.1}\%$ for stars of approximately a solar mass [179, 207, 235] and $24^{+18}_{-8}\%$ for lower-mass M-dwarf stars [73]. The result for solar mass stars is less robust due to lack of identified Earth-like planets at high stellar masses. Owing to the large measurement uncertainties, we assume a constant η_{Earth} within the range of stellar masses under consideration. The specific constant value of η_{Earth} drops out of the calculation due to the normalisation factor N.

There is scope for considerable refinement in the choice of the second factor $p(\text{life}|\text{HZ})$. One could suppose that the probability of life evolving on a planet increases with the amount of time that the planet exists or that increasing the surface area of the planet should increase the likelihood of life beginning. However, given our ignorance, we will set this probability factor to a constant, an assumption which can be improved upon by statistical data from future searches for biosignatures in the molecular composition of the atmospheres of habitable planets [151, 162, 222, 231]. In our simplified treatment, this constant value has no effect on $dP(t)/dt$ since its contribution is also cancelled by the normalisation factor N.

12.6.3 Results

The top and bottom panels in Figure 12.8 show the probability per log time interval $t\,dP(t)/dt = dP/d\ln t$ and the cumulative probability $P(<t) = \int_0^t [dP(t')/dt']dt'$ based on Eq. 12.28 for different choices of the low-mass cut-off in the distribution of host stars for life-hosting planets (equally spaced in $\ln m$). The upper stellar mass cut-off has a negligible influence on $dP/d\ln t$, due to the short lifetime and low abundance of massive stars. In general, $dP/d\ln t$ cuts off roughly at the lifetime of the longest-lived stars in each case, as indicated by the upper axis labels. For the full range of hydrogen-burning stars, $dP(t)/d\ln t$ peaks around the lifetime of the lowest mass stars $t \sim 10^{13}$ yr with a probability value that is a thousand times larger than for the Sun, implying that life around low-mass stars in the distant future is much more likely than terrestrial life around the Sun today.

12.6.4 Section Summary and Implications

Figure 12.8 implies that the probability for life per logarithm interval of cosmic time, $dP(t)/d\ln t$, has a broad distribution in $\ln t$ and is peaked in the future, as long as life is likely around low-mass stars. High-mass stars are shorter lived and less abundant and hence make a relatively small contribution to the probability distribution.

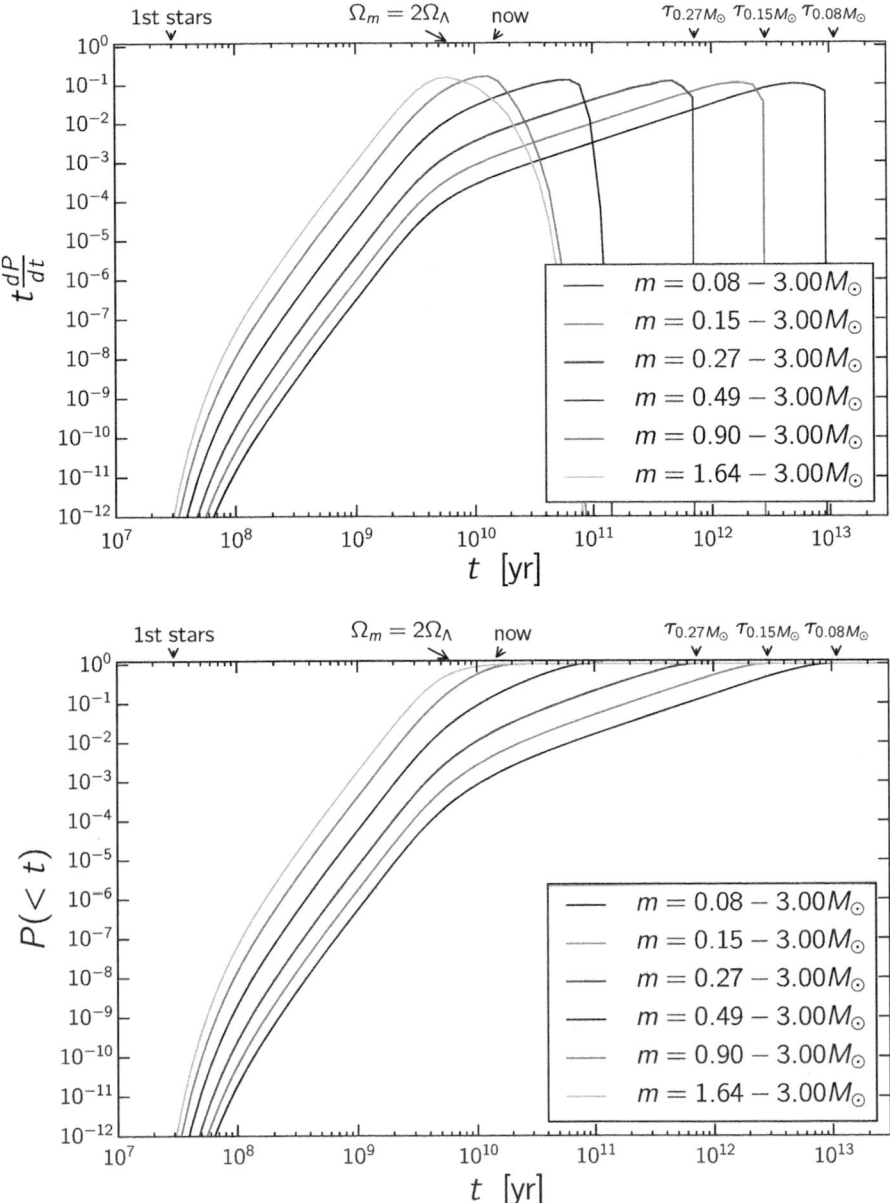

Figure 12.8 Probability distribution for the emergence of life within a fixed co-moving volume of the Universe as a function of cosmic time. We show the probability per log time, $t\,dP/dt$ (top panel) as well as the cumulative probability up to a time t, $P(<t)$ (bottom panel), for different choices of the minimum stellar mass, equally spaced in $\log m$ between 0.08 $M_\odot$ and 3 $M_\odot$. The contribution of stars above 3 $M_\odot$ to $dP(t)/dt$ is ignored due to their short lifetimes and low abundances. The labels on the top axis indicate the formation time of the first stars, the time when the cosmic expansion started accelerating (i.e., when the density parameter of matter, Ω_m, was twice that of the vacuum, Ω_Λ), the present time (now) and the lifetimes of stars with masses of 0.08 $M_\odot$, 0.15 $M_\odot$, and 0.27 $M_\odot$. The lines correspond to the masses indicated in the legend, from right to left.

Future searches for molecular biosignatures (such as O_2 combined with CH_4) in the atmospheres of planets around low-mass stars [162, 222, 231] could inform us whether life will exist at late cosmic times [137]. If we were to insist that life near the Sun is typical and not premature i.e., require that the peak in $dP(t)/d\ln t$ would coincide with the lifetime of Sun-like stars at the present time – then we must conclude that the physical environments of low-mass stars are hazardous to life. This outcome could be the result of a more conventional line of reasoning [217]). It could be a natural outcome, for example, from the enhanced UV emission and flaring activity of young low-mass stars during their extended ($\sim$ Gyr long) pre-main-sequence phase, which is capable of stripping rocky planets of their atmospheres [200]. The habitable zone is much closer in around low-mass stars, leading to tidal locking and extreme temperature contrast due between the permanent day and night sides [140] as well as an increased impact of the stellar wind and enhanced stellar variability in stripping the atmospheres of rocky planets [4, 17, 31, 168]. The recent discoveries of a habitable planet around the nearest, $0.12\ M_\odot$ star Proxima Centauri [9] and of three transiting habitable planets around the $0.08\ M_\odot$ star TRAPPIST-1 [97] offer new prospects for testing the conjecture. Empirical testing is crucial since self-selection arguments, such as the one associated with the peak in $dP(t)/d\ln t$, are known to be subject to statistical caveats [36, 101, 149]. Moreover, Figure 12.8 makes the implicit assumption that the emergence of life does not correlate with the early stages of planet formation, providing an equal probability for appearing early or late. In particular, water loss from the surface of a planet could be substantial at late times, suppressing the survival of life over trillions of years.

Values of the cosmological constant below the observed one should not affect the probability distribution, as they would introduce only mild changes to the star formation history due to the modified formation history of galaxies [44, 190]. However, much larger values of the cosmological constant would suppress galaxy formation and reduce the total number of stars per co-moving volume [158], hence limiting the overall likelihood for life altogether [264].

Our results provide a new perspective on the so-called coincidence problem, *why do we observe $\Omega_m \sim \Omega_\Lambda$* [46]? The answer comes naturally if we consider the history of Sun-like star formation, as the number of habitable planets peaks around present time for $m \sim 1\ M_\odot$. We note that for the majority of stars, this coincidence will not exist as $dP(t)/dt$ peaks in the future, where $\Omega_m \ll \Omega_\Lambda$. The question is, then, why do we find ourselves orbiting a star like the Sun now rather than a lower-mass star in the future?

We derived our numerical results based on a conservative set of assumptions and guided by the latest empirical data for the various components of Eq. (12.28). However, the emergence of life may be sensitive to additional factors that were not included in our formulation, such as the existence of a moon to stabilise the climate on an Earth-like planet [145], the existence of asteroid belts [181], the orbital structure of the host planetary system (e.g., the existence of nearby giant planets or orbital eccentricity), the effects of a binary star companion [106], the location of the planetary system within the host galaxy [154], and the detailed properties of the host galaxy (e.g., galaxy type [59] or metallicity [127, 156]), including the environmental effects of quasars, gamma ray bursts [210], or the hot gas in clusters of galaxies. These additional factors are highly uncertain and complicated to model and were ignored for simplicity in our analysis.

The probability distribution $dP(t)/d\ln t$ is of particular importance for studies attempting to gauge the level of fine-tuning required for the cosmological or fundamental physics parameters that would allow life to emerge in our Universe.

Acknowledgements

I thank my collaborators on the work described in this chapter, namely Rafael Batista, Shmuel Bialy, Laura Kreidberg, Gonzalo Gonzalez, James Guillochon, Henry Lin, Dani Maoz, Natalie Mashian, David Sloan, Amiel Sternberg, Ed Turner, and Matias Zaldarriaga.

References

[1] Adams, F. C., and Spergel, D. N. 2005. 'Lithopanspermia in Star-Forming Clusters'. *Astrobiology*, **5**(Aug.), 497–514.

[2] Agol, E. 2011. 'Transit Surveys for Earths in the Habitable Zones of White Dwarfs'. *The Astrophysical Journal Letters*, **731**(Apr.), L31.

[3] Albrow, M., *et al.* 1998. 'The 1995 Pilot Campaign of PLANET: Searching for Microlensing Anomalies through Precise, Rapid, Round-the-Clock Monitoring'. *Astrophysical Journal*, **509**(Dec.), 687–702.

[4] Alibert, Y., and Benz, W. 2017. 'Formation and Composition of Planets around Very Low Mass Stars'. *Astronomy and Astrophysics*, **598**(Jan.), L5.

[5] Anderson, L., *et al.* 2014. 'The Clustering of Galaxies in the SDSS-III Baryon Oscillation Spectroscopic Survey: Baryon Acoustic Oscillations in the Data Releases 10 and 11 Galaxy Samples'. *Monthly Notices of the Royal Astronomical Society*, **441**(June), 24–62.

[6] Andrews, S. M., and Williams, J. P. 2005. 'Circumstellar Dust Disks in Taurus-Auriga: The Submillimeter Perspective'. *Astrophysical Journal*, **631**(Oct.), 1134–1160.

[7] Andrews, S. M., *et al.* 2009. 'Protoplanetary Disk Structures in Ophiuchus'. *Astrophysical Journal*, **700**(Aug.), 1502–1523.

[8] Andrews, S. M., *et al.* 2010. 'Protoplanetary Disk Structures in Ophiuchus: II. Extension to Fainter Sources'. *Astrophysical Journal*, **723**(Nov.), 1241–1254.

[9] Anglada-Escudé, G., *et al.* 2016. 'A Terrestrial Planet Candidate in a Temperate Orbit around Proxima Centauri'. *Nature*, **536**(Aug.), 437–440.

[10] Aoki, W. 2010 (Mar.). 'Carbon-Enhanced Metal-Poor (CEMP) Stars'. Cunha, K., Spite, M., and Barbuy, B. eds. *Chemical Abundances in the Universe: Connecting First Stars to Planets*. IAU Symposium, vol. 265 Pages 111–116.

[11] Aoki, W., *et al.* 2007. 'Carbon-Enhanced Metal-Poor Stars: I. Chemical Compositions of 26 Stars'. *Astrophysical Journal*, **655**(Jan.), 492–521.

[12] Armitage, P. J. 2010. *Astrophysics of Planet Formation*. Cambridge University Press.

[13] Ashok, S. K., and Douglas, M. R. 2004. 'Counting Flux Vacua'. *Journal of High Energy Physics*, **1**(Jan.), 060.

[14] Asplund, M., *et al.* 2009. 'The Chemical Composition of the Sun'. *Annual Reviews of Astronomy and Astrophysics*, **47**(Sept.), 481–522.

[15] Baltz, E. A., and Gondolo, P. 2001. 'Binary Events and Extragalactic Planets in Pixel Microlensing'. *Astrophysical Journal*, **559**(Sept.), 41–52.

[16] Barnes, L., *et al.* 2005. 'The Influence of Evolving Dark Energy on Cosmology'. *Publications of the Astronomical Society of Australia*, **22**, 315–325.

[17] Barnes, R., *et al.* 2009. 'Tidal Limits to Planetary Habitability'. *Astrophysical Journal Letters*, **700**(July), L30–L33.

[18] Barrow, J. D., and Tipler, F. J. 1986. *The Anthropic Cosmological Principle*. Oxford University Press.

[19] Bayet, E., *et al.* 2009. 'Molecular Tracers of PDR-Dominated Galaxies'. *Astrophysical Journal*, **696**(May), 1466–1477.

[20] Beaulieu, J. P., Bennett, D. P. Fouqué, P., *et al.* 2006. 'Discovery of a Cool Planet of 5.5 Earth Masses through Gravitational Microlensing'. *Nature*, **439**(Jan.), 437–440.

[21] Beckwith, S. V. W., Henning, T., and Nakagawa, Y. 2000. 'Dust Properties and Assembly of Large Particles in Protoplanetary Disks'. *Protostars and Planets IV*, May, 533.

[22] Beers, T. C., and Christlieb, N. 2005. 'The Discovery and Analysis of Very Metal-Poor Stars in the Galaxy'. *Annual Reviews of Astronomy and Astrophysics*, **43**(Sept.), 531–580.

[23] Beers, T. C., Preston, G. W., and Shectman, S. A. 1985. 'A Search for Stars of Very Low Metal Abundance: I'. *The Astronomical Journal*, **90**(Oct.), 2089–2102.

[24] Beers, T. C., *et al.* 2007. 'Near-Infrared Spectroscopy of Carbon-Enhanced Metal-Poor Stars: I. A SOAR/OSIRIS Pilot Study'. *The Astronomical Journal*, **133**(Mar.), 1193–1203.

[25] Behroozi, P., and Peeples, M. S. 2015. 'On the History and Future of Cosmic Planet Formation'. *Monthly Notices of the Royal Astronomical Sociey*, **454**(Dec.), 1811–1817.

[26] Beichman, C. A., Greene, T., and Krist, J. 2009 (Feb.). 'Future Observations of Transits and Light Curves from Space'. Pont, F., Sasselov, D., and Holman, M. J., eds. *Transiting Planets*. IAU Symposium, vol. 253. Pages 391–328.

[27] Bell, T. A., *et al.* 2006. 'Molecular Line Intensities as Measures of Cloud Masses: I. Sensitivity of CO Emissions to Physical Parameter Variations'. *Monthly Notices of the Royal Astronomical Society*, **371**(Oct.), 1865–1872.

[28] Bialy, S., and Sternberg, A. 2015. 'CO/H_2, C/CO, OH/CO, and OH/O_2 in Dense Interstellar Gas: From High Ionization to Low Metallicity'. *Monthly Notices of the Royal Astronomical Society*, **450**(July), 4424–4445.

[29] Bialy, S., Sternberg, A., and Loeb, A. 2015. 'Water Formation during the Epoch of First Metal Enrichment'. *The Astrophysical Journal Letters*, **804**(May), L29.

[30] Birch, F. 1947. 'Finite Elastic Strain of Cubic Crystals'. *Physical Review*, **71**(June), 809–824.

[31] Bolmont, E., *et al.* 2017. 'Water Loss from Terrestrial Planets Orbiting Ultracool Dwarfs: Implications for the Planets of TRAPPIST-1'. *Monthly Notices of the Royal Astronomical Society*, **464**(Jan.), 3728–3741.

[32] Bond, I. A., *et al.* 2002a. 'Improving the Prospects for Detecting Extrasolar Planets in Gravitational Microlensing Events in 2002'. *Monthly Notices of the Royal Astronomical Society*, **331**(Mar.), L19–L23.

[33] Bond, I. A., Rattenbury, N. J., and Skuljan, J., *et al.* 2002b. 'Study by MOA of Extrasolar Planets in Gravitational Microlensing Events of High Magnification'. *Monthly Notices of the Royal Astronomical Society*, **333**(June), 71–83.

[34] Bond, J. C., O'Brien, D. P., and Lauretta, D. S. 2010. 'The Compositional Diversity of Extrasolar Terrestrial Planets: I. In Situ Simulations'. *Astrophysical Journal*, **715**(June), 1050–1070.

[35] Bonifacio, P., *et al.* 2015. 'TOPoS: II. On the Bimodality of Carbon Abundance in CEMP Stars Implications on the Early Chemical Evolution of Galaxies'. *Astronomy and Astrophysics*, **579**(July), A28.

[36] Bostrom, N. 2002. *Anthropic Bias: Observation Selection Effects in Science and Philosophy*. Routledge.

[37] Bouchy, F., *et al.* 2004. 'Two New 'Very hot Jupiters' among the OGLE Transiting Candidates. *Astronomy and Astrophysics*, **421**(July), L13–L16.

[38] Bousso, R., and Polchinski, J. 2000. 'Quantization of Four-Form Fluxes and Dynamical Neutralization of the Cosmological Constant'. *Journal of High Energy Physics*, **6**(June), 006.

[39] Bouwens, R. J., *et al.* 2015a. 'Reionization after Planck: The Derived Growth of the Cosmic Ionizing Emissivity Now Matches the Growth of the Galaxy UV Luminosity Density'. *Astrophysical Journal*, **811**(Oct.), 140.

[40] Bouwens, R. J., Illingworth, G. D., Oesch, P. A., *et al.* 2015b. 'UV Luminosity Functions at Redshifts z = 4 to z = 10: 10,000 Galaxies from HST Legacy Fields'. *Astrophysical Journal*, **803**(Apr.), 34.

[41] Bromm, V., and Larson, R. B. 2004. 'The First Stars'. *Annual Reviews of Astronomy and Astrophysics*, **42**(Sept.), 79–118.

[42] Bromm, V., and Loeb, A. 2003. 'The Formation of the First Low-Mass Stars from Gas with Low Carbon and Oxygen Abundances'. *Nature*, **425**(Oct.), 812–814.

[43] Bromm, V., Kudritzki, R. P., and Loeb, A. 2001. 'Generic Spectrum and Ionization Efficiency of a Heavy Initial Mass Function for the First Stars'. *Astrophysical Journal*, **552**(May), 464–472.

[44] Busha, M. T., Evrard, A. E., and Adams, F. C. 2007. 'The Asymptotic Form of Cosmic Structure: Small-Scale Power and Accretion History'. *Astrophysical Journal*, **665**(Aug.), 1–13.

[45] Carollo, D., *et al.* 2012. 'Carbon-Enhanced Metal-Poor Stars in the Inner and Outer Halo Components of the Milky Way'. *Astrophysical Journal*, **744**(Jan.), 195.

[46] Carter, B. 1983. 'The Anthropic Principle and Its Implications for Biological Evolution'. *Philosophical Transactions of the Royal Society of London Series A*, **310**(Dec.), 347–363.

[47] Carter-Bond, J. C., *et al.* 2012a. 'Low Mg/Si Planetary Host Stars and Their Mg-depleted Terrestrial Planets'. *Astrophysical Journal*, **747**(Mar.), L2.

[48] Carter-Bond, J. C., O'Brien, D. P., and Raymond, S. N. 2012b. 'The Compositional Diversity of Extrasolar Terrestrial Planets: II. Migration Simulations'. *Astrophysical Journal*, **760**(Nov.), 44.

[49] Cazaux, S., and Tielens, A. G. G. M. 2002. 'Molecular Hydrogen Formation in the Interstellar Medium'. *Astrophysical Journal*, **575**(Aug.), L29–L32.

[50] Cen, R., and Ostriker, J. P. 1999. 'Where Are the Baryons?'. *Astrophysical Journal*, **514**(Mar.), 1–6.

[51] Chabrier, G. 2003. 'Galactic Stellar and Substellar Initial Mass Function'. *Publications of the Astronomical Society of Australia*, **115**(July), 763–795.

[91] Galametz, M., *et al.* 2011. 'Probing the Dust Properties of Galaxies Up to SubMillimetre Wavelengths: II. Dust-to-Gas Mass Ratio Trends with Metallicity and the Submm Excess in Dwarf Galaxies'. *Astronomy and Astrophysics*, **532**(Aug.), A56.

[92] Garriga, J., and Vilenkin, A. 2003. 'Testable Anthropic Predictions for Dark Energy'. *Physical Review D*, **67**(4), 043503.

[93] Garriga, J., Livio, M., and Vilenkin, A. 2000. 'Cosmological Constant and the Time of its Dominance'. *Physical Review D*, **61**(2), 023503.

[94] Gaudi, B. S., and Gould, A. 1997. 'Planet Parameters in Microlensing Events'. *Astrophysical Journal*, **486**(Sept.), 85–99.

[95] Giddings, S. B., Kachru, S., and Polchinski, J. 2002. 'Hierarchies from Fluxes in String Compactifications'. *Physical Review D*, **66**(10), 106006.

[96] Gilliland, R. L., Brown, T. M., and Guhathakurta, P., *et al.* 2000. 'A Lack of Planets in 47 Tucanae from a Hubble Space Telescope Search'. *Astrophysical Journal*, **545**(Dec.), L47–L51.

[97] Gillon, M., *et al.* 2017. 'Seven Temperate Terrestrial Planets around the Nearby Ultracool Dwarf Star TRAPPIST-1'. *Nature*, **542**(Feb.), 456–460.

[98] Glover, S. C. O., and Clark, P. C. 2014. Molecular Cooling in the Diffuse Interstellar Medium'. *Monthly Notices of the Royal Astronomical Society*, **437**(Jan.), 9–20.

[99] Gonzalez, G., Brownlee, D., and Ward, P. 2001. 'The Galactic Habitable Zone: Galactic Chemical Evolution'. *Icarus*, **152**(July), 185–200.

[100] Gorti, U., and Hollenbach, D. 2009. 'Photoevaporation of Circumstellar Disks By Far-Ultraviolet, Extreme-Ultraviolet and X-Ray Radiation from the Central Star'. *Astrophysical Journal*, **690**(Jan.), 1539–1552.

[101] Gott, III, J. R. 1993. 'Implications of the Copernican Principle for Our Future Prospects'. *Nature*, **363**(May), 315–319.

[102] Gould, A., *et al.* 2006. 'Microlens OGLE-2005-BLG-169 Implies That Cool Neptune-Like Planets Are Common'. *Astrophysical Journal*, **644**(June), L37–L40.

[103] Guillochon, J., and Loeb, A. 2015a. 'SETI via Leakage from Light Sails in Exoplanetary Systems'. *The Astrophysical Journal Letters*, **811**(Oct.), L20.

[104] Guillochon, J., and Loeb, A. 2015b. 'The Fastest Unbound Stars in the Universe'. *Astrophysical Journal*, **806**(June), 124.

[105] Gyuk, G., Dalal, N., and Griest, K. 2000. 'Self-Lensing Models of the Large Magellanic Cloud'. *Astrophysical Journal*, **535**(May), 90–103.

[106] Haghighipour, N., and Kaltenegger, L. 2013. 'Calculating the Habitable Zone of Binary Star Systems: II. P-Type Binaries'. *Astrophysical Journal*, **777**(Nov.), 166.

[107] Hahn, J. M., and Malhotra, R. 1999. 'Orbital Evolution of Planets Embedded in a Planetesimal Disk'. *The Astronomical Journal*, **117**(June), 3041–3053.

[108] Haiman, Z., Thoul, A. A., and Loeb, A. 1996. 'Cosmological Formation of Low-Mass Objects'. *Astrophysical Journal*, **464**(June), 523.

[109] Hanfland, M., Beister, H., and Syassen, K. 1989. 'Graphite under Pressure: Equation of State and First-Order Raman Modes'. *Physical Review B*, **39**(June), 12598–12603.

[110] Heger, A., and Woosley, S.E. 2002. 'The Nucleosynthetic Signature of Population III'. *Astrophysical Journal*, **567**(Mar.), 532–543.

[111] Herbst, E., and Klemperer, W. 1973. 'The Formation and Depletion of Molecules in Dense Interstellar Clouds'. *Astrophysical Journal*, **185**(Oct.), 505–534.

[112] Herrera-Camus, R., *et al.* 2012. 'Dust-to-Gas Ratio in the Extremely Metal-poor Galaxy I Zw 18'. *Astrophysical Journal*, **752**(June), 112.

[113] Hirata, C. M., and Padmanabhan, N. 2006. 'Cosmological Production of H_2 before the Formation of the First Galaxies'. *Monthly Notices of the Royal Astronomical Society*, **372**(Nov.), 1175–1186.

[114] Hollenbach, D., *et al.* 2009. 'Water, O2, and Ice in Molecular Clouds'. *Astrophysical Journal*, **690**(Jan.), 1497–1521.

[115] Hollenbach, D., *et al.* 2012. 'The Chemistry of Interstellar OH^+, H_2O^+, and H_3O^+: Inferring the Cosmic-Ray Ionization Rates from Observations of Molecular Ions'. *Astrophysical Journal*, **754**(Aug.), 105.

[116] Hollenbach, D. J., Werner, M. W., and Salpeter, E. E. 1971. 'Molecular Hydrogen in H i Regions'. *Astrophysical Journal*, **163**(Jan.), 165.

[117] Hubbard, W. B. *Planetary Interiors* Van Nostrand Reinhold.

[118] Huxor, A. P., *et al.* 2005. 'A New Population of Extended, Luminous Star Clusters in the Halo of M31'. *Monthly Notices of the Royal Astronomical Society*, **360**(July), 1007–1012.

[119] Indebetouw, R., *et al.* 2014. 'Dust Production and Particle Acceleration in Supernova 1987A Revealed with ALMA'. *Astrophysical Journal*, **782**(Feb.), L2.

[120] Indriolo, N., and McCall, B. J., 2012. 'Investigating the Cosmic-Ray Ionization Rate in the Galactic Diffuse Interstellar Medium through Observations of H_3^+'. *Astrophysical Journal*, **745**(Jan.), 91.

[121] Ishigaki, M. N., *et al.* 2014. 'Faint Population III Supernovae as the Origin of the Most Iron-Poor Stars'. *Astrophysical Journal*, **792**(Sept.), L32.

[122] Iwamoto, N., *et al.* 2005. 'The First Chemical Enrichment in the Universe and the Formation of Hyper Metal-Poor Stars'. *Science*, **309**(July), 451–453.

[123] Janson, M., *et al.* 2011. 'High-Contrast Imaging Search for Planets and Brown Dwarfs around the Most Massive Stars in the Solar Neighborhood'. *Astrophysical Journal*, **736**(Aug.), 89.

[124] Ji, A. P., Frebel, A., and Bromm, V. 2014. 'The Chemical Imprint of Silicate Dust on the Most Metal-Poor Stars'. *Astrophysical Journal*, **782**(Feb.), 95.

[125] Joggerst, C. C., Woosley, S. E., and Heger, A. 2009. 'Mixing in Zero- and Solar-Metallicity Supernovae'. *Astrophysical Journal*, **693**(Mar.), 1780–1802.

[126] Johansen, A., Youdin, A., and Mac Low, M. M. 2009. 'Particle Clumping and Planetesimal Formation Depend Strongly on Metallicity'. *Astrophysical Journal*, **704**(Oct.), L75–L79.

[127] Johnson, J. L., and Li, H. 2012. 'The First Planets: The Critical Metallicity for Planet Formation'. *Astrophysical Journal*, **751**(June), 81.

[128] Jura, M. 1974. 'Formation and Destruction Rates of Interstellar H_2'. *Astrophysical Journal*, **191**(July), 375–379.

[129] Kachru, S., *et al.* 2003. 'De Sitter Vacua in String Theory'. *Physical Review D*, **68**(4), 046005.

[130] Kasting, J. 2010. *How to Find a Habitable Planet*. Princeton University Press.

[131] Kasting, J. F., Whitmire, D. P., and Reynolds, R. T. 1993. 'Habitable Zones around Main Sequence Stars'. *Icarus*, **101**(Jan.), 108–128.

[132] Kaufman, M. J., and Neufeld, D. A. 1996. 'Far-Infrared Water Emission from Magnetohydrodynamic Shock Waves'. *Astrophysical Journal*, **456**(Jan.), 611.

[133] Keller, S. C., *et al.* 2014. 'A Single Low-Energy, Iron-Poor Supernova as the Source of Metals in the Star SMSS J031300.36-670839.3'. *Nature*, **506**(Feb.), 463–466.

[134] Kneib, J. P., *et al.* 2004. 'A Probable z~7 Galaxy Strongly Lensed by the Rich Cluster A2218: Exploring the Dark Ages'. *Astrophysical Journal*, **607**(June), 697–703.

[135] Kobayashi, H., *et al.* 2011. 'Sublimation Temperature of Circumstellar Dust Particles and Its Importance for Dust Ring Formation'. *Earth, Planets, and Space*, **63**(Oct.), 1067–1075.

[136] Konacki, M., *et al.* 2003. 'An Extrasolar Planet That Transits the Disk of Its Parent Star'. *Nature*, **421**(Jan.), 507–509.

[137] Kopparapu, R. K., *et al.* 2013. 'Habitable Zones around Main-Sequence Stars: New Estimates'. *Astrophysical Journal*, **765**(Mar.), 131.

[138] Kornet, K., *et al.* 2005. 'Formation of Giant Planets in Disks with Different Metallicities'. *Astronomy and Astrophysics*, **430**(Feb.), 1133–1138.

[139] Kozasa, T., Hasegawa, H., and Nomoto, K. 1989. 'Formation of Dust Grains in the Ejecta of SN 1987A'. *Astrophysical Journal*, **344**(Sept.), 325–331.

[140] Kreidberg, L., and Loeb, A. 2016. 'Prospects for Characterizing the Atmosphere of Proxima Centauri b'. *The Astrophysical Journal Letters*, **832**(Nov.), L12.

[141] Kress, M. E., and Tielens, A. G. G. M. 2001. 'The Role of Fischer-Tropsch Catalysis in Solar Nebula Chemistry'. *Meteoritics and Planetary Science*, **36**(Jan.), 75–92.

[142] Krisciunas, K., Garnavich, P. M., Challis, P., *et al.* 2005. 'Hubble Space Telescope Observations of Nine High-Redshift ESSENCE Supernovae1'. *The Astronomical Journal*, **130**(Dec.), 2453–2472.

[143] Kuchner, M. J., and Seager, S. 2005. 'Extrasolar Carbon Planets'. *ArXiv Astrophysics e-prints*, Apr.

[144] Larimer, J. W., 1975. 'The Effect of C/O Ratio on the Condensation of Planetary Material'. *Geochimica et Cosmochimica Acta*, **39**(Mar.), 389–392.

[145] Laskar, J., Joutel, F., and Robutel, P. 1993. 'Stabilization of the Earth's Obliquity by the Moon'. *Nature*, **361**(Feb.), 615–617.

[146] Laughlin, G., Bodenheimer, P., and Adams, F. C. 1997. 'The End of the Main Sequence'. *Astrophysical Journal*, **482**(June), 420–432.

[147] Le Bourlot, J., Pineau des Forêts, G., and Flower, D. R. 1999. 'The Cooling of Astrophysical Media by H_2'. *Monthly Notices of the Royal Astronomical Society*, **305**(May), 802–810.

[148] Léger, A., *et al.* 2004. 'A New Family of Planets? "Ocean-Planets"'. *Icarus*, **169**(June), 499–504.

[149] Leslie, J. 1996. *The End of the World: The Science and Ethics of Human Extinction* Routledge.

[150] Lin, H. W., and Loeb, A. 2014. 'Finding Rocky Asteroids around White Dwarfs by Their Periodic Thermal Emission'. *The Astrophysical Journal Letters*, **793**(Oct.), L43.

[151] Lin, H. W., and Loeb, A. 2015. 'Statistical Signatures of Panspermia in Exoplanet Surveys'. *The Astrophysical Journal Letters*, **810**(Sept.), L3.

[152] Lin, H. W., Gonzalez Abad, G., and Loeb, A. 2014. 'Detecting Industrial Pollution in the Atmospheres of Earth-Like Exoplanets'. *The Astrophysical Journal Letters*, **792**(Sept.), L7.

[153] Lineweaver, C. H. 2001. 'An Estimate of the Age Distribution of Terrestrial Planets in the Universe: Quantifying Metallicity as a Selection Effect'. *Icarus*, **151**(June), 307–313.

[154] Lineweaver, C. H., Fenner, Y., and Gibson, B. K. 2004. 'The Galactic Habitable Zone and the Age Distribution of Complex Life in the Milky Way'. *Science*, **303**(Jan.), 59–62.

[155] Lissauer, J. J. 1993. 'Planet Formation'. *Annual Reviews of Astronomy and Astrophysics*, **31**, 129–174.

[156] Livio, M. 1999. 'How Rare Are Extraterrestrial Civilizations, and When Did They Emerge?' *Astrophysical Journal*, **511**(Jan.), 429–431.

[157] Loeb, A. 2002. 'Long-Term Future of Extragalactic Astronomy'. *Physical Review D*, **65**(4), 047301.

[158] Loeb, A. 2006. 'An Observational Test for the Anthropic Origin of the Cosmological Constant'. *Journal of Cosmology and Astroparticle Physics*, **5**(May), 009.

[159] Loeb, A. 2014. 'The Habitable Epoch of the Early Universe'. *International Journal of Astrobiology*, **13**(Sept.), 337–339.

[160] Loeb, A., and Furlanetto, S. R. 2013. *The First Galaxies in the Universe*. Princeton University Press.

[161] Loeb, A., and Guillochon, J. 2014. 'Observational Cosmology with Semi-Relativistic Stars'. *ArXiv e-prints*, Nov.

[162] Loeb, A., and Maoz, D. 2013. 'Detecting Biomarkers in Habitable-Zone Earths Transiting White Dwarfs'. *Monthly Notices of the Royal Astronomical Society*, **432**(May), 11–15.

[163] Loeb, A., and Turner, E. L. 2012. 'Detection Technique for Artificially Illuminated Objects in the Outer Solar System and Beyond'. *Astrobiology*, **12**(Apr.), 290–294.

[164] Loeb, A., and Zaldarriaga, M. 2007. 'Eavesdropping on Radio Broadcasts from Galactic Civilizations with Upcoming Observatories for Redshifted 21 cm Radiation'. *Journal of Cosmology and AstroParticle Physics*, **1**(Jan.), 020.

[165] Loeb, A., Batista, R. A., and Sloan, D. 2016. 'Relative Likelihood for Life as a Function of Cosmic Time'. *Journal of Cosmology and Astroparticle Physics*, **2016**(Aug.), 040.

[166] LoVerde, M., and Smith, K. M. 2011. 'The Non-Gaussian Halo Mass Function with f_{NL}, g_{NL} and τ_{NL}'. *Journal of Cosmology and Astroparticle Physics*, **8**(Aug.), 003.

[167] Lucy, L. B., *et al.* 1989. 'Dust Condensation in the Ejecta of SN 1987 A'. Tenorio-Tagle, G., Moles, M., and Melnick, J. eds., *IAU Colloquium 120: Structure and Dynamics of the Interstellar Medium*. Lecture Notes in Physics, Berlin Springer Verlag, vol. 350 Page 164.

[168] Luger, R., and Barnes, R. 2015. 'Extreme Water Loss and Abiotic O_2 Buildup on Planets throughout the Habitable Zones of M Dwarfs'. *Astrobiology*, **15**(Feb.), 119–143.

[169] Madau, P., and Dickinson, M. 2014. 'Cosmic Star-Formation History'. *Annual Reviews of Astronomy and Astrophysics*, **52**(Aug.), 415–486.

[170] Madhusudhan, N., Harrington, J., Stevenson, K. B., *et al.* 2011. 'A High C/O Ratio and Weak Thermal Inversion in the Atmosphere of Exoplanet WASP-12b'. *Nature*, **469**(Jan.), 64–67.

[171] Madhusudhan, N., Lee, K. K. M., and Mousis, O. 2012. 'A Possible Carbon-Rich Interior in Super-Earth 55 Cancri e'. *Astrophysical Journal*, **759**(Nov.), L40.

[172] Maio, U., *et al.* 2012. 'Counts of High-Redshift GRBs as Probes of Primordial Non-Gaussianities'. *Monthly Notices of the Royal Astronomical Society*, **426**(Nov.), 2078–2088.

[173] Maldacena, J. 2003. 'Non-Gaussian Features of Primordial Fluctuations in Single Field Inflationary Models'. *Journal of High Energy Physics*, **5**(May), 013.

[174] Mallén-Ornelas, G., *et al.* 2003. 'The EXPLORE Project: I. A Deep Search for Transiting Extrasolar Planets'. *Astrophysical Journal*, **582**(Jan.), 1123–1140.

[175] Maloney, A., Silverstein, E., and Strominger, A. 2002. 'De Sitter Space in Non-Critical String Theory'. *NASA STI/Recon Technical Report N*, **3**(June).

[176] Mao, S., and Paczynski, B. 1991. 'Gravitational Microlensing by Double Stars and Planetary Systems'. *Astrophysical Journal*, **374**(June), L37–L40.

[177] Marassi, S., *et al.* 2014. 'The Origin of the Most Iron-Poor Star'. *Astrophysical Journal*, **794**(Oct.), 100.

[178] Marassi, S., *et al.* 2015. 'The Metal and Dust Yields of the First Massive Stars'. *Monthly Notices of the Royal Astronomical Society*, **454**(Dec.), 4250–4266.

[179] Marcy, G. W., *et al.* 2014. 'Occurrence and Core-Envelope Structure of 1-4 Earth-Size Planets around Sun-Like Stars'. *Proceedings of the National Academy of Science*, **111**(Sept.), 12655–12660.

[180] Martel, H., Shapiro, P. R., and Weinberg, S. 1998. 'Likely Values of the Cosmological Constant'. *Astrophysical Journal*, **492**(Jan.), 29–40.

[181] Martin, R. G., and Livio, M. 2013. 'On the Formation and Evolution of Asteroid Belts and Their Potential Significance for Life'. *Monthly Notices of the Royal Astronomical Society*, **428**(Jan.), L11–L15.

[182] Mashian, N., and Loeb, A. 2016. 'CEMP Stars: Possible Hosts to Carbon Planets in the Early Universe'. *Monthly Notices of the Royal Astronomical Society*, May.

[183] McElroy, D., *et al.* 2013. 'The UMIST Database for Astrochemistry 2012'. *Astronomy and Astrophysics*, **550**(Feb.), A36.

[184] Meylan, G., and Heggie, D. C. 1997. 'Internal Dynamics of Globular Clusters'. *Astronomy and Astrophysicsr*, **8**, 1–143.

[185] Mojzsis, S. J., *et al.* 1996. 'Evidence for Life on Earth before 3,800 Million Years Ago'. *Nature*, **384**(Nov.), 55–59.

[186] Monje, R. R., *et al.* 2011. 'Herschel/HIFI Observations of Hydrogen Fluoride Toward Sagittarius B2(M)'. *Astrophysical Journal*, **734**(June), L23.

[187] Moore, B., *et al.* 2006. 'Globular Clusters, Satellite Galaxies and Stellar Haloes from Early Dark Matter Peaks'. *Monthly Notices of the Royal Astronomical Society*, **368**(May), 563–570.

[188] Morrison, H. L., *et al.* 2003. 'Andromeda VIII: A New Tidally Distorted Satellite of M31'. *Astrophysical Journal*, **596**(Oct.), L183–L186.

[189] Musso, M., and Sheth, R. K., 2014. 'The Excursion Set Approach in Non-Gaussian Random Fields'. *Monthly Notices of the Royal Astronomical Society*, **439**(Apr.), 3051–3063.

[190] Nagamine, K., and Loeb, A. 2004. 'Future Evolution of the Intergalactic Medium in a Universe Dominated by a Cosmological Constant'. *New Astronomy*, **9**(Oct.), 573–583.

[191] Nagasawa, M., *et al.* 2007. 'The Diverse Origins of Terrestrial-Planet Systems'. *Protostars and Planets V*, 639–654.

[192] Naka, S., *et al.* 1976. 'Direct Conversion of Graphite to Diamond under Static Pressure'. *Nature*, **259**(Jan.), 38–39.

[193] Naoz, S., Noter, S., and Barkana, R. 2006. 'The First Stars in the Universe'. *Monthly Notices of the Royal Astronomical Society*, **373**(Nov.), L98–L102.

[194] Niedzielski, A., *et al.* 2009. 'Substellar-Mass Companions to the K-Dwarf BD+14 4559 and the K-Giants HD 240210 and BD+20 2457'. *Astrophysical Journal*, **707**(Dec.), 768–777.

[195] Norris, J. E., Ryan, S. G., and Beers, T. C. 1997. 'Extremely Metal-Poor Stars: The Carbon-Rich, Neutron Capture Element–Poor Object CS 22957-027'. *Astrophysical Journal*, **489**(Nov.), L169.

[196] Norris, J. E., *et al.* 2013. 'The Most Metal-Poor Stars: IV. The Two Populations with [Fe/H] $<^\sim -3.0$'. *Astrophysical Journal*, **762**(Jan.), 28.

[197] Nozawa, T., *et al.* 2003. 'Dust in the Early Universe: Dust Formation in the Ejecta of Population III Supernovae'. *Astrophysical Journal*, **598**(Dec.), 785–803.

[198] Ober, W. W., El Eid, M. F., and Fricke, K. J. 1983. 'Evolution of Massive Pregalactic Stars: Part Two; Nucleosynthesis in Pair Creation Supernovae and Pregalactic Enrichment'. *Astronomy and Astrophysics*, **119**(Mar.), 61.

[199] Omukai, K., *et al.* 2005. 'Thermal and Fragmentation Properties of Star-Forming Clouds in Low-Metallicity Environments'. *Astrophysical Journal*, **626**(June), 627–643.

[200] Owen, J. E., and Mohanty, S. 2016. 'Habitability of Terrestrial-Mass Planets in the HZ of M Dwarfs: I. H/He-Dominated Atmospheres'. *Monthly Notices of the Royal Astronomical Society*, **459**(July), 4088–4108.

[201] Papaloizou, J. C. B., and Terquem, C. 2006. 'Planet Formation and Migration'. *Reports on Progress in Physics*, **69**(Jan.), 119–180.

[202] Papoular, R., *et al.* 1996. 'A Comparison of Solid-State Carbonaceous Models of Cosmic Dust'. *Astronomy and Astrophysics*, **315**(Nov.), 222–236.

[203] Park, B. G., *et al.* 2006. 'Microlensing Sensitivity to Earth-Mass Planets in the Habitable Zone'. *Astrophysical Journal*, **643**(June), 1233–1238.

[204] Penteado, E. M., Cuppen, H. M., and Rocha-Pinto, H. J. 2014. 'Modelling the Chemical Evolution of Molecular Clouds as a Function of Metallicity'. *Monthly Notices of the Royal Astronomical Society*, **439**(Apr.), 3616–3629.

[205] Pepper, J., and Gaudi, B. S. 2005. 'Searching for Transiting Planets in Stellar Systems'. *Astrophysical Journal*, **631**(Sept.), 581–596.

[206] Perlmutter, S. 2005. 'Studying Dark Energy with Supernovae: Now, Soon, and the Not-Too-Distant Future'. *Physica Scripta Volume T*, **117**(Jan.), 17–28.

[207] Petigura, E. A., Howard, A. W., and Marcy, G. W. 2013. 'Prevalence of Earth-Size Planets Orbiting Sun-Like Stars'. *Proceedings of the National Academy of Science*, **110**(Nov.), 19273–19278.

[208] Pierrehumbert, R., and Gaidos, E. 2011. 'Hydrogen Greenhouse Planets beyond the Habitable Zone'. *The Astrophysical Journal Letters*, **734**(June), L13.

[209] Pinotti, R., *et al.* 2005. 'A link between the Semimajor Axis of Extrasolar Gas Giant Planets and Stellar Metallicity'. *Monthly Notices of the Royal Astronomical Society*, **364**(Nov.), 29–36.

[210] Piran, T., and Jimenez, R. 2014. 'Possible Role of Gamma Ray Bursts on Life Extinction in the Universe'. *Physical Review Letters*, **113**(23), 231102.

[211] Planck Collaboration *et al.* 2015a. 'Planck 2015 Results: XIII. Cosmological Parameters'. *ArXiv e-prints*, Feb.

[212] Planck Collaboration *et al.* 2015b. 'Planck 2015 Results: XVII. Constraints on Primordial Non-Gaussianity'. *ArXiv e-prints*, Feb.

[213] Plez, B., and Cohen, J. G. 2005. 'Analysis of the Carbon-Rich Very Metal-Poor Dwarf G77-61'. *Astronomy and Astrophysics*, **434**(May), 1117–1124.

[214] Polchinski, J. 2006. 'The Cosmological Constant and the String Landscape'. *ArXiv High Energy Physics: Theory e-prints*, Mar.

[215] Pritchard, J. R., and Loeb, A. 2012. '21 cm Cosmology in the 21st Century'. *Reports on Progress in Physics*, **75**(8), 086901.

[216] Ray, A., and Loeb, A. 2015. 'Inferring the Composition of Super-Jupiter Mass Companions of Pulsars with Radio Line Spectroscopy'. *ArXiv e-prints*, Oct.

[217] Raymond, S. N., Scalo, J., and Meadows, V. S. 2007. 'A Decreased Probability of Habitable Planet Formation around Low-Mass Stars'. *Astrophysical Journal*, **669**(Nov.), 606–614.

[218] Rhoads, J. E., and Malhotra, S. 2001. 'Lyα Emitters at Redshift z = 5.7'. *Astrophysical Journal*, **563**(Dec.), L5–L9.

[219] Rhoads, J. E., *et al.* 2003. 'Spectroscopic Confirmation of Three Redshift z~5.7 Lyα Emitters from the Large-Area Lyman Alpha Survey'. *The Astronomical Journal*, **125**(Mar.), 1006–1013.

[220] Ricotti, M. 2002. 'Did Globular Clusters Reionize the Universe?'. *Monthly Notices of the Royal Astronomical Society*, **336**(Oct.), L33–L37.

[221] Ricotti, M., and Gnedin, N. Y., 2005. 'Formation Histories of Dwarf Galaxies in the Local Group'. *Astrophysical Journal*, **629**(Aug.), 259–267.

[222] Rodler, F., and López-Morales, M. 2014. 'Feasibility Studies for the Detection of O_2 in an Earth-Like Exoplanet'. *Astrophysical Journal*, **781**(Jan.), 54.

[223] Roederer, I. U., *et al.* 2014. 'A Search for Stars of Very Low Metal Abundance: VI. Detailed Abundances of 313 Metal-Poor Stars'. *The Astronomical Journal*, **147**(June), 136.

[224] Rushby, A. J., *et al.* 2013. 'Habitable Zone Lifetimes of Exoplanets around Main Sequence Stars'. *Astrobiology*, **13**(Sept.), 833–849.

[225] Sahu, K. C. 1994. 'Stars within the Large Magellanic Cloud as Potential Lenses for Observed Microlensing Events'. *Nature*, **370**(July), 275.

[226] Salaris, M., and Cassisi, S. 2006. *Evolution of Stars and Stellar Populations*. Wiley.

[227] Salpeter, E. E., and Zapolsky, H. S. 1967. 'Theoretical High-Pressure Equations of State Including Correlation Energy'. *Physical Review*, **158**(June), 876–886.

[228] Santos, M. R., *et al.* 2004. 'The Abundance of Low-Luminosity Lyα Emitters at High Redshift'. *Astrophysical Journal*, **606**(May), 683–701.

[229] Schaller, E. L., and Brown, M. E. 2007. 'Volatile Loss and Retention on Kuiper Belt Objects'. *The Astrophysical Journal Letters*, **659**(Apr.), L61–L64.

[230] Seager, S., *et al.* 2007. 'Mass-Radius Relationships for Solid Exoplanets'. *Astrophysical Journal*, **669**(Nov.), 1279–1297.

[231] Seager, S., Bains, W., and Hu, R. 2013. 'A Biomass-Based Model to Estimate the Plausibility of Exoplanet Biosignature Gases'. *Astrophysical Journal*, **775**(Oct.), 104.

[232] Shandera, S., *et al.* 2013. 'X-Ray Cluster Constraints on Non-Gaussianity'. *Journal of Cosmology and Astroparticle Physics*, **8**(Aug.), 004.

[233] Sheth, R. K., and Tormen, G. 1999. 'Large-Scale Bias and the Peak Background Split'. *Monthly Notices of the Royal Astronomical Society*, **308**(Sept.), 119–126.

[234] Sigurdsson, S., *et al.* 2003. 'A Young White Dwarf Companion to Pulsar B1620-26: Evidence for Early Planet Formation'. *Science*, **301**(July), 193–196.

[235] Silburt, A., Gaidos, E., and Wu, Y. 2015. 'A Statistical Reconstruction of the Planet Population around Kepler Solar-Type Stars'. *Astrophysical Journal*, **799**(Feb.), 180.

[236] Sloan, D., Alves Batista, R., and Loeb, A. 2017. 'The Resilience of Life to Astrophysical Events'. *Scientific Reports*, **7**(July), 5419.

[237] Sonnentrucker, P., Neufeld, D. A., Phillips, T. G., *et al.* 2010. 'Detection of Hydrogen Fluoride Absorption in Diffuse Molecular Clouds with Herschel/HIFI: An Ubiquitous Tracer of Molecular Gas'. *Astronomy and Astrophysics*, **521**(Oct.), L12.

[238] Sonnentrucker, P., *et al.* 2013. 'Herschel Observations Reveal Anomalous Molecular Abundances toward the Galactic Center'. *Astrophysical Journal*, **763**(Jan.), L19.

[239] Stark, D. P., and Ellis, R. S., 2006. 'Searching for the Sources Responsible for Cosmic Reionization: Probing the Redshift Range 7 < z < 10 and Beyond'. *New Astronomy*, **50**(Mar.), 46–52.

[240] Sternberg, A., *et al.* 2014. 'H I-to-H_2 Transitions and H I Column Densities in Galaxy Star-Forming Regions'. *Astrophysical Journal*, **790**(July), 10.

[241] Stevenson, D. J. 1999. 'Life-Sustaining Planets in Interstellar Space?'. *Nature*, **400**(July), 32.

[242] Susskind, L. 2003 (Mar.). 'The Anthropic Landscape of String Theory'. In *The Davis Meeting on Cosmic Inflation*. Page 26.

[243] Takai, K., *et al.* 2003. Structure and Electronic Properties of a Nongraphitic Disordered Carbon System and Its Heat-Treatment Effects'. *Physical Review B*, **67**(21), 214202.

[244] Tegmark, M., *et al.* 1997. 'How Small Were the First Cosmological Objects?.' *Astrophysical Journal*, **474**(Jan.), 1.

[245] Tegmark, M., *et al.* 2006. 'Dimensionless Constants, Cosmology, and Other Dark Matters'. *Physical Review D*, **73**(2), 023505.

[246] Tielens, A. G. G. M., *et al.* 1987. 'Shock Processing of Interstellar Dust: Diamonds in the Sky'. *Astrophysical Journal*, **319**(Aug.), L109–L113.

[247] Todini, P., and Ferrara, A. 2001. 'Dust Formation in Primordial Type II Supernovae'. *Monthly Notices of the Royal Astronomical Society*, **325**(Aug.), 726–736.

[248] Tominaga, N., Iwamoto, N., and Nomoto, K. 2014. 'Abundance Profiling of Extremely Metal-Poor Stars and Supernova Properties in the Early Universe'. *Astrophysical Journal*, **785**(Apr.), 98.

[249] Traub, W. A. 2015. 'Steps towards Eta-Earth, from Kepler Data'. *International Journal of Astrobiology*, **14**(July), 359–363.

[250] Tseliakhovich, D., Barkana, R., and Hirata, C. M. 2011. 'Suppression and Spatial Variation of Early Galaxies and Minihaloes'. *Monthly Notices of the Royal Astronomical Society*, **418**(Dec.), 906–915.

[251] Udalski, A. 2003. 'The Optical Gravitational Lensing Experiment: Real Time Data Analysis Systems in the OGLE-III Survey'. *Acta Astronomica*, **53**(Dec.), 291–305.

[252] Udalski, A., *et al.* 2002. 'The Optical Gravitational Lensing Experiment: Search for Planetary and Low-Luminosity Object Transits in the Galactic Disk; Results of 2001 Campaign'. *Acta Astronomica*, **52**(Mar.), 1–37.

[253] Umeda, H., and Nomoto, K. 2003. 'First-Generation Black-Hole-Forming Supernovae and the Metal Abundance Pattern of a Very Iron-Poor Star'. *Nature*, **422**(Apr.), 871–873.

[254] Umeda, H., and Nomoto, K. 2005. 'Variations in the Abundance Pattern of Extremely Metal-Poor Stars and Nucleosynthesis in Population III Supernovae'. *Astrophysical Journal*, **619**(Jan.), 427–445.

[255] Urakawa, S., *et al.* 2006. 'Extrasolar Transiting Planet Search with Subaru Suprime-Cam'. *ArXiv Astrophysics e-prints*, Mar.

[256] Valencia, D., O'Connell, R. J., and Sasselov, D. 2006. 'Internal Structure of Massive Terrestrial Planets'. *Icarus*, **181**(Apr.), 545–554.

[257] Valley, J. W., *et al.* 2002. 'A Cool Early Earth'. *Geology*, **30**(Apr.), 351.

[258] van Dishoeck, E. F., Herbst, E., and Neufeld, D. A. 2013. 'Interstellar Water Chemistry: From Laboratory to Observations'. *Chemical Reviews*, **113**(Dec.), 9043–9085.

[259] Vanderburg, A., *et al.* 2015. 'A Disintegrating Minor Planet Transiting a White Dwarf'. *Nature*, **526**(Oct.), 546–549.

[260] Veras, D., Crepp, J. R., and Ford, E. B. 2009. 'Formation, Survival, and Detectability of Planets beyond 100 AU'. *Astrophysical Journal*, **696**(May), 1600–1611.

[261] Vilenkin, A. 1995. 'Predictions from Quantum Cosmology'. *Physical Review Letters*, **74**(Feb.), 846–849.

[262] Vinet, P., *et al.* 1987. 'Compressibility of Solids'. *Journal of Geophysical Research*, **92**(Aug.), 9319–9325.

[263] Vinet, P., *et al.* 1989. 'Universal Features of the Equation of State of Solids'. *Journal of Physics Condensed Matter*, **1**(Mar.), 1941–1963.

[264] Weinberg, S. 1987. 'Anthropic Bound on the Cosmological Constant'. *Physical Review Letters*, **59**(Nov.), 2607–2610.

[265] Weinberg, S. 2000. 'A Priori Probability Distribution of the Cosmological Constant'. *Physical Review D*, **61**(10), 103505.

[266] Weinberg, S. 2005. 'Living in the Multiverse'. *ArXiv High Energy Physics: Theory e-prints*, Nov.

[267] Winn, J. N., and Fabrycky, D. C. 2015. 'The Occurrence and Architecture of Exoplanetary Systems'. *Annual Reviews of Astronomy and Astrophysics*, **53**(Aug.), 409–447.

[268] Wisotzki, L., *et al.* 1996. 'The Hamburg/ESO survey for Bright QSOs: I. Survey Design and Candidate Selection Procedure'. *Astronomy and Astrophysics Supplement Series*, **115**(Feb.), 227.

[269] Wyse, R. F. G., and Gilmore, G. 2006. 'Stellar Populations with ELTs'. Whitelock, P., Dennefeld, M., and Leibundgut, B., eds. *The Scientific Requirements for Extremely Large Telescopes*. IAU Symposium, vol. 232 Pages 140–148.

[270] Yanny, B., *et al.* 2009. 'SEGUE: A Spectroscopic Survey of 240,000 Stars with g = 14–20'. *The Astronomical Journal*, **137**(May), 4377–4399.

[271] Yasui, C., *et al.* 2009. 'The Lifetime of Protoplanetary Disks in a Low-metallicity Environment'. *Astrophysical Journal*, **705**(Nov.), 54–63.

[272] Yong, D., *et al.* 2013. The Most Metal-Poor Stars: III. The Metallicity Distribution Function and Carbon-Enhanced Metal-Poor Fraction'. *Astrophysical Journal*, **762**(Jan.), 27.

[273] Yoo, J., *et al.* 2004. 'Constraints on Planetary Companions in the Magnification A = 256 Microlensing Event OGLE-2003-BLG-423'. *Astrophysical Journal*, **616**(Dec.), 1204–1214.

[274] York, D. G., Adelman, J., Anderson, J. E., *et al.* 2000. 'The Sloan Digital Sky Survey: Technical Summary'. *The Astronomical Journal*, **120**(Sept.), 1579–1587.

[275] Zackrisson, E., *et al.* 2016. 'Terrestrial Planets across Space and Time'. *ArXiv e-prints*, Feb.

[276] Zahnle, K. J., and Catling, D. C. 2017. 'The Cosmic Shoreline: The Evidence That Escape Determines Which Planets Have Atmospheres, and What This May Mean for Proxima Centauri b'. *ArXiv e-prints*, Feb.

[277] Zapolsky, H. S., and Salpeter, E. E. 1969. 'The Mass-Radius Relation for Cold Spheres of Low Mass'. *Astrophysical Journal*, **158**(Nov.), 809.

[278] Zinner, E. 1998. 'Leonard Award Address: Trends in the Study of Presolar Dust Grains from Primitive Meteorites'. *Meteoritics and Planetary Science*, **33**(July), 549–564.

13

Climbing Up the Theories of Nature: Fine-Tuning and Biological Molecules

GEORGE ELLIS, JEAN-PHILIPPE UZAN, AND DAVID SLOAN

Abstract

The understanding of the world around us requires the use of various theories representing different levels of emergent structure. While they enjoy a decoupling, they are not completely independent. It has been acknowledged that the possibility for complexity to emerge sets constraints, indeed a fine-tuning, on the fundamental constants of nature. It turns out that at each level of emergent complexity, new such constraints appear. It is not clear whether they are more stringent or obviously satisfied in view of constraints relevant for lower levels. We discuss the connection between fine-tuning of fundamental physics, atomic physics, chemistry, and biology to highlight the importance of the analysis of the connection between these different theories in relation to fine-tuning.

13.1 Introduction

13.1.1 The Basic Idea

Fine-tuning relates the existence of life to the values of physical constants. Only certain values of the fundamental constants will allow any life whatsoever to exist [2, 16, 23, 24]. This is because these parameters underlie existence of galaxies, stars, and planets with suitable heavy elements out of which organic molecules can be constructed. However, they also underlie the values of energies and angles in molecules and, hence, also govern the organic chemistry and molecular biology that underlie all living systems, so these, too, will only be possible for a restricted range of these constants. It is a fundamental problem to find out what these relations and limitations are.

Fundamental constants of physics are parameters in the theory that cannot be reduced to other parameters [30, 33]. The following are two key points:

- They are dependent on the theoretical framework used.
- They should be dimensionless to be physically meaningful.

Examples: the fine-structure constant $\alpha = e^2/(\hbar c)$ and electron-proton mass ratio $\mu = m_e/m_p$ are key dimensionless fundamental constants. The speed of light c, Planck's constant $\hbar$, and gravitational constant G are not fundamental constants because they are not dimensionless. You can make their value whatever you want by your choice of units (e.g., you can always set $c = 1$ by choice of time and length units [10]).

There is no agreement on the number of constants or what they are; a comprehensive survey has been given in Reference [27] as well as Chapter 2 of this volume. As mentioned earlier, only a limited range of values of these constants is compatible with existence of life of any kind [2, 16, 23, 24].

Fine-tuning discussions conventionally involve only issues to do with

- Existence of galaxies
- Existence of key elements out of which organic molecules can be made: C, N, O, P, S (as a result of stellar nucleosynthesis followed by explosions of first generation stars)
- Existence of second-generation stars with planets, some of which have an atmosphere and water, perhaps with a moon.

While all these are necessary, they are not sufficient. They do not touch the nature of life itself. That is what we touch on here.

Biologically Important Molecules

One needs also the existence of essential organic molecules and aqueous solutions, thus

- Water with a suitable dipole [3]
- Nucleic acids: suitable DNA [4] and RNA
- Organic molecules: suitable proteins [22]

One also needs lipids and carbohydrates, but their structure is not so crucial. Proteins are the key molecules that do the needed work in all sorts of ways [22], e.g.,

- *Catalysis:* speeding up reactions by a huge amount (enzymes)
- Controlling *gene expression* (gene transcription factors, gene corrections)
- *Logical operations*: controlling flow of ions into and out of axons (e.g., voltage gated ion channels [12])

DNA is important only because it creates proteins at the right time and place [6] (because of gene transcription networks). However, that role is, of course, crucial. RNA is, in effect, at the present time a support actor for DNA (although it may have played a different more role at earlier epochs).

The reader should note that what is presented here is very much a work in progress. We explore a variety of approaches and obtain some preliminary results, but there is much work to be done. In effect, this chapter is an outline of a research program rather than a presentation of results from a completed such program.

13.1.2 The Hierarchy of Structure

Nature confronts us with structures of different scales, complexities, and properties – from fundamental particles to molecules and cells to planets, stars, and galaxies. Each of them can be described by a scientific theory, with its own ontology and structures. The structures form a hierarchy of theories organised in modules in interaction (see References [8, 9, 11] for a discussion). Indeed, those theories are not independent. Higher levels are built on more fundamental theories. And higher-level theories set the context in which the dynamics of the lower-level theories develop. This is related to both bottom-up action and top-down causation [11].

The fact that we can understand the Universe and its laws has deep implications for this structure of theories. At each step in our construction of physical theories, we have been dealing with phenomena below a typical energy scale, mostly because of technological constraints, and it has turned out (empirically) that we always have been able to design a consistent theory that is valid in such a restricted regime. This is not expected in general and is deeply rooted in the mathematical structure of the theories that describe nature. They have to enjoy a *scale decoupling principle* in the sense that there exist energy scales below which effective theories are sufficient to understand a set of physical phenomena that can be observed. *Effective theories* are then the most fundamental concepts in the scientific approach to the understanding of nature, and they always come with a domain of validity inside which they are efficient to describe all related phenomena [28, 29]. They are a successful explanation at a given level of complexity based on the concepts of that particular level.

This implies that the structure of the theories is such that there is a kind of stability and independence of higher levels with respect to more fundamental ones. It follows that various disciplines have developed independently in almost quasi-autonomous domains, each of them having its own ontology and dynamics that are independent of our ability to formulate a theory explaining these concepts in lower-level terms. In each case, we can hope to relate the concepts and constants of a given level to those of an underlying level. For instance, we understand that the proton is a composite structure of three quarks, and we may try to determine its physical characteristic (charge, mass, gyromagnetic factor, quantum numbers) in terms of those of these more fundamental entities [19]. However, we know that this can only be achieved for some structures, since there exist emergent phenomena (information, life, consciousness) that cannot be reduced to the concepts of a lower level.

13.1.3 Fundamental Physics

Today, gravitation is well described by general relativity, and the most fundamental (experimentally tested) theory of matter is the Standard Model of particle physics. This is a theoretical construction based on an action and many choices such as the mathematical description of the matter fields (this is not completely arbitrary and is based on representations of the Poincaré group, which allows one to define scalar, spinor, vector, etc., structures, but we decide to identify each kind of particle with a particular mathematical structure [31]), symmetries (such as $SU(2)$ or $SU(3)$), and constants. None of them can be explained by the theory at hand. In particular, it is important for the constants to be measurable [21].

Table 13.1 *List of the fundamental constants of our Standard Model. See Reference [21] for further details of the measurements.*

Constant	Symbol	Value
Speed of light	c	299,792,458 m s^{-1}
Planck constant (reduced)	$\hbar$	$1.054\,571\,628(53) \times 10^{-34}$ J s
Newton constant	G	$6.674\,28(67) \times 10^{-11}$ m^2 kg^{-1} s^{-2}
Weak coupling constant (at m_Z)	$g_2(m_Z)$	0.6520 ± 0.0001
Strong coupling constant (at m_Z)	$g_3(m_Z)$	1.221 ± 0.022
Weinberg angle	$\sin^2\theta_W(91.2\ \text{GeV})_{\overline{\text{MS}}}$	0.23120 ± 0.00015
Electron Yukawa coupling	h_e	2.94×10^{-6}
Muon Yukawa coupling	h_μ	0.000607
Tauon Yukawa coupling	h_τ	0.0102156
Up Yukawa coupling	h_u	0.000016 ± 0.000007
Down Yukawa coupling	h_d	0.00003 ± 0.00002
Charm Yukawa coupling	h_c	0.0072 ± 0.0006
Strange Yukawa coupling	h_s	0.0006 ± 0.0002
Top Yukawa coupling	h_t	1.002 ± 0.029
Bottom Yukawa·coupling	h_b	0.026 ± 0.003
Quark CKM matrix angle	$\sin\theta_{12}$	0.2243 ± 0.0016
	$\sin\theta_{23}$	0.0413 ± 0.0015
	$\sin\theta_{13}$	0.0037 ± 0.0005
Quark CKM matrix phase	δ_{CKM}	1.05 ± 0.24
Higgs potential quadratic coefficient	$\hat{\mu}^2$	$-(250.6 \pm 1.2)$ GeV^2
Higgs potential quartic coefficient	λ	1.015 ± 0.05
QCD vacuum phase	θ_{QCD}	10^{-9}

For this model, they actually are and have been; see Table 13.1. There is no way to express them in terms of more fundamental quantities (else they would not be fundamental constants), and there is no equation for them. Testing their constancy reveals that the hypothesis that they are constant is a good hypothesis, at least to the level of accuracy and timescale over which the experiment is conducted [27]. In case of disagreement, one can promote them to a dynamical field but then would have to explain why they are almost frozen [26, 27].

It is also important to remember that any measurement is just a comparison between two physical systems, one usually defining a system of units. It follows that only dimensionless constants can be measured [10], and only the change of such constants would change the physics. We will consider only these parameters. Given a list of N constants, one can always pick three of them to define units so that one is left with $N - 3$ fundamental parameters that affect the magnitude of any physical process.

Coming back to the Standard Model of particle physics, we have assumed it is our fundamental theory, even though it does not incorporate massive neutrinos and dark matter, so that we know it calls for an extension. This theory offers the possibility of space for higher levels of complexity to emerge. Changing the value of the fundamental constants may result in the technical impossibility for nuclei to be stable – which is an example of fine-tuning. The issue is, how fine-tuned is the Universe? This is what we shall now illustrate, keeping in mind that we would like to estimate how far up the chain of physical theories of higher complexity levels these fine-tunings propagate.

13.1.4 Towards Higher Complexity

Scaling Up to Nuclear Physics

The first level to consider above fundamental particle physics is nuclear physics. There, one needs to determine how cross sections, binding energies, lifetimes of unstable nuclei (or of the neutron), or simply characteristics such as the mass of the protons depends on the fundamental constants listed in Table 13.1. This has been intensively investigated, and we refer to References [26, 27] for reviews. They are of huge importance for the description of Big Bang nucleosynthesis and for stellar nucleosynthesis. As examples, one can compute the binding energy of the deuterium or the lifetime of the proton.

The dependence of the Hoyle state on the nuclear parameters and the fine-structure constant was studied in Reference [7]. It was shown there that the requirement that one forms both carbon-12 and oxygen-16 in population III stars set a constraint on α of the order of 10^{-3}. The dependence of the mass of the proton and gyromagnetic factors are described in Reference [19]. It is important to realise that the connection from QCD to nuclear physics is difficult and intricate so that the accuracy of these scaling is lively debated.

Scaling Up to Atomic Physics

The goal of this chapter is to describe the effect of a change in the value of the fundamental constants on the structure of molecules, where the most importat relevant atoms are H, C, N, O, P, and S. It will thus focus on the standard Schrödinger equation,

$$i\hbar\partial_t \psi(\boldsymbol{r},t) = H\psi(\boldsymbol{r},t), \qquad (13.1)$$

where H is the Hamiltonian. As far as atomic physics and chemistry are concerned, this Hamiltonian depends on a few constants: the fine-structure constant α, the masses of the nuclei m_I, the mass of the electron m_e, and the gyromagnetic factors of the nuclei and electrons g_I and g_e. The observables are of two types – either dimensionless, such as geometric conformations and angles, or dimensional, such as bond lengths, energy levels, and magnetic and electric moments. These quantities affect the properties of matter. For example, a change in the electric and magnetic properties of the water molecule can impact those of water in a drastic way – e.g., if it changes the solid-liquid phase boundary in the phase diagram, which has a negative slope [3], contrary to most substances. This feature is important for the development of life as we know it.

Hypothesis: *If one changes fundamental constants, such as α and μ, chemistry will not work the same in the case of larger molecules. It will change conformation and so change biological function.*

A delicate issue is whether a change that alters all scales simultaneously would affect biological function or not or, rather, just result in the same function but rescaled in size. It would affect relevant energies, which are, of course, crucial to biology; this might or might not affect function because, for example, the photoelectric effect is not determined by the Schrödinger equation and would scale differently. Presumably, this would alter some functions – for example, those related to photosynthesis and vision – crucial ways.

The text is organised as followed. Section 13.2 describes the Schrödinger equation, focusing on the issue of units and dimensions. We then detail how to compute the effects of a variation of the constants on the properties of molecule in Section 13.3 and apply this to some simple molecules in Section 13.4. The issue of conformal scaling is considered in Section 13.5. We then draw some implications in Section 13.6 in particular, in evaluating whether the fine-tuning required from chemistry is greater or less than that arising from successful carbon production in stars.

13.2 Quantum Physics and Observables

The link from physics to chemistry is via the Schrödinger equation (13.1).

13.2.1 Schrödinger Equation in Dimensionless Form

As discussed previously, one can start by rewriting it in dimensionless form. For instance, among the constants of the problem we can pick $(m_e, \hbar, c)$ to construct our standard units in the form

$$\ell_a = \frac{\hbar}{m_e c} \qquad t_a = \frac{\hbar}{m_e c^2} \qquad E_a = m_e c^2. \tag{13.2}$$

and then perform the rescalings

$$\tau = t/t_a, \qquad \rho = \frac{r}{\ell_a}, \qquad \mathcal{E} = \frac{E}{E_a}, \qquad \mu_i \equiv \frac{m_i}{m_e}. \tag{13.3}$$

It follows that Eq. (13.1) takes the form

$$i \partial_\tau \psi(\rho, \tau) = - \left(\frac{1}{2} \frac{m_e}{m} \Delta + \mathcal{V} \right) \psi(\rho, \tau), \tag{13.4}$$

where Δ is Laplacian associated to ρ and $\mathcal{V} = V/m_e c^2$ is a dimensionless potential. For instance, the ionisation energy of hydrogen $-E_I = \frac{1}{2} m_e c^2 \alpha^2 = 13.60580 \, \text{eV}$ in standard units reduces to $-E_I = \frac{1}{2}\alpha^2$ in these atomic units. What is important is that the ratios of any energies will be independent of the choice of the units. The Bohr radius $a_0 = \hbar/m_e c \alpha = 5.29 \times 10^{-11}$ m in standard units reduces to $a_0 = 1/\alpha$ in these atomic units.

This means that we are working with the Hamiltonian

$$\mathcal{H} = \sum_{\text{electrons}} \frac{1}{2} P_i^2 + \sum_{\text{nuclei}} \frac{1}{2\mu_I^2} P_I^2 + V \equiv \sum_{\text{electrons}} \mathcal{T}_e + \sum_{\text{nuclei}} \mathcal{T}_N(\mu_i) + \mathcal{V}(\alpha, \mu_i, g_i) \quad (13.5)$$

so that one needs to solve

$$\mathcal{H} |\psi\rangle = \mathcal{E} |\psi\rangle. \qquad (13.6)$$

13.2.2 Example: Hydrogen Atom

For the hydrogen atom including fine and hyperfine structure, we get for the electron

$$i\partial_\tau \psi_e(\rho, \tau) = -\left(\frac{1}{2}\frac{m_e}{m_p}\Delta + \frac{\alpha}{\rho} + \frac{W}{m_e c^2}\right)\psi_e(\rho, \tau), \qquad (13.7)$$

where the term W includes the fine and hyperfine structure. The dimensionless potential $\mathcal{V}$ is the last two terms in (13.7) and is no longer a function of (α/ρ). It is explicitly given by

$$\mathcal{V} = \frac{\alpha}{\rho} + \frac{\alpha}{2\rho^3}\frac{g_e}{2}\frac{\mathbf{L}}{\hbar}\cdot\frac{\mathbf{S}}{\hbar} - \frac{\Delta^2}{8} + \pi\alpha\delta(\rho)$$

$$+ \alpha\frac{g_p}{2}\frac{m_e}{m_p}\left\{-\frac{1}{2\rho^3}\frac{\mathbf{L}}{\hbar}\cdot\frac{\mathbf{I}}{\hbar} + \frac{2\pi}{3}\frac{g_e}{2}\frac{\mathbf{I}}{\hbar}\cdot\frac{\mathbf{S}}{\hbar}\delta(\rho) + \frac{1}{4\rho^3}\frac{g_e}{2}\left[3\left(\frac{\mathbf{S}}{\hbar}\cdot\mathbf{n}\right)\left(\frac{\mathbf{I}}{\hbar}\cdot\mathbf{n}\right) - \frac{\mathbf{I}}{\hbar}\cdot\frac{\mathbf{S}}{\hbar}\right]\right\},$$

$$(13.8)$$

where the first line is the non-relativistic term, the second line is the the fine-structure terms (discussed further in Section 13.5.2), and the final one is the hyperfine-structure terms. Indeed, it involves only dimensionless quantities: $\alpha, \mu := m_e/m_p$, g_e and g_p. So solving this equation will result in dimensionless energies – i.e., energies in units of $m_e c^2$: $\mathcal{E} = E/m_e c^2$. The optical lines are given by

$$\mathcal{E}_n = \frac{\alpha^2}{2n^2}(1+\mu)^{-1} \qquad (13.9)$$

and the fine structure by

$$\mathcal{E}_{nlJ} = 1 + \frac{\alpha^2}{2n^2} - \frac{\alpha^4}{2n^4}\left(\frac{n}{J+1/2} - \frac{3}{4}\right) + \cdots \qquad (13.10)$$

For a full solution, of course, we also need to include the nucleus – or nuclei, in the case of heavier atoms – as in Eq. (13.5).

13.2.3 Metre Dependence on Fundamental Constants

In the international system of units, the second and the metre are defined by a hyperfine transition in ^{133}Cs. This implies that when the constants are varied, the physical systems to which the system of study is compared are also modified.

In general, the hyperfine frequency in a given electronic state of an alkali-like atom – such as ^{133}Cs, ^{87}Rb, ^{199}Hg$^+$ – is given by

$$\nu_{\text{hfs}} \simeq R_\infty c \times A_{\text{hfs}} \times g_i \times \alpha^2 \times \bar{\mu} \times F_{\text{hfs}}(\alpha), \tag{13.11}$$

where $g_i = \mu_i/\mu_N$ is the nuclear g factor. A_{hfs} is a numerical factor depending on each particular atom, and we have set $F_{\text{rel}} = F_{\text{hfs}}(\alpha)$. The Rydberg constant R_∞ is given by $R_\infty = \alpha/4\pi a_0$ – i.e., $\alpha^2/4\pi$ in atomic units. Similarly, the frequency of an electronic transition is well approximated by

$$\nu_{\text{elec}} \simeq R_\infty c \times A_{\text{elec}} \times F_{\text{elec}}(Z, \alpha), \tag{13.12}$$

where – as before – A_{elec} is a numerical factor depending on each particular atom and F_{elec} is the function accounting for relativistic effects, spin-orbit couplings, and many-body effects. Even though an electronic transition should also include a contribution from the hyperfine interaction, it is generally only a small fraction of the transition energy and, thus, should not carry any significant sensitivity to a variation of the fundamental constants.

It follows that the metre, defined as $L_{1m} \equiv c/\nu_{\text{Cs}}$, scales as

$$L_{1m}^{-1} = R_\infty \times A_{\text{hfs}} \times g_i \times \alpha^2 \times \bar{\mu} \times F_{\text{Cs}}(\alpha). \tag{13.13}$$

The relativistic corrections can be characterised by introducing the sensitivity of the relativistic factors to a variation of α,

$$\kappa_\alpha \equiv \frac{\partial \ln F}{\partial \ln \alpha}. \tag{13.14}$$

Table 13.2 summarises the values of some of them. Indeed, a reliable knowledge of these coefficients at the 1%–10% level is required to deduce limits to a possible variation of the constants. The interpretation of the spectra in this context relies, from a theoretical point of view, only on quantum electrodynamics (QED) so that we can safely obtain constraints on (α, μ, g_i), still keeping in mind that the computation of the sensitivity factors requires numerical N-body simulations.

Table 13.2 *Sensitivity of various transitions on a variation of the fine-structure constant.*

Atom	Transition	Sensitivity κ_α
^{1}H	$1s - 2s$	0.00
^{87}Rb	hf	0.34
^{133}Cs	$^2S_{1/2}(F = 2) - (F = 3)$	0.83
^{171}Yb$^+$	$^2S_{1/2} - {}^2D_{3/2}$	0.9
^{199}Hg$^+$	$^2S_{1/2} - {}^2D_{5/2}$	-3.2
^{87}Sr	$^1S_0 - {}^3P_0$	0.06
^{27}Al$^+$	$^1S_0 - {}^3P_0$	0.008

13.3 Influence on the Structure of Molecules

In this section, we set up a general procedure for finding the effect of a variation of fundamental constants on chemical structure.

13.3.1 Evaluating the Effect of a Change in the Constants

The starting point of the analysis is the Schrödinger equation (13.1). In most problems, it can only be solved numerically or by means of perturbation theory. The main idea is then that the full Hamiltonian H can be split as $H = H_0 + W$, where H_0 has eigenstates that can be computed, and W is a perturbation.

The existing constraints on the time variation of the fundamental constants teach us that we are looking for small effects so that perturbation theory is a well-suited approach. The Hamiltonian depends on a set of constants, c_a, the value of which is c_{a0} in the laboratory today. It can be decomposed as

$$H(c_i) = H_0 + W \tag{13.15}$$

with

$$H_0 \equiv H(c_a = c_{a0}) \tag{13.16}$$

and

$$W \equiv \sum_a \left(\left. \frac{\delta H}{\delta \ln c_a} \right|_{c_a = c_{a0}} \right) \delta \ln c_a \tag{13.17}$$

since we assume that $\delta c_a / c_a \ll 1$. Given the general structure of the Hamiltonian (see Eqs. (13.5), and (13.6)), one concludes that the dependences on the fine-structure constant and gyromagnetic factor appear only in the potential while the masses appear in both the potential and kinetic terms. Hence, rescaling the perturbation potential as $\mathcal{W} = W/m_e c^2$, we get

$$\mathcal{W} = \sum_a \left(\left. \frac{\delta \mathcal{H}}{\delta \ln c_a} \right|_{c_a = c_{a0}} \right) \delta \ln c_a \equiv \sum_a f_a \delta \ln c_a, \tag{13.18}$$

which defines f_a.

We shall now perform a perturbation on the value of the fundamental constants. We work with the rescaled adimensional quantities. Indeed, solving the Schrödinger equation with the Hamiltonian $\mathcal{H}_0$ may require us to use another perturbation expansion in terms of the parameter of the interaction potential. Let us assume this leads to a set of eigenmodes $\left| \varphi_p^i \right\rangle$ associated to energies $\mathcal{E}_p^0$ so that i labels degenerate states of same energy:

$$\mathcal{H}_0 \left| \varphi_p^i \right\rangle = \mathcal{E}_p^0 \left| \varphi_p^i \right\rangle. \tag{13.19}$$

As usual, the eigenstates form an orthonormal basis of the space of states,

$$\left\langle \varphi_p^i | \varphi_q^j \right\rangle = \delta_{pq}\delta^{ij}, \qquad \sum_{p,i} \left| \varphi_p^i \right\rangle \left\langle \varphi_p^i \right| = 1. \tag{13.20}$$

Perturbation theory allows one to determine the perturbed energy levels and associated eigenstates. The derivation can be found in many textbooks. In a case of a non-degenerate state, one gets, at first order,

$$\mathcal{E}_n = \mathcal{E}_n^0 + \left\langle \varphi_p | \mathcal{W} | \varphi_p \right\rangle \tag{13.21}$$

for the energy and

$$\left| \psi_p \right\rangle = \left| \varphi_p \right\rangle + \sum_{n \neq p} \sum_i \frac{\left\langle \varphi_n^i | \mathcal{W} | \varphi_p \right\rangle}{\mathcal{E}_p^0 - \mathcal{E}_n^0} \left| \varphi_n^i \right\rangle \tag{13.22}$$

for the eigenstate.

13.3.2 *Effect on Bond Length*

A molecule with N atoms and P electrons has a state that is specified by $\vec{R}_I$ for the position of the atoms and $\vec{r}_i$ for the position of the electrons so that we are looking for a wave function $\psi(\vec{R}_I, \vec{r}_i; c_a)$. The bond length is then given by

$$\ell_{IJ} \equiv \left\langle \psi | R_{IJ} | \psi \right\rangle \tag{13.23}$$

for a given bond between nuclei I and J with $\vec{R}_{IJ} \equiv \vec{R}_J - \vec{R}_I$. However, this has units. We can define two dimensionless quantities, either the bound length compared to the metre (the value of which depends on the value of the constants; see Section 13.2.3, or in units of the Bohr radius a_0,

$$\tilde{\ell}_{IJ} \equiv \frac{\ell_{IJ}}{L_{1m}}, \qquad \lambda_{IJ} \equiv \frac{\ell_{IJ}}{\ell_a}. \tag{13.24}$$

Both are dimensionless quantities, the variation of which characterises a change in the geometry of the molecule. Another quantity that satisfies this criteria is the ratio between different bond lengths,

$$\alpha_{IJ;KL} \equiv \frac{\ell_{IJ}}{\ell_{KL}}. \tag{13.25}$$

Given the preceding description, we get that

$$\delta\lambda_{IJ}^{(p)} = \sum_a \left(\sum_{n \neq p} \sum_i \frac{\left\langle \varphi_n^i | f_a | \varphi_p \right\rangle}{\mathcal{E}_p^0 - \mathcal{E}_n^0} \left\langle \varphi_p | R_{IJ} | \varphi_n^i \right\rangle + c.c. \right) \delta \ln c_a \tag{13.26}$$

for the state p. What is usually called the bound length refers to the fundamental state. It follows that the sensitivity of the bond length to the variation of the constant c_a is

$$\frac{\delta \lambda_{IJ}}{\delta \ln c_a} = \sum_{n \neq p} \sum_i \frac{\langle \varphi_n^i | f_a | \varphi_p \rangle}{\mathcal{E}_p^0 - \mathcal{E}_n^0} \langle \varphi_p | R_{IJ} | \varphi_n^i \rangle + c.c., \tag{13.27}$$

where p is the fundamental state. This expression depends only on the energies and eigenstates of the standard molecule so that it can be computed from existing codes.

13.3.3 Effect on the Geometry

Angles are dimensionless so that a good indicator of the change of the geometry is

$$\beta_{\widehat{IJK}} \equiv \left\langle \psi \left| \frac{\vec{R}_{IJ}.\vec{R}_{IK}}{R_{IJ} R_{IK}} \right| \psi \right\rangle. \tag{13.28}$$

As before, we deduce the sensitivity of the angle to the constant c_a as

$$\frac{\delta \beta_{\widehat{IJK}}}{\delta \ln c_a} = \sum_{n \neq p} \sum_i \frac{\langle \varphi_n^i | f_a | \varphi_p \rangle}{\mathcal{E}_p^0 - \mathcal{E}_n^0} \left\langle \varphi_p \left| \frac{\vec{R}_{IJ}.\vec{R}_{IK}}{R_{IJ} R_{IK}} \right| \varphi_n^i \right\rangle + c.c. \tag{13.29}$$

13.4 Applications

The simplest applications to test the effect of changing the fine-structure constant on molecular structure are the hydrogen atom (Section 13.4.1) and hydrogen molecule (Section 13.4.2). We thank Peter Atkins for discussions and help with this section. Atkins's book *Molecular Quantum Mechanics* [1] (MQM) is a great resource as regards concepts and calculational methods. The next simplest application is the water molecule (Section 13.4.3), which is of considerable biological importance.

13.4.1 Hydrogen Atom

The constants needed in what follows, chosen in order to simplify the equations, are

$$\alpha = e^2/4\pi\epsilon_0\hbar, \ \beta = \hbar/(2m_e c), \ \gamma = c\hbar, \tag{13.30}$$

which imply

$$e^2/4\pi\epsilon_0 = \alpha c\hbar = \alpha\gamma, \ \hbar^2/(2m_e) = \beta\hbar c = \beta\gamma. \tag{13.31}$$

The energies (see [1], p. 89; $Z = 1$) are

$$E_n = -\frac{m_e e^4}{32\pi^2\epsilon_0^2\hbar^2}\cdot\frac{1}{n^2} = -\frac{m_e}{2\hbar^2}\left(\frac{e^2}{4\pi\epsilon_0}\right)^2\cdot\frac{1}{n^2} = -\frac{\gamma\alpha^2}{4\beta}\cdot\frac{1}{n^2}. \tag{13.32}$$

The orbital of lowest energy ($1s$; $n = 1, l = 0, m_l = 0$) is ([1], p. 88 and Y from p. 78)

$$\Psi_{1s}(r,\theta,\phi) = \frac{1}{\pi^{1/2}}\left(\frac{1}{a_0}\right)^{3/2} e^{-r/a_0} = \frac{1}{\pi^{1/2}}\left(\frac{\alpha}{2\beta}\right)^{3/2} e^{-2\beta r/\alpha}. \tag{13.33}$$

The most probable distance of the electron from the nucleus is the Bohr radius a_0:

$$a_0 = (4\pi\epsilon_0\hbar^2)/(m_e e^2) = \frac{2\beta}{\alpha} \Rightarrow \frac{\partial a_0}{\partial \alpha} = -\frac{2\beta}{\alpha^2} = -\frac{a_0}{\alpha}. \tag{13.34}$$

That is,

$$\partial \ln a_0/\partial\alpha = -1/\alpha = -1/137 = -0.0007. \tag{13.35}$$

As expected, an increase in α (an increase in charge) decreases the size of the atom.

13.4.2 Hydrogen Molecule

The Hamiltonian for the hydrogen molecule has terms for the two protons and the two electrons, all interacting though a Coulomb potential,

$$\mathcal{H} = -\frac{1}{2}\sum_{i=1,2} P_i^2 - \frac{1}{2\mu_p}\sum_{I=1,2} P_I^2 + \frac{\alpha}{R_{12}} + \frac{\alpha}{r_{12}} - \sum_{I,i}\frac{\alpha}{r_{Ii}}, \tag{13.36}$$

from which we extract the dependence on α and μ_p separately:

$$f_\alpha = \frac{\alpha}{R_{12}} + \frac{\alpha}{r_{12}} - \sum_{I,i}\frac{\alpha}{r_{Ii}}, \qquad f_{\mu_p} = +\frac{1}{2\mu_p}\sum_{I=1,2} P_I^2. \tag{13.37}$$

To determine specific values, we follow Peter Atkins thus: consider the hydrogen molecule ion (H_2^+). From [1], p. 265 in the LCAO approximation, for an internuclear separation R, with $s = R/a_0 = \alpha R/2\beta$, one has

$$E_+ - E_{1s} = \frac{\alpha\gamma}{R} - \frac{j' + k'}{1 + S} \tag{13.38}$$

$$j' = \frac{j_0}{R}\{1 - (1 + s)e^{-2s}\} = \frac{\alpha\gamma}{R}\left[1 - \left(1 + \frac{\alpha R}{2\beta}\right)e^{-\alpha R/\beta}\right] \tag{13.39}$$

$$k' = \frac{j_0}{a_0}1 + se^{-s} = \frac{2\alpha^2}{\beta}\left[1 + \frac{\alpha R}{2\beta}\right]e^{-\alpha R/2\beta} \tag{13.40}$$

$$S = \left(1 + s + \frac{1}{3}s^2\right)e^{-s} = \left[1 + \frac{\alpha R}{2\beta} + \frac{1}{3}\left(\frac{\alpha R}{2\beta}\right)^2\right]e^{-\alpha R/2\beta}. \tag{13.41}$$

Exploring numerically how the energy varies with r for different values of α results in Figures 13.1–13.3. The key outcome is the result,

Variation of the fine-structure constant by about 6% gives a change in the bond length of the hydrogen ion molecule by about 6%.

This agrees with the above result (Eq. (13.35)) for the hydrogen atom.

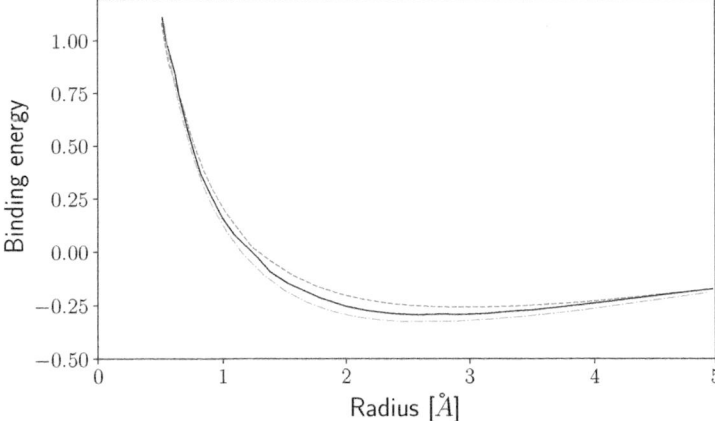

Figure 13.1 Graph of energies of the system as a function of radius. Units: Angstroms (but note that the metre depends on α). Binding will correspond to a minimum. Various values of α are shown. The middle one is the physical value of α. The upper and lower are $\pm 6\%$ change in the binding radius.

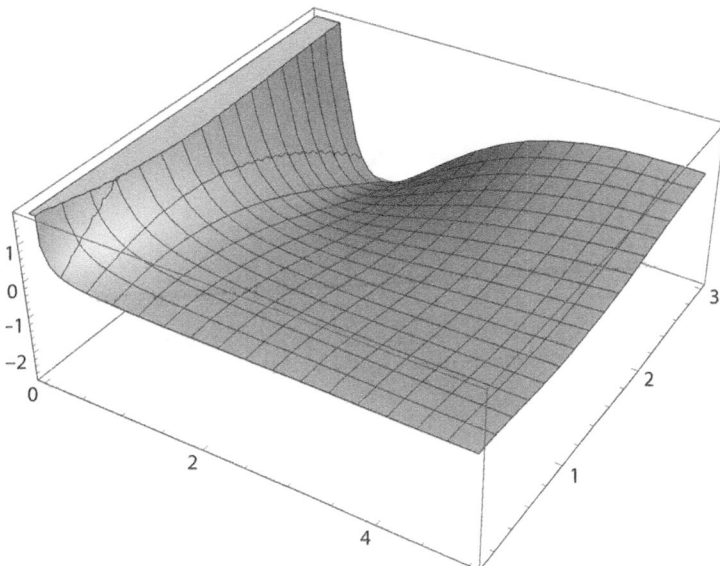

Figure 13.2 Variation of energy with radius (on the x-axis marked 0–4) against change in the fine-structure constant α on the y-axis. For each value of α, the minimum energy gives the binding radius.

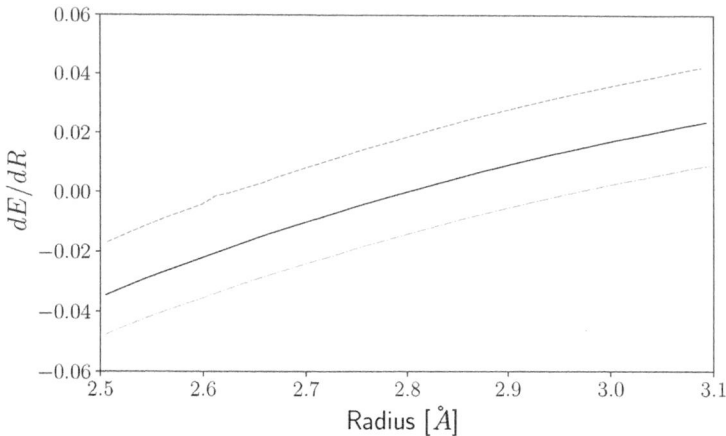

Figure 13.3 Finding the minimum. Graph of dE/dr as a function of radius. Various values of α are shown. The middle line is the physical value of α, the lowest curve is α reduced by 5.6%, and the highest curve is α increased by 6.4%. Both give $\pm6\%$ change in the binding radius, which may be the anthropic bound (Section 13.6).

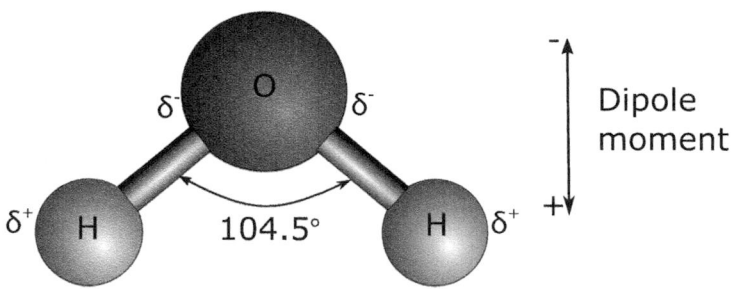

Figure 13.4 The water molecule has a crucial dipole moment because of its H-O-H angle of 104.5°. From [18], Figure 2.5.

13.4.3 Water Molecule

The water molecule H_2O [3] is a significant biological molecule because it is an excellent solvent that is liquid at room temperature, but freezes from the top because ice floats (frozen water is less dense than liquid water).

Water forms a liquid at room temperature because of its H-O-H angle of 104°, resulting in a dipole moment (the positive charges on the hydrogen atoms and negative charge on the oxygen atom are not aligned; Figure 13.4). Electrical attraction between water molecules due to this dipole reduces intermolecular distances, increasing the interaction energy and so raising its boiling point. Water is also a good solvent, because this polarity enables it to be strongly attracted to other polar molecules such as NaCl and, indeed, to disrupt them into their component atoms.

Hamiltonian

The Hamiltonian for the water molecule is given by the general formula for any molecule. Let i label the electrons in a molecule and I the nuclei, with masses m_i and M_I; then the non-relativistic equation is

$$\mathcal{H}\Psi = E\Psi, \tag{13.42}$$

where

$$\mathcal{H} = -\frac{\hbar^2}{2m_e}\sum_i \nabla_i^2 - \frac{\hbar^2}{2m_e}\sum_i \nabla_I^2 + \frac{e^2}{4\pi\epsilon_0}\sum_{(I,J)}\frac{Z_I Z_J}{r_{IJ}}$$

$$- \frac{e^2}{4\pi\epsilon_0}\sum_{(i,I)}\frac{Z_I}{r_{iI}} + \frac{1}{2}\frac{e^2}{4\pi\epsilon_0}\sum_{(i,j)\neq}\frac{1}{r_{ij}}, \tag{13.43}$$

where all positions are determined with respect to the oxygen nucleus. r_{IJ} are the internuclear distances, r_{iI} the electron to nucleus distances, and R_{ij} the inter-electron distances.

In the case of the water molecule with the most abundant isotope of oxygen, there are two hydrogen nuclei (protons of atomic mass 1), one oxygen nucleus (atomic mass 16), and 18 electrons.

Standard Lore

The general geometric form of molecules of the form $AX_m E_n$ – where A is the central atom, X is the other atom, and E is the free electron pairs – is depicted in Figure 13.5. The sum $m + n$ determines the geometry. Water, and all its isotopes, have $m + n = 4$ and are of form f in Figure 13.5. Because of the attraction by the nuclei of the X atom, a bonding pair is always further away from the central atom than a free electron pair (i.e., a non-bonding pair). The electronic interaction, always repulsive, is then always stronger between two non-bonding pairs than between two bonding pairs. Hence, the interactions can be ordered as *non-bonding/non-bonding > non-bonding/bonding > bonding-bonding*. Hence, in practice, the existence of a non-bonding doublet results in the deformation of the angle of the molecule because of the competition between the different replsive actions of the doublets. This induces a dipolar moment, which is important for the interaction between the molecules. It follows that

A free doublet leads to the opening of the E-A-X angles and the closure of the angle X-A-X.

Water enjoys a tetrahedral symmetry but with only two summits occupied by a hydrogen atom. It follows that its opening angle is $104.45°$ instead of $109.47°$, which it would be if the symmetry were exact (and as it actually is for methane CH_4). To compare, NH_3 is similar but has an opening angle of $107°$ (see Figure 13.6). This shows that while the angle

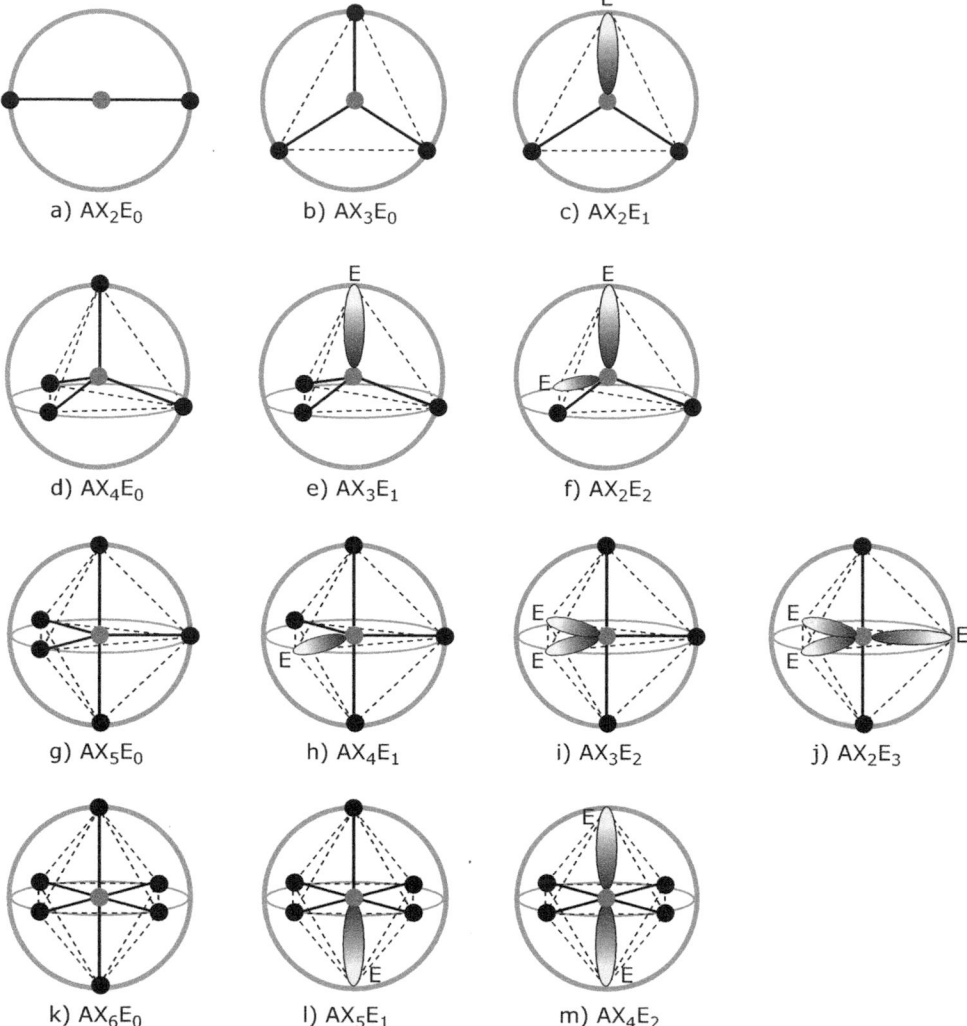

Figure 13.5 Configuration of molecules of the type $AX_m E_n$.

is determined in the first approximation by pure geometry, it also depends on the electric forces and the competition between bonding and non-bonding pairs. The more non-bonding the pairs, the smaller the angle.

13.4.4 Effect of the Variation of the Constants

On the numerical side, King *et al.* [15] have already investigated the effect of variation of the fundamental constants on water in detail via massive computer calculations. They

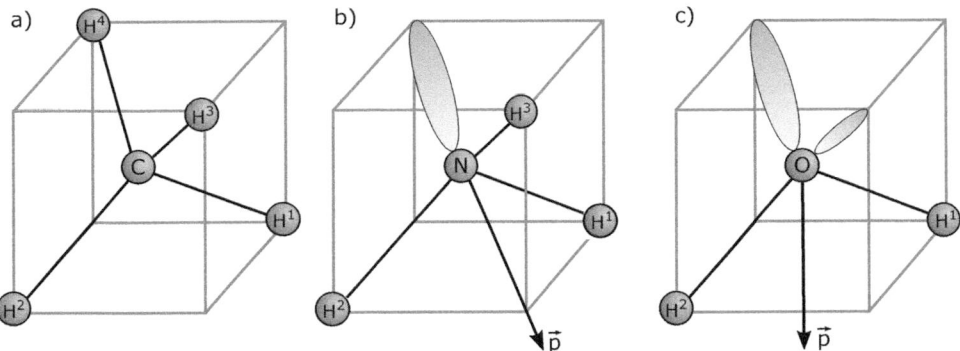

Figure 13.6 Tetragonal molecules. The opening angle changes when an atom is replaced by an electronic pair. It is perfect for methane (a), $\theta = 109.47°$, and descreases for amonia (b), $\theta = 107°$, and water (c), $\theta = 104.47°$.

state, 'The dramatic advances of recent decades in electronic structure methods, numerical algorithms, and raw computing power permit the determination of solutions very close to the ab initio limit for molecular systems of reasonable size'.

Their method is as follows. For non-relativistic chemistry and the Schrödinger equation, the relevant equation is (13.43). To solve it, one can use the Born-Oppenheimer approximation, whereby the electronic part of the Schröedinger equatin is first solved with clamped nuclei [15]. Then,

$$\hat{\mathcal{H}}_0 \Psi_e = \left[-\frac{\hbar^2}{2m_e} \sum_i \nabla_i^2 + \frac{e^2}{4\pi\epsilon_0} \sum_{(I,J)} \frac{Z_I Z_J}{r_{IJ}} - \frac{e^2}{4\pi\epsilon_0} \sum_{(i,I)} \frac{Z_I}{r_{iI}} + \frac{1}{2} \frac{e^2}{4\pi\epsilon_0} \sum_{(i,j\neq)} \frac{1}{r_{ij}} \right] \Psi_e \tag{13.44}$$

$$= \epsilon_e(\rho_{nuc}) \Psi_e, \tag{13.45}$$

yielding a potential energy surface (ρ_{nuc}) for the motion of the nuclei as a function of the scaled nuclei positions ρ_{nuc}. The resulting nuclear wave equation is then solved for the final rovibronic energy levels:

$$\left[-\frac{1}{2} \sum_I \frac{\beta}{\mu_I} \nabla_I^2 + \epsilon_e(\rho_{muc}) \right] \Psi_{nuc} = \epsilon \Psi_{nuc}. \tag{13.46}$$

The topography of the surfaces $\epsilon_e(\rho_{nuc})$ provides the basis for ascribing geometrical structure to molecules [15]. The local minima of this multi-dimensional wave surface correspond to the three-dimensional molecular structures of chemistry, allowing bond lengths and angles to be estimated. The results depend on basis sets chosen.

Figure 13.7 The water molecule bond length (Angstroms) as a function of the fine-structure constant α, computed for various basis sets.

Figure 13.8 The water molecule angle (degrees) as a function of the fine-structure constant α, calculated using various basis sets.

Figure 13.7 shows the water bond length as a function of α computed for various basis sets (for details, see [15]). Figure 13.8 shows the water molecule angle as a function of α, and Figure 13.9 shows the resulting variation of the dipole moment as a function of α. Finally, Figure 13.10 shows the bond length and angle as a function of the electron-proton mass ration μ.

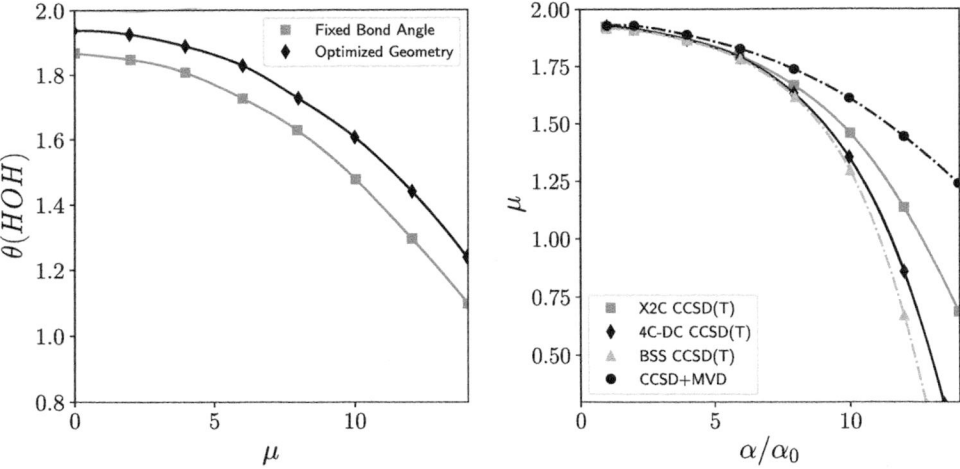

Figure 13.9 The water molecule dipole moment in Debyes as a function of the fine-structure constant α, calculated using various basis sets.

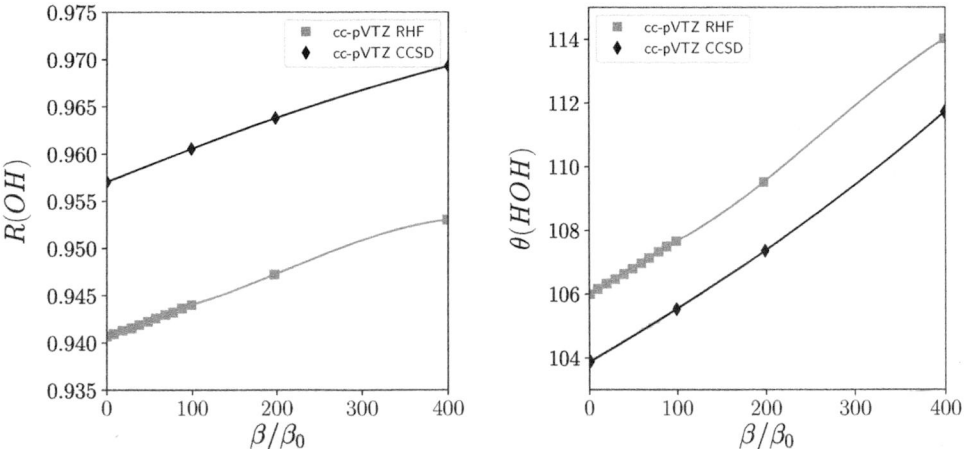

Figure 13.10 The water molecule bond length in Angstroms (left) and angle in degrees (right) as a function of the electron-proton mass ratio μ.

These interesting figures show that, indeed, there is an effect on key molecules if the variation in α and μ is large enough. Note that unlike the H_2 molecule, the radius of separation for water appears to increase as the coupling, through α, increases. This

13.5 The Issue of Scaling

However a key issue now arises: what if what we have been working out is just a similarity transformation, that leaves biological function unchanged? (Section 13.5.1). That leads to a consideration of relativistic effects in biology (Section 13.5.2).

13.5.1 What about Conformal Invariance?

Maybe a variation of α is just a change in size or, equivalently, of units (cf. Section 13.2.3). To explore this, consider a simple quantum mechanical system which has only a single force law acting, with a homogeneous potential term $V(\lambda r) = \lambda^n V(r)$, which has a linear dependence on some coupling constant C – e.g., $V(r) = C|r_1 - r_2|^n$.

Use this in the Schrödinger equation for a system of particles:

$$\frac{-\hbar^2}{2m}\nabla^2\psi(r) = (V(r) + E)\psi(r), \tag{13.47}$$

If we change C, we can find a second solution to our equations by defining a second wave function $\xi(r) = \psi(\lambda r)$, which will solve the Schrödinger equation with a different eigenvalue (energy level) but the same shape; altering C does not affect the angular modes that come into our solution.

To see this, suppose we look at a Coulomb potential, $V(r) \propto C/r$, and take a solution to the Schrödinger equation $\psi(r)$. If change our coordinate system to $R = 2r$, and let $\xi(R) = \psi(2r)$, then in this system, $\nabla_R^2 = \nabla_r^2/4$, and $\hat{V}\xi = \hat{V}\psi/2$. Thus, if we reduce C by a factor of two and call the new potential V, we find that

$$\frac{-\hbar^2}{2m}\nabla^2\xi(R) - V'(R)\xi(R)+ = \frac{1}{4}\left(\frac{-\hbar^2}{2m}\nabla^2\psi(r) - V(r)\psi(r)\right) = \frac{E}{4}\psi(r) = \frac{E}{4}\xi(R).$$
$$\tag{13.48}$$

So in reducing the coupling by a factor of two, we find that there is a solution of exactly the same shape as before, but with all distances doubled and one quarter of the energy. This leads us to consider conformal changes.

Conformal change: By rescaling the potential by C, it just changes all lengths together: it's just a larger molecule. Perhaps its biological function is unchanged?

But firstly, energies matter also. It is not clear whether one would obtain a rescaled system with the same conformational properties and energies rescaled to allow the same biological function but with different size systems or not. Specifically, some key effects related, for example, to chlorophyll and rhodopsin depend on photon energies $E = h\nu$ that are not related to the Schrödinger equation and will not scale with α. Secondly, there are spin effects in biology that are not conformally invariant. This has to account for the changes in angles shown in the results of (13.8) of King *et al.* [15], which clearly demonstrate that the molecular changes with variation of α do not just result in a change of scale. QED and relativistic effects matter (Figure 13.11).

13.5.2 Relativistic Effects

According to Likhtenshtein [17], 'The pivotal role of electron spin interactions in Nature cannot be overestimated'. It can be shown, for example, to have significant effects on brain function [25].

Figure 13.11 QED and relativistic effects on the bond angle (degrees) of the water molecule as a function of the fine-structure constant [15].

The relativistic terms for the hydrogen atom including fine and hyperfine structure were given in (13.8); we now discuss the fine-structure terms given there. This structure can be determined using perturbation theory; that is, by setting

$$H = H_0 + W. \tag{13.49}$$

The spin-orbit interaction is described in dimensionless form by

$$\frac{W_{S,O}}{m_e c^2} = \frac{\alpha}{2\rho^3} \frac{g_e}{a} \left(\frac{\mathbf{L}}{\hbar} \cdot \frac{\mathbf{S}}{\hbar} \right), \tag{13.50}$$

where g_e is the electron gyromagnetic factor. The second correction arises from the $(v/c)^2$ relativistic terms and is of the form

$$W_{rel} = -\frac{\mathbf{P}^4}{8m_e^3 c^2}. \tag{13.51}$$

The third and last correction is known as the Darwin term and arises from the fact that in the Dirac equation, the interaction between the electron and the Coulomb field is local. But the non-relativistic approximation leads to a non-local equation for the electron spinor that is sensitive to the field in a zone of order the Compton wavelength centred on $\mathbf{r}$. It follows that

$$W_D = \frac{\pi \hbar^2 q^2}{m_e^2 c^2} \delta(\mathbf{r}). \tag{13.52}$$

The average in an atomic state is of order

$$\langle W_D \rangle = \pi \hbar^2 q^2 / (2m_e^2 c^2) |\psi(\mathbf{0})|^2 \sim m_e c^2 \alpha^4 \sim \alpha^2 H_0. \tag{13.53}$$

Its dimensionless version is (not forgetting that the Dirac distribution has a dimension)

$$\frac{W_D}{m_e c^2} = \pi \alpha \delta(\rho). \tag{13.54}$$

These interactions will no longer be scale invariant under change of α.

In the calculations of King *et al.* [15], they state the following: the dimensionless ratios that have consequences for chemistry are the fine-structure constant $\alpha = e^2/(\hbar c)$ and the electron-proton mass ratio $\mu = m_e/m_p$. In conventional, non-relativistic quantum chemistry within the Born-Oppenheimer approximation, it is assumed that both ratios are negligibly small. The most important relativistic effects in chemistry can be investigated by means of the Cowan-Griffin Hamiltonian in which $\hat{H}_0$ is augmented with one-electron mass-velocity and Darwin terms:

$$\hat{H}_1 = \alpha^2 \left[-\frac{1}{8} \sum_i \nabla_i^4 + \frac{\pi}{2} \sum_{I,j} \delta(\rho_{Ii}) \right]. \tag{13.55}$$

The consequences of finite m_e/m_p ratios on chemical systems can be probed by means of the diagonal Born-Oppenheimer correction (DBOC)

$$E_{DBOC} = -\frac{\mu}{2} \sum_I \frac{1}{\mu_I} \left\langle \Psi_e | \nabla_I^2 | \Psi_e \right\rangle. \tag{13.56}$$

This is what is represented directly in Figure 13.11 and indirectly through the (Figures 13.8 and 13.9), where the angle and dipole moment depend on the fundamental constants.

13.6 Extrapolations and Conclusions

Can we extend these results to key organic molecules?

13.6.1 Scaling to Biology: Organic Molecules

The structure of DNA has very tight constraints in order for it to function. In order for the DNA helix structure to exist, one needs the equality between the lengths of $A + T$ and $G + C$ to be equal. According to Reference [4], 'The distance between adjacent sugars or phosphates in the DNA chain is 6 Angstroms. It must be between 5.5 and 6.5 Angstroms for it to work'. This means that we require

$$\delta \left(\frac{\ell_{A+T}}{\ell_{G+C}} \right) < 10\%. \tag{13.57}$$

How this relates to a constraint on the fine-structure constant is unclear without detailed study using the methods of King *et al.* [15]. This may not be as hopeless as it sounds because of the very regular structure of the molecule in its double-helix form, so there will be symmetries that can be exploited. However, one might hope that the results will be indicated by those for the water molecule, given that the underlying equation is the same. This constraint on a dimensionless parameter that specifies the geometry and thus the possibility to build a helix should be of the order of between 6% and 10% variation in α. However, this rough estimate could be completely wrong.

13.6.2 Proteins: Voltage Gated Ion Channels

As mentioned at the beginning of this chapter, what one really wants is to study the viability of molecules such as voltage gated ion channels, which play a key role in biology [12], under variation of the fundamental constants. This is much more difficult than DNA because their tertiary and quaternary structure is so much more complex [5, 20]. However, it may not be completely hopeless because they are made of subunits such as α and β helices that have a more regular form. This may be hopeless, but it is worth stating as a long-term goal.

13.6.3 A General Theorem?

Peter Atkins has provided the following argument for the general case. Let i label the electrons in a molecule and I, the nuclei. Using constants as in (13.30), the Hamiltonian can be written as

$$H = -\beta\gamma \sum_i \nabla_i^2 - \alpha\gamma \sum_{(i,I)} \frac{Z_I}{r_{iI}} + \frac{1}{2}\alpha\gamma \sum_{(i,j?)} \frac{1}{r_{ij}} + \frac{1}{2}\alpha\gamma \sum_{I,J\neq I} \frac{Z_I Z_J}{r_{IJ}}. \tag{13.58}$$

Taking partial derivatives,

$$\frac{\partial H}{\partial \alpha} = -\gamma \sum_{(i,I)} \frac{Z_I}{r_{iI}} + \frac{1}{2}\gamma \sum_{(i,j)} \frac{1}{r_{ij}} + \frac{1}{2}\gamma \sum_{I,J\neq I} \frac{Z_I Z_J}{r_{IJ}} = \frac{H}{\alpha} + \frac{\beta\gamma}{\alpha} \sum_i \nabla_i^2. \tag{13.59}$$

$$\frac{\partial H}{\partial \beta} = -\gamma \sum_i \nabla_i^2. \tag{13.60}$$

From the first of these expressions,

$$\left\langle \frac{\partial H}{\partial \alpha} \right\rangle = \frac{1}{\alpha}\langle H \rangle + \frac{\beta\gamma}{\alpha}\left\langle \sum_i \nabla_i^2 \right\rangle. \tag{13.61}$$

And then from the second,

$$\left\langle \frac{\partial H}{\partial \alpha} \right\rangle = \frac{1}{\alpha}\langle H \rangle - \frac{\beta}{\alpha}\left\langle \frac{\partial H}{\partial \beta} \right\rangle. \tag{13.62}$$

From the Hellmann-Feynman theorem [13, 14], which states that for a parameter P and exact, normalised wave function,

$$\left\langle \frac{\partial H}{\partial P} \right\rangle = \frac{d}{dP}, \tag{13.63}$$

this expression, with $\langle H \rangle = E$, becomes

$$\frac{dE}{d\alpha} = \frac{E}{\alpha} - \frac{\beta}{\alpha}\frac{dE}{d\beta} \qquad (13.64)$$

or

$$\frac{\alpha}{E}\frac{dE}{d\alpha} + \frac{E}{\beta}\frac{dE}{d\beta} = 1 \iff \frac{d\ln E}{d\ln\alpha} + \frac{d\ln E}{d\ln\beta} = 1, \qquad (13.65)$$

This equation with $f = \ln E$, $x = \ln\alpha$, and $y = \ln\beta$ – has the form $df/dx + df/dy = 1$ and, therefore, has the solution (with a constant of integration $\ln k$) $f = ax + by + \ln k$, with $a + b = 1$. That is,

$$\ln kE = a\ln\alpha + b\ln\beta + \ln k = \ln k\alpha^a\beta^b. \qquad (13.66)$$

It then follows that, regardless of the identity of the molecule,

$$E = k\alpha^a\beta^b \text{ with } a + b = 1. \qquad (13.67)$$

This is consistent with the H atom, where $E_n \propto \alpha^2/\beta$, for which $a = 2, b = -1$.

Dimensional analysis helps to take this further. Write $k = \gamma\kappa$, with $\gamma = c\hbar$ (as in (13.30)). Then,

$$E = \kappa\gamma\alpha^a\beta^b. \qquad (13.68)$$

The dimensions are as follows:

$$[E] = [\kappa][\gamma][\alpha^a][\beta^b] : [ML^2T^{-2}] = [\kappa][ML^3T^{-2}][1]^a[L]^b = [\kappa][ML^{3+b}T^{-2}], \qquad (13.69)$$

which implies that

$$[\kappa] = L^{-(b+1)}. \qquad (13.70)$$

One can then try the following. The only combination of fundamental constants in the Hamiltonian that has dimensions of length form the combination β. Therefore, κ must be proportional to $\beta^{-(b+1)}$. However, κ is independent of β (it is proportional to the constant k). Therefore, the only value of b that is allowed is $b = -1$, with the implication that $[\kappa] = [1]$ and $a = 2$. Therefore,

$$E = \kappa\gamma\alpha^2\beta^{-1}. \qquad (13.71)$$

for some constant κ, which is

$$E = 2m_ec^2\kappa\alpha^2, \iff \mathcal{E} = \frac{E}{m_ec^2} = \kappa\alpha^2, \qquad (13.72)$$

which is what we get for the hydrogen atom. And if the argument is valid, this is the case for all atoms and molecules, regardless of their identity (solids too, which are just big molecules). A puzzle is the result for H_2^+, where this is not the case, but this may be because the LCAO treatment of H^+ there is an approximation, whereas this result (depending as it does on the Hellmann-Feynman theorem) is for exact energies and wave functions.

If an argument like this works, it will suggest how energies scale with α (and other constants), which is key to one of the major themes of biology: metabolism [32]. As mentioned before, there cannot just be a scaling of metabolic effects with α because some, such as photosynthesis, are independent of α.

13.6.4 Other Scales

Finally, we return to considering the relation of this work to other scales.

- Nuclear scale: stable nuclei possible.
- Stellar scale: existence of stable nuclei due to stellar processes. Requires the Hoyle state to exist.
- Galactic to stellar scale: existence of galaxies, stars, and planets due to cosmological and astrophysical processes. Requires restricted cosmological parameters.

The fine-structure constant α and mass ratio μ are not the only relevant constants for these effects to be favourable; thus, there, in principle, can be many physical situations where life cannot occur (e.g., because carbon does not come into being), even though the organic molecules would functon fine, were they to come into existence.

13.6.5 Fine-Tuning Relations

The effect of fine-tuning on biomolecules has importance in the understanding of the fine-tuning of the Uxniverse. We can ask a series of questions:

1. Are constraints from physics (stability of nuclei) stronger than those of biology?
2. Are constraints propagating?
3. Can we argue that if the constraints for the formation of carbon-12, etc., are satisfied, then conditions for life (at least on those parameters) are satisfied? Or do they set further non-trivial constraints?

We can ask, *which sets the tighter limits: physics/astrophysics or biology?*

It appears from the preceding that it is likely that the physics constraints are indeed tighter than the constraints from water and biomolecules. As regards α:

- Stellar nucleosynthesis limits, holding everything else fixed, give limits $\simeq 10^{-3}$.
- This seems much tighter than the likely limits from water.
- This may or may not be tighter than limits from DNA and proteins, but an initial estimate based on the preceding argument is that it will indeed be tighter.

As regards μ: there are only combinations of limits. Limts on μ per se are unclear.

If physics uniquely leads to molecules compatible with life in this way, then in some sense, physics foresees the existence of life.

Why? This is a conundrum: *physics seems fine-tuned to expect life.*

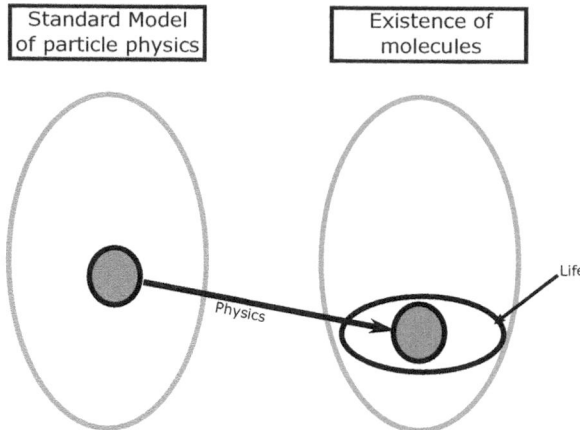

Figure 13.12 Fundamental physics and biology: the Standard Model of particles physics might have been implemented with a variety of fundamental constants (oval on left). The set that actually occurred (small circle on left) is in the small subset of these possibilities that leads to possible existence of molecules compatible with life (open circle on right).

13.7 Conclusion

It is a legitimate and important question to quantify the dependence of higher levels of complexity on the parameters of a fundamental theory. In particular, it allows one to discuss the level at which the fine-tuning on the fundamental constant is the more stringent. This can be compared to Darwinian selection, which takes place simultaneously but with different timescales, on the scale of the gene, the cell, the organ, the animal, and the population.

Two final issues remain. First, what about other forms of life? Is it too restrictive to only allow the consideration of water and organic molecules? Our viewpoint is that it is only organic molecules, and specifically proteins [22], that have the capacity to produce the complexity of living systems that respond to the environment in an adaptive way and allow biology to emerge from physics [12]. We do not believe life could come into existence without being based in carbon and its huge variety of organic molecules that allow extraordinary complexity of function, as characterised by the possibility spaces of molecular biology [32].

Second, we have not considered the relation to a more fundamental theory that might underlie the Standard Model of particle physics. We do not believe enough is known to make reliable estimates in this regard.

References

[1] Atkins, P., and Friedman, R. 1997. *Molecular Quantum Mechanics*. Oxford University Press.
[2] Barrow, J. D., and Tipler, F. J. 1986. *The Anthropic Cosmological Principle*.
[3] Cabane, B., and Vuilleumier, R. 2005. 'The Physics of Liquid Water'. *Comptes Rendus Geoscience*, **337**(Feb.), 159–171.

[4] Calladine, C. R., and Drew, H. R. 1992. *Understanding DNA: The Molecule & How It Works*. Academic Press.

[5] Catterall, W. A. 1993. 'Structure and Function of Voltage-Gated Ion Channels'. *Trends in Neurosciences*, **16**(12), 500–506.

[6] Watson, J. D. 2004. *Molecular Biology of the Gene*. Pearson Education.

[7] Ekström, S., *et al.* 2010. 'Effects of the Variation of Fundamental Constants on Population III Stellar Evolution'. *Astronomy and Astrophysics*, **514**(May), A62.

[8] Ellis, G. F. R. 2005a. 'Physics and the Real World'. *Physics Today*, **58**(7), 49–54.

[9] Ellis, G. F. R. 2005b. 'Physics, Complexity and Causality'. *Nature*, **435**(June), 743.

[10] Ellis, G. F. R., and Uzan, J. P. 2005. '*c Is the Speed of Light, Isn't It?*.' *American Journal of Physics*, **73**(Mar.), 240–247.

[11] Ellis, G. 2016. *How Can Physics Underlie the Mind?: Top-Down Causation in the Human Context*. The Frontiers Collection. Springer.

[12] Ellis, G. F. R., and Kopel, J. 2018. *Wandering towards a Goal: The Key Role of Biomolecules*. Springer International Publishing. Pages 227–243.

[13] Feynman, R. P. 1939. 'Forces in Molecules'. *Physical Review*, **56**(Aug.), 340–343.

[14] Hellmann, H. 1937. *Einführung in Die Quantenchemie: Texte Imprimé*. F. Deuticke.

[15] King, R. A., *et al.* 2010. 'Chemistry as a Function of the Fine-Structure Constant and the Electron-Proton Mass Ratio'. *Physical Review A: General Physics*, **81**(4), 042523.

[16] Lewis, G. F., and Barnes, L. A. 2016. *A Fortunate Universe*. Cambridge University Press.

[17] Likhtenshtein, G. 2014. *Electron Spin Interactions in Chemistry and Biology: Fundamentals, Methods, Reactions Mechanisms, Magnetic Phenomena, Structure Investigation*. Biological and Medical Physics, Biomedical Engineering. World Scientific Publishing Company Incorporated.

[18] Lodish, H. 2008. *Molecular Cell Biology*. 6th edn. W. H. Freeman and Company.

[19] Luo, F., Olive, K. A., and Uzan, J. P. 2011. 'Gyromagnetic Factors and Atomic Clock Constraints on the Variation of Fundamental Constants'. *Physical Review D*, **84**(9), 096004.

[20] Magleby, K. L. 2017. 'Structural Biology: Ion-Channel Mechanisms Revealed'. *Nature*, **541**(7635), 33.

[21] Nakamura, K., and Particle Data Group. 2010. 'Review of Particle Physics'. *Journal of Physics G Nuclear Physics*, **37**(7), 075021.

[22] Petsko, G. A., and Ringe, D. 2004. *Protein Structure and Function*. Vol. Primers in Biology. New Science.

[23] Rees, M. 2000. *Just Six Numbers: The Deep Forces That Shape the Universe*. Basic Books.

[24] Rees, M. 2003. *Our Cosmic Habitat*. Princeton University Press.

[25] Turin, L., Skoulakis, E. M. C., and Horsfield, A. P. 2014. 'Electron Spin Changes during General Anesthesia in Drosophila'. *Proceedings of the National Academy of Science*, **111**(Aug.), E3524–E3533.

[26] Uzan, J. P. 2003. 'The Fundamental Constants and Their Variation: Observational and Theoretical Status'. *Reviews of Modern Physics*, **75**(Apr.), 403–455.

[27] Uzan, J. P. 2011. 'Varying Constants, Gravitation and Cosmology'. *Living Reviews in Relativity*, **14**(Mar.), 2.

[28] Uzan, J. P. 2013. 'Models of the Cosmos and the Emergence of Complexity. In Ellis, G. F. R., Heller, M., and Pabjan, T., eds., *The Causal Universe*. Copernicus Center Press.

[29] Uzan, J. P. 2017. *Emergent Structures of Effective Field Theories*. Cambridge University Press.

[30] Uzan, J., and Lehoucq, R. 2005. *Les constantes fondamentales*. Belin Sup. Belin.

[31] Villani, C., Uzan, J., and Moncorgé, V. 2017. *The House of Mathematics*. Le Cherche Midi.

[32] Wagner, A. 2014. *Arrival of the Fittest: Solving Evolution's Greatest Puzzle*. Penguin.

[33] Wilczek, F. 2007. 'Fundamental Constants'. *arXiv e-prints*, Aug.

Index